Development, Characterization, Application and Recycling of Novel Construction Materials

Development, Characterization, Application and Recycling of Novel Construction Materials

Mouhamadou Amar
Nor Edine Abriak

Basel • Beijing • Wuhan • Barcelona • Belgrade • Novi Sad • Cluj • Manchester

Mouhamadou Amar
Research Center for Materials
and Processes
Institut Mines-Télécom
IMT Nord Europe
France

Nor Edine Abriak
Research Center for Materials
and Processes
Institut Mines-Télécom
IMT Nord Europe
France

Editorial Office
MDPI AG
Grosspeteranlage 5
4052 Basel, Switzerland

This is a reprint of articles from the Special Issue published online in the open access journal *Applied Sciences* (ISSN 2076-3417) (available at: www.mdpi.com/journal/applsci/special_issues/4UKS4SN5D2).

For citation purposes, cite each article independently as indicated on the article page online and using the guide below:

Lastname, A.A.; Lastname, B.B. Article Title. *Journal Name* **Year**, *Volume Number*, Page Range.

ISBN 978-3-7258-2146-4 (Hbk)
ISBN 978-3-7258-2145-7 (PDF)
https://doi.org/10.3390/books978-3-7258-2145-7

Cover image courtesy of Mouhamadou Amar

© 2024 by the authors. Articles in this book are Open Access and distributed under the Creative Commons Attribution (CC BY) license. The book as a whole is distributed by MDPI under the terms and conditions of the Creative Commons Attribution-NonCommercial-NoDerivs (CC BY-NC-ND) license.

Contents

About the Editors vii

Preface ix

Mouhamadou Amar and Nor-Edine Abriak
Development, Characterization, Application and Recycling of Novel Construction Materials
Reprinted from: *Appl. Sci.* 2024, 14, 6951, doi:10.3390/app14166951 1

Ivana Perná, Martina Havelcová, Monika Šupová, Margit Žaloudková and Olga Bičáková
The Synthesis and Characterization of Geopolymers Based on Metakaolin and on Automotive Glass Waste
Reprinted from: *Appl. Sci.* 2024, 14, 3439, doi:10.3390/app14083439 5

Pablo Plaza, Isabel Fuencisla Sáez del Bosque, Javier Sánchez and César Medina
Recycled Eco-Concretes Containing Fine and/or Coarse Concrete Aggregates. Mechanical Performance
Reprinted from: *Appl. Sci.* 2024, 14, 3995, doi:10.3390/app14103995 24

Girum Mindaye Mengistu and Rita Nemes
Recycling 3D Printed Concrete Waste for Normal Strength Concrete Production
Reprinted from: *Appl. Sci.* 2024, 14, 1142, doi:10.3390/app14031142 45

Matija Zvonarić, Mirta Benšić, Ivana Barišić and Tihomir Dokšanović
Prediction Models for Mechanical Properties of Cement-Bound Aggregate with Waste Rubber
Reprinted from: *Appl. Sci.* 2024, 14, 470, doi:10.3390/app14010470 59

Joanna Julia Sokołowska, Paweł Łukowski and Alicja Baczek
Mortars with Polypropylene Fibers Modified with Tannic Acid to Increase Their Adhesion to Cement Matrices
Reprinted from: *Appl. Sci.* 2024, 14, 2677, doi:10.3390/app14072677 74

Haris Brevet, Rose-Marie Dheilly, Nicolas Montrelay, Koffi Justin Houessou, Emmanuel Petit and Adeline Goullieux
Effects of Flaxseed Mucilage Admixture on Ordinary Portland Cement Fresh and Hardened States
Reprinted from: *Appl. Sci.* 2024, 14, 3862, doi:10.3390/app14093862 86

Yosra Rmili, Khadim Ndiaye, Lionel Plancher, Zine El Abidine Tahar, Annelise Cousture and Yannick Melinge
Properties and Durability of Cementitious Composites Incorporating Solid-Solid Phase Change Materials
Reprinted from: *Appl. Sci.* 2024, 14, 2040, doi:10.3390/app14052040 106

Anita Gojević, Ivanka Netinger Grubeša, Sandra Juradin and Ivana Banjad Pečur
Resistance of Concrete with Crystalline Hydrophilic Additives to Freeze–Thaw Cycles
Reprinted from: *Appl. Sci.* 2024, 14, 2303, doi:10.3390/app14062303 127

François El Inaty, Bugra Aydin, Maryam Houhou, Mario Marchetti, Marc Quiertant and Othman Omikrine Metalssi
Long-Term Effects of External Sulfate Attack on Low-Carbon Cementitious Materials at Early Age
Reprinted from: *Appl. Sci.* 2024, 14, 2831, doi:10.3390/app14072831 139

Shenqiu Lin, Ping Chen, Weiheng Xiang, Cheng Hu, Fangbin Li and Jun Liu et al.
Exploring the Effect of Moisture on CO_2 Diffusion and Particle Cementation in Carbonated Steel Slag
Reprinted from: *Appl. Sci.* **2024**, *14*, 3631, doi:10.3390/app14093631 . . . . **156**

Lluís Gil, Luis Mercedes, Virginia Mendizabal and Ernest Bernat-Maso
Shear Performance of the Interface of Sandwich Specimens with Fabric-Reinforced Cementitious Matrix Vegetal Fabric Skins
Reprinted from: *Appl. Sci.* **2024**, *14*, 883, doi:10.3390/app14020883 . . . . **171**

Sung-Won Cho, Sung Eun Cho and Alexander S. Brand
A Meta-Analysis of the Effect of Moisture Content of Recycled Concrete Aggregate on the Compressive Strength of Concrete
Reprinted from: *Appl. Sci.* **2024**, *14*, 3512, doi:10.3390/app14083512 . . . . **184**

About the Editors

Mouhamadou Amar

Assoc. Prof. Mouhamadou Amar focuses on research in material reuse and design, low-carbon concrete, durable materials, geopolymers and alkali-activated materials (AAMs), waste treatment, and numerical modeling, with major contributions in the field.

Nor Edine Abriak

Prof. Nor-edine Abriak has over 30 years of expertise in mechanical engineering, chemistry, soil science, and building materials. His work has significantly advanced these interdisciplinary fields, enhancing their applications and knowledge.

Preface

In an era where sustainability and innovation drive scientific progress, materials science plays a crucial role in shaping our future. With growing environmental concerns and the need for resource optimization, research in sustainable materials, cement, and waste reuse is more vital than ever. This reprint offers a platform for the latest findings in these areas. It covers key topics such as the formulation of eco-friendly materials, advanced techniques and characterization methods for understanding material properties, and innovations in alternative binders like geopolymers. Additionally, it explores waste reuse through materials like recycled aggregates and fibers in construction. By bringing together leading research, this reprint aims to be an essential resource for those committed to advancing sustainable materials in construction.

Mouhamadou Amar and Nor Edine Abriak
Editors

Editorial

Development, Characterization, Application and Recycling of Novel Construction Materials

Mouhamadou Amar [1,*] and Nor-Edine Abriak [2]

1 Centre for Materials and Processes, Institut Mines-Telecom, Institut Mines Telecom Nord Europe, F-59508 Douai, France
2 Laboratoire de Génie Civil et Géo-Environnement, ULR 4515—LGCgE, Institut Mines-Télécom, University Lille, F-59000 Lille, France; nor-edine.abriak@imt-nord-europe.fr
* Correspondence: mouhamadou.amar@imt-nord-europe.fr

Citation: Amar, M.; Abriak, N.-E. Development, Characterization, Application and Recycling of Novel Construction Materials. *Appl. Sci.* **2024**, *14*, 6951. https://doi.org/10.3390/app14166951

Received: 31 July 2024
Accepted: 2 August 2024
Published: 8 August 2024

Copyright: © 2024 by the authors. Licensee MDPI, Basel, Switzerland. This article is an open access article distributed under the terms and conditions of the Creative Commons Attribution (CC BY) license (https://creativecommons.org/licenses/by/4.0/).

1. Introduction

The construction industry ranks among the most polluting sectors globally. Efforts are underway, guided by international forums and conference resolutions, to reduce CO_2 emissions. Since the onset of the industrial revolution, the growing impact of human activity has become increasingly pronounced, exacerbated by one key indicator: carbon footprint. This footprint is considered a major factor in climate change and is a phenomenon that continues to intensify. According to the Intergovernmental Panel on Climate Change (IPCC), post-2030 temperature increases are projected to range between +1.1 °C and +6.4 °C. In its latest 2023 report, the IPCC proposed mitigation measures such as emission reduction or resource consumption proportioning to limit global warming to +1.5 °C by 2050 [1–3]. Thus, at the beginning of the 21st century, humanity faces an unprecedented challenge: How can we ensure sustainability by protecting nature and biodiversity for future generations while simultaneously meeting the current growing demands for energy, materials, and resources? More recently, between 2020 and 2022, the EU launched its new climate action plan, unveiling ambitious initiatives throughout the entire lifecycle of products. This plan includes product design, promotes circular economy processes, and encourages sustainable consumption. Known as the "climate package", this approach aims to minimize waste proliferation and conserve resources for as long as possible [4]. Resource efficiency and environmental protection have become major concerns in addressing climate issues. Global economic growth and the rapid development of cities, which are responsible for approximately 80% of global CO_2 emissions [5], significantly increase the demand for materials, natural resources, and energy. The need for resources such as water, land, energy, and minerals has never been higher. There is an urgent need to enhance resource efficiency and increasingly reuse building materials [6–8] by adopting a circular economy approach. The UN sustainable cities program, aligned with the Paris Agreement (COP 21, 2015), proposes an action plan that emphasizes integrating environmental concerns into urban planning and management: "Mainstreaming environmental concerns in urban planning and management" [1,9,10]. Finding solutions for wastes, i.e., secondary materials generated from industries, infrastructure, and construction activities, has become imperative.

Concrete, predominantly made from ordinary Portland cement (OPC), is the most widely used building material. Each constituent in OPC production impacts the environment, leading to sustainability concerns. The manufacture and use of cement and concrete significantly affect the environment, driven by infrastructure development, building operations, and CO_2 emissions, which constitute 7–8% of global emissions [11]. Consequently, the construction industry faces pressure to develop eco-friendly alternatives. As the environmental issues associated with OPC have become evident, numerous studies are seeking new binding materials that can match the cost and performance of currently used construction materials. Also, currently, alternative binders based on sustainable materials, e.g., geopolymers, seem to be preferred to conventional cementitious materials. Scientific studies have highlighted several key reasons why geopolymers are considered promising alternatives to traditional construction materials [12]. The concrete production process is notably energy-intensive, with studies by Muhamad et al. [13] and Shirkani et al. [14] highlighting the cement sector's significant contribution to global carbon dioxide emissions and climate change.

Hence, the utilization of binders made from byproducts and alternative materials, like metakaolin, fly ash, sludge ash, blast furnace slag, silica fume, fiber glass, and waste rubber, is strongly encouraged, as their effectiveness is well documented [15–18]. This approach addresses the previously mentioned environmental concerns and enhances the durability of structures exposed to harsh conditions [16]. For this purpose, several solutions have been envisaged in the construction sector involving the use of greener and more innovative materials, for example, using alternative fuels for material combustion in cement kilns and enhancing decarbonization through replacing some raw materials (such as calcium carbonate) with other products that are already decarbonated and confer the same chemical properties.

2. An Overview of Published Articles

Recent advancements in the field of cementitious materials have focused on enhancing mechanical properties, sustainability, and durability through innovative approaches, such as incorporating waste materials, innovative additives, and recycled aggregates. One notable approach involves using waste rubber in cement-bound aggregates (CBAs) to reduce their traditionally high stiffness. By developing prediction models for compressive strength and the modulus of elasticity based on non-destructive ultrasonic pulse velocity (UPV) tests, it was found that incorporating rubber not only reduces stiffness but also simplifies the prediction of mechanical properties, especially compressive strength, which does not significantly depend on the curing period [19].

Additionally, the incorporation of solid–solid phase change materials (SS-PCMs) into cementitious composites was studied due to their heat storage capabilities. While increasing porosity and reducing mechanical strength, SS-PCMs enhanced thermal insulation and shrinkage resistance, demonstrating superior stability over multiple thermal cycles despite their fast carbonation kinetics due to their high porosity. Research on concrete with crystalline hydrophilic additives (CAs) has focused on enhancing resistance to freeze–thaw cycles. Standard air-entraining agents have been found to be effective, but CAs, especially at a 1% dosage, can improve internal damage resistance.

Another innovative study investigated the role of moisture in CO_2 diffusion and particle cementation in carbonated steel slag. Optimal moisture content was crucial for balancing CO_2 diffusion and particle cementation, enhancing the compressive strength and carbon sequestration capacity of the slag, and thus contributing to the development of effective carbon sequestration materials. These researchers also analyzed the mechanical performance of eco-concrete using recycled concrete aggregates. They found minimal strength decreases and increased permeability with higher concentrations of recycled content, demonstrating the viability of incorporating recycled aggregates into eco-concretes. In examining low-carbon cementitious materials, the effects of external sulfate attack (ESA) were assessed. It was found that a mix of CEM I, slag, and metakaolin exhibited the highest resistance to sulfate attack, while a 100% CEM I mix deteriorated significantly. This highlights the importance of using supplementary cementitious materials to increase durability in sulfate-rich environments [20].

The use of flaxseed mucilage (FM) as a bio-admixture in ordinary Portland cement was explored. FM delayed the time taken for the cement to set but did not hinder its hydration properties. It increased the cement's porosity and carbonation while reducing its bulk density and thermal conductivity. FM's hygroscopic properties and controlled water release improved the mechanical properties of the cement over time, suggesting that it may confer potential self-healing capabilities. In another study, vegetal fabric-reinforced cementitious matrix (FRCM) sandwich panels were investigated for their sustainable advantages. When using vegetal fabrics like hemp and sisal in combination with extruded polystyrene cores, these panels showed competitive strength compared to synthetic fibers, with steel connectors providing enhanced stiffness and shear strength [21].

The recycling of 3D-printed concrete waste as aggregate in new concrete mixtures was also explored. It was found that replacing conventional aggregates with recycled ones from 3D-printed waste at a ratio of up to 67% did not significantly reduce the compressive strength of the concrete and sometimes even improved it, particularly in higher-strength classes like C40/50.

These studies collectively highlight significant advancements in the field of cementitious materials, with the goal of enhancing mechanical properties, sustainability, and durability.

3. Conclusions

By exploring the incorporation of waste materials, innovative additives, and recycled aggregates into construction materials, these research efforts contribute to the development of more resilient and environmentally friendly construction materials. The sustainable alternatives considered include waste rubber, vegetal fabrics, and recycled aggregates, which could all be used to enhance materials. Innovations like solid–solid phase change materials, crystalline hydrophilic additives, and low-carbon cementitious materials improve the durability and performance of construction materials. Overall, these advancements aim to facilitate the creation of eco-friendly construction materials.

Author Contributions: M.A.: writing—original draft preparation; N.-E.A.: writing—review and editing. All authors have read and agreed to the published version of the manuscript.

Funding: This research received no external funding.

Conflicts of Interest: The authors declare no conflicts of interest.

References

1. Ghezloun, A.; Saidane, A.; Merabet, H. The COP 22 New commitments in support of the Paris Agreement. *Energy Procedia* **2017**, *119*, 10–16. [CrossRef]
2. Bernstein, L.; Bosch, P.; Canziani, O.; Chen, Z.; Christ, R.; Davidson, O.; Hare, W.; Huq, S.; Karoly, D.; Kattsov, V.; et al. *Climate Change 2007: Synthesis Report: An Assessment of the Intergovernmental Panel on Climate Change*; IPCC: Geneva, Switzerland, 2008.
3. Scrivener, K.L.; John, V.M.; Gartner, E.M. Eco-efficient cements: Potential economically viable solutions for a low-CO_2 cement-based materials industry. *Cem. Concr. Res.* **2018**, *114*, 2–26. [CrossRef]
4. Directorate-General for Communication (European Commission). *Circular Economy Action Plan: For a Cleaner and More Competitive Europe*; Publications Office of the European Union: Luxembourg, 2020. [CrossRef]
5. Heinonen, J.; Junnila, S. Case study on the carbon consumption of two metropolitan cities. *Int. J. Life Cycle Assess.* **2011**, *16*, 569–579. [CrossRef]
6. Blengini, G.A.; Garbarino, E. Resources and waste management in Turin (Italy): The role of recycled aggregates in the sustainable supply mix. *J. Clean. Prod.* **2010**, *18*, 1021–1030. [CrossRef]
7. Gangolells, M.; Casals, M.; Forcada, N.; Macarulla, M. Analysis of the implementation of effective waste management practices in construction projects and sites. *Resour. Conserv. Recycl.* **2014**, *93*, 99–111. [CrossRef]
8. Eras, J.J.C.; Gutiérrez, A.S.; Capote, D.H.; Hens, L.; Vandecasteele, C. Improving the environmental performance of an earthwork project using cleaner production strategies. *J. Clean. Prod.* **2013**, *47*, 368–376. [CrossRef]
9. Fabre, P.; Prévot, A.-C.; Semal, L. Le Grand Paris, ville durable? Limites pour la biodiversité urbaine dans un projet de métropolisation emblématiqueThe Greater Paris, a sustainable city? Limits for urban biodiversity in emblematic metropolis plan. *Dev. Durable Territ.* **2016**, *7*, 1. [CrossRef]
10. Blau, J. The Paris Agreement. In *The Paris Agreement: Climate Change, Solidarity, and Human Rights*; Springer Nature: Dordrecht, The Netherlands, 2017. [CrossRef]
11. Pławecka, K.; Bazan, P.; Lin, W.-T.; Korniejenko, K.; Sitarz, M.; Nykiel, M. Development of Geopolymers Based on Fly Ashes from Different Combustion Processes. *Polymers* **2022**, *14*, 1954. [CrossRef] [PubMed]
12. Palmero, P.; Formia, A.; Tulliani, J.-M.; Antonaci, P. Processing and applications of geopolymers as sustainable alternative to traditional cement. In Proceedings of the 5th International Conference on Development, Energy, Environment, Economics (DEEE '14), Florence, Italy, 22–24 November 2014.
13. Mohamad, N.; Muthusamy, K.; Embong, R.; Kusbiantoro, A.; Hashim, M.H. Environmental impact of cement production and Solutions: A review. *Mater. Today Proc.* **2022**, *48*, 741–746. [CrossRef]

14. Shirkhani, A.; Kouchaki-Penchah, H.; Azmoodeh-Mishamandani, A. Environmental and exergetic impacts of cement production: A case study. *Environ. Prog. Sustain. Energy* **2018**, *37*, 2042–2049. [CrossRef]
15. Payá, J.; Borrachero, M.; Monzó, J.; Peris-Mora, E.; Amahjour, F. Enhanced conductivity measurement techniques for evaluation of fly ash pozzolanic activity. *Cem. Concr. Res.* **2001**, *31*, 41–49. [CrossRef]
16. Gastaldini, A.; Hengen, M.; Gastaldini, M.; Amaral, F.D.; Antolini, M.; Coletto, T. The use of water treatment plant sludge ash as a mineral addition. *Constr. Build. Mater.* **2015**, *94*, 513–520. [CrossRef]
17. Schöler, A.; Lothenbach, B.; Winnefeld, F.; Zajac, M. Hydration of quaternary Portland cement blends containing blast-furnace slag, siliceous fly ash and limestone powder. *Cem. Concr. Compos.* **2015**, *55*, 374–382. [CrossRef]
18. Nicolas, R.S.; Cyr, M.; Escadeillas, G. Characteristics and applications of flash metakaolins. *Appl. Clay. Sci.* **2013**, *83–84*, 253–262. [CrossRef]
19. Zvonarić, M.; Benšić, M.; Barišić, I.; Dokšanović, T. Prediction Models for Mechanical Properties of Cement-Bound Aggregate with Waste Rubber. *Appl. Sci.* **2024**, *14*, 470. [CrossRef]
20. El Inaty, F.; Aydin, B.; Houhou, M.; Marchetti, M.; Quiertant, M.; Metalssi, O.O. Long-Term Effects of External Sulfate Attack on Low-Carbon Cementitious Materials at Early Age. *Appl. Sci.* **2024**, *14*, 2831. [CrossRef]
21. Gil, L.; Mercedes, L.; Mendizabal, V.; Bernat-Maso, E. Shear Performance of the Interface of Sandwich Specimens with Fabric-Reinforced Cementitious Matrix Vegetal Fabric Skins. *Appl. Sci.* **2024**, *14*, 883. [CrossRef]

Disclaimer/Publisher's Note: The statements, opinions and data contained in all publications are solely those of the individual author(s) and contributor(s) and not of MDPI and/or the editor(s). MDPI and/or the editor(s) disclaim responsibility for any injury to people or property resulting from any ideas, methods, instructions or products referred to in the content.

Article

The Synthesis and Characterization of Geopolymers Based on Metakaolin and on Automotive Glass Waste

Ivana Perná *[ ], Martina Havelcová, Monika Šupová [ ], Margit Žaloudková and Olga Bičáková [ ]

Institute of Rock Structure and Mechanics, Czech Academy of Sciences, V Holešovičkách 41, 18209 Prague, Czech Republic; havelcova@irsm.cas.cz (M.H.); supova@irsm.cas.cz (M.Š.); zaloudkova@irsm.cas.cz (M.Ž.); bicakova@irsm.cas.cz (O.B.)
* Correspondence: perna@irsm.cas.cz; Tel.: +420-266-009-253

Abstract: The presented article studies a metakaolin-based geopolymer matrix for which two types of non-recyclable automotive glass waste (AGW) have been used as an alternative aggregate. Their composition and character, as well as their influence on the properties and structure of geopolymer composites (AGW-Gs), have been investigated by means of X-ray fluorescence and X-ray diffraction analyses, scanning electron microscopy, Fourier transform infrared spectrometry and gas chromatography/mass spectrometry. Infrared analysis has proven that the use of AGW does not affect the formation of geopolymer bonds. GC/MS analysis has revealed the presence of triethylene glycol bis(2-ethylhexanoate) in AGW and geopolymers, whose concentration varied according to the size of the fractions used. Preliminary compressive-strength tests have shown the promising potential of AGW-Gs. From the presented results, based on the study of two types of automotive glass waste, it is possible to assume that automotive glass will generally behave in the same or a similar manner in metakaolin-based geopolymer matrices and can be considered as potential alternative aggregates. The result is promising for the current search for new sources of raw materials, for ensuring resource security, for the promotion of sustainability and innovation and for meeting the needs of the growing world population while reducing dependence on limited resources.

Keywords: geopolymer; automotive glass waste; alternative aggregate; characterization

Citation: Perná, I.; Havelcová, M.; Šupová, M.; Žaloudková, M.; Bičáková, O. The Synthesis and Characterization of Geopolymers Based on Metakaolin and on Automotive Glass Waste. *Appl. Sci.* **2024**, *14*, 3439. https://doi.org/10.3390/app14083439

Academic Editors: Mouhamadou Amar and Nor Edine Abriak

Received: 14 March 2024
Revised: 11 April 2024
Accepted: 16 April 2024
Published: 18 April 2024

Copyright: © 2024 by the authors. Licensee MDPI, Basel, Switzerland. This article is an open access article distributed under the terms and conditions of the Creative Commons Attribution (CC BY) license (https:// creativecommons.org/licenses/by/ 4.0/).

1. Introduction

In the European Union (EU), more than 250,000 tons of glass waste (GW) per month were produced in 2021 [1]. Non-recyclable or hardly recyclable glass, such as automotive, TV-screen, monitor, mirror and solar-panel (photovoltaic) glass, is a type of waste material related to developments in a wide range of industries. Currently, the EU environmental policy is focused on waste recycling, although the process of waste recovery is sometimes complicated in some cases. The main objectives are to reduce landfill waste and increase recycling [1–3]. The inseparable additives that impede common glass recycling include resin protective films or tint foils (automotive glass), luminescence substances (TV-screen and monitor glass) and metal coatings (mirror, TV-screen, monitor and solar-panel glass) [4,5]. For example, automotive glass (usually windshield) could contain tint or acoustic foils or safety films (polyester—PE, polycarbonate—PC, polyvinyl butyral—PVB); in some cases, it also has heating elements incorporated (copper (Cu), silver (Ag) and contains conductive foils from metal oxides such as $BaSnO_3$, TiO_2, SnO_2, ZnO and ZrO_2) [4–6].

The eco-friendly solution concerning the reduction, reuse and recycling of non-recyclable or hardly recyclable GW is the use of geopolymer materials [7–12]. Geopolymers have received significant attention for their potential environmental benefits, including the possibility of using industrial by-products, their lower carbon-dioxide emissions during production and their increased resistance to certain types of degradation (e.g., chemical attack, thermal stress, freeze–thaw cycles) [13–18].

Geopolymers are typically formed through the chemical reaction of aluminosilicate materials with an alkaline activator (a solution of sodium, potassium or calcium silicates) [10,19]. A prerequisite is the presence of tetra- or penta-coordinated aluminum in the aluminosilicate precursor [7,19]. The most extensively utilized and well-examined aluminosilicate precursors are metakaolin [20,21], fly ash [22], different types of slag [23,24] and various secondary or waste materials [10,11,25]. The precursor used significantly impacts the resulting microstructure, and obtaining the typical microstructure of the geopolymer is also associated with a precursor with a low calcium content [26–28].

The properties of geopolymers, including mechanical properties, durability, porosity, thermal insulation and others, can easily be modified by judicious selection of the matrix and/or aggregate used [11,29–31]. In addition, geopolymers are capable of stabilizing many hazardous substances, including heavy metals or organics, in their structure [25,32,33]. On the other hand, the addition of organic additives to hybrid geopolymer-organic composites can help to improve their flexibility, tensile strength and overall durability [34].

Glass waste, a byproduct of various industries and municipal recycling programs, can be incorporated into geopolymer formulations as a partial or complete replacement for traditional precursors or aggregates [35,36]. Moreover, GW can be used in the preparation of alkaline activators [36,37].

The finely ground glass particles contribute supplementary sources of silica and alumina, enhancing the geopolymerization process. It has been demonstrated that GW serves as a feasible alternative to the commercial sodium silicate hydrates (commonly known as water glass) typically utilized in the activation of aluminosilicate materials like fly ash for geopolymer preparation [38]. The study by Tchakouté et al. has proven that sodium water glass derived from glass waste and rice husk ash can serve as viable alternative alkaline solutions in the production of metakaolin-based geopolymer binders [39]. Puertas and Torres-Carrasco have reached similar conclusions regarding alkali-activated slag (AAS) [40]. They found that solutions obtained by processing GW lead to the formation of compounds and microstructures similar to those observed for AAS prepared using water glass [40].

In the preparation of geopolymers based only on waste glass, it has been found that waste glass is suitable as a precursor because it produces geopolymer materials with suitable mechanical strength, but whose development is dependent on the curing conditions, namely relative humidity [38]. In the case of metakaolin-based geopolymer, El-Naggar and El-Dessouky have determined that substituting 3% of metakaolin with GW (finer than 38 µm) resulted in a 2% enhancement in the 28-day compressive strength of the reference sample, reaching 82.36 MPa [41]. However, additional incorporation of GW had a negative impact on strength [41].

Fine glass waste was confirmed by Hajimohammadi et al. as a suitable alternative to fine sand for use in fly ash/slag-based geopolymer concrete because fine glass particles increase the alkalinity of the matrix, which promotes a greater range of dissolution and reaction in the vicinity of aggregates [42]. However, it was found by Tahwia et al. that when GW was used as a partial replacement for fine sand aggregate, on the one hand, the flowability of the mixture improved with the glass content, but on the other hand, when 22.5% natural sand was replaced by waste glass, a slight decrease in strength from 126 MPa to 121 MPa was observed [43]. The research revealed that AGW can serve as a replacement for fine aggregate in the production of fly ash-based geopolymer mortar, offering outstanding thermal insulation and fire protection characteristics [44].

The utilization of waste glass as a coarse aggregate has been explored in research such as that conducted by Kuri et al., who investigated its impact on the properties of both portland cement (PC) concrete and geopolymer concrete [45]. It was observed that the 28-day compressive and tensile strengths of both PC concrete and geopolymer concrete decrease with an increase in the percentage of recycled glass coarse aggregate [45]. In contrast, a study by Srivastava et al. has demonstrated that GW can be effectively used as a replacement for coarse aggregate in concrete (up to 50%) without a significant change in strength [46].

Although it seems that the use of glass waste for the preparation of geopolymer materials has already been investigated, studies have mainly focused on the use of GW in the powdered state. Automotive glass waste is non- or hardly recyclable and, in contrast to ordinary broken glass, contains safety films and, in some cases, heating elements [47]. AGW materials have only been tentatively tested as a resource for silica in the production of ferrosilicon [48,49], silicon carbide [50] or ceramic [51]. However, there has been almost no study of automotive glass waste in the context of geopolymer materials. The present paper deals with the innovative use of automotive glass waste (AGW) as an aggregate in a metakaolin-based geopolymer matrix, and the characterization of leachable organic substances from AGW, and their behavior concerning the particle size applied. Two different AGW materials have been characterized, and the influence of alkaline solutions on pure glass waste has been examined. The effect of AGW in different granulometric fractions on the properties and microstructure of geopolymer composites (AGW-Gs) has been systematically investigated. The quality of the aluminosilicate network in geopolymer solids has been verified by Fourier-transform infrared spectrometry (FTIR). The release of organics from AGW and subsequently from AGW-G has been monitored by gas chromatography/mass spectrometry (GC/MS). The study has been complemented by scanning electron microscopy (SEM) images identifying the AGW, its microstructure affected by alkaline solutions and the AGW-G structure. Additionally, the indicative compressive strength has been measured.

2. Materials and Methods

2.1. Materials

The clay-based geopolymer matrix was prepared from the clay material, industrially supplied by ČLUZ a.s. (Nové Strašecí, Czech Republic) under the trade name L05, an aqueous solution of potassium silicate (Vodní sklo, a.s., Prague, Czech Republic) and potassium hydroxide (Penta, Prague, Czech Republic). The clay material L05 was calcined at 750 °C for four hours before being used to obtain metakaolin. Based on X-ray diffraction (XRD) analysis (not presented), the clay material L05 is primarily composed of amorphous phases, with quartz (SiO_2) and anatase (TiO_2) present in notable quantities, alongside minor amounts of muscovite ($KAl_2(AlSi_3O_{10})(OH)_2$) and hematite ($Fe_2O_3$). The chemical composition determined by X-ray fluorescence (XRF) analysis, including a loss on ignition (LOI), and the particle size distribution of the clay material are listed in Tables 1 and 2, respectively.

Table 1. The chemical composition of the clay material used.

Oxides	SiO_2	Al_2O_3	MgO	CaO	K_2O	Fe_2O_3	LOI	Residues
L05	50.28	41.99	<0.02	0.14	0.9	1.03	3.65	2.01

Table 2. The particle size distribution of the clay material used (% of total).

Particle Size	2	5	8	10 [μm]	15	25	d10	d50 [μm]	d90
L05	27.03	57.98	77.70	88.49	99.14	100.00	1.00	3.94	10.43

Automotive glass waste from two different car windshields was used as an aggregate. The first glass (AGW1), contained only a safety film, and the second one (AGW2) additionally contained heating elements (Figure S1 of the Supplementary Files). The collected car windshields were cut into smaller pieces with a chopping saw (Figure S2 of the Supplementary Files), which were further fragmented into finer particles using a jaw crusher. A representative sample of crushed AGW (<0.2 mm) was taken from the obtained glass fragments by quartering for subsequent analyses. The residue was separated by sieving into different fractions, which were evaluated in terms of particle-size distribution.

2.2. Geopolymer Sample Preparation

The geopolymer matrix was prepared by mixing the calcined L05 (100 g) with an alkaline solution (100 g; molar rates: $SiO_2/Al_2O_3 = 2.62$, $K_2O/SiO_2 = 0.19$ and $H_2O/K_2O = 12.43$) for 20 min using a shaft mixer. The resulting material was transferred to a planetary mixer, where, after the addition of a defined fraction of crushed AGW, the mixture was mixed for another five minutes. The amount of AGW materials incorporated was 40 g per 100 g of the geopolymer matrix. The geopolymer mixture was poured into molds, vibrated on a vibrating table to remove air bubbles and covered until the next day to prevent water evaporation. After demolding, the geopolymer samples were stored in a plastic bag for 28 days and then left uncovered under laboratory conditions.

In order to be able to compare the influence of the different AGW fractions on the behavior of leachable organic substances in geopolymer solids, the ratio of AGW to the geopolymer matrix was kept at 40:100 (35 wt.% of AGW per dry L05) for all the samples prepared (AGW-G), even though the filling was not exactly ideal. In the case of larger fractions, the mixture was fluid, and there was partial sedimentation of the glass waste. Concerning smaller fractions, the consistency of the mixture was thicker to slightly pasty.

2.3. Methods

Chemical analyses of pure metakaolin, AGWs and AGW-G composites were performed using a Spectro IQ X-ray fluorescence (XRF) analyzer (SPECTRO Analytical Instruments GmbH., Kleve, Germany), which allows the analysis of major and trace elements of solid, powder and liquid materials. This instrument has a target made of palladium; the target angle was 90° from the central beam, and the focal size was 1 mm × 1 mm. The measurements were carried out in an inert helium atmosphere. The data obtained were evaluated using the computer program XLabPro, which enables automatic recalculation of the elemental content in the sample to the oxide form. The tested specimens were produced by the pressed-pellet method: 4.0 g of the material (particle size: 15–20 μm) were mixed for 10 min with 0.9 g of the binder (HWC Hoechst wax, FLUXANA® GmbH & Co., KG, Bedburg-Hau, Germany). The compaction pressure was 80 kN.

The samples were analyzed by X-ray diffraction (XRD) on a Bruker D8 Advance powder X-ray diffractometer in a Bragg–Brentano geometry using a Lynx Eye XE detector and CuKα radiation (Bruker AXS, Karlsruhe, Germany). The powder sample was mounted on a planar substrate (diffraction-free silicon). Diffraction was recorded in the angular range of 4–80° 2Θ with a step of 0.015° and a readout time of 0.8 s per step. The acquired X-ray diffraction patterns were qualitatively evaluated in the Diffrac.Eva 4.1. software (Bruker AXS, Karlsruhe, Germany, 2015) using the ICDD PDF-2 database (ICDD 2018).

The organic substances in the AGW and subsequently in the AGW-G composites were determined on gas chromatography/mass spectrometry (GC/MS) equipment, consisting of a Trace 1310 gas chromatograph coupled with an ISQ single quadrupole system (ThermoScientific, Waltham, MA, USA) equipped with a CP5 capillary column (30 m × 0.25 mm × 25 μm). Data were acquired and integrated using the Chromeleon system (ThermoScientific, Waltham, MA, USA). Dichloromethane and methanol (HPLC grade) were purchased from Sigma-Aldrich (St. Luis, MO, USA), and ethylene glycol from Penta (Prague, Czech Republic).

The ground samples (<0.2 mm) were extracted with a mixture of dichloromethane and methanol (93:7, v/v) using an accelerated solvent extractor (ASE 150, Dionex, Sunnyvale, CA, USA). The extracts were filtered through a glass-fiber filter and concentrated using a Christ RVC 2–18 rotator evaporator. The total extracts were re-dissolved in 1.0 mL of dichloromethane/methanol (97:3) and analyzed.

GC/MS analysis was performed using helium as the carrier gas with a constant flow rate of 1.5 mL/min. The extract solution (1 μL) was injected into the GC system in splitless mode for each analysis. The injection port and detector temperatures were set at 250 °C; the oven temperature program started at 40 °C and was maintained for 1 min, after which it increased to 120 °C at 15 °C/min, then to 250 °C at 6 °C/min and finally to 300 °C at 12 °C/min, being maintained for 5 min. The analyte detected was identified by comparing the spectrum with the NIST mass spectral library and data from the literature. Each sample was analyzed three times, and the mean and the relative standard deviation (RSD) were calculated.

The compound identified was quantified using ethylene glycol as a standard after its derivatization with bis(trimethylsilyl)trifluoroacetamide (BSTFA) in pyridine. Ethylene–glycol stock solutions were prepared in dichloromethane to achieve the five concentration levels of the calibrators: 1, 5, 10, 25 and 100 µg/mL. The derivatization reagent (BSTFA, 25 µL) and the catalytic reagent (pyridine, 25 µL) were added to an aliquot (200 µL) of the ethylene–glycol solutions in a vial, and the mixture was placed in a laboratory dryer at 40 °C for 20 min. After cooling to room temperature and the evaporation of excessive BSTFA, the resulting trimethylsilyl derivatives were diluted in dichloromethane (50 µL) and analyzed by GC/MS.

The effect of the alkaline activator on the waste glass was determined only for selected fractions of both types of AGW (0.16–0.25 mm, 0.4–0.63 mm, 1.25–2 mm and 2–5 mm), which were exposed to the alkaline solution in a ratio of 1:10 (glass:alkaline activator). The samples were stirred on a shaker once a day for 1 h. After 28 days, the waste glass was filtered, washed with demi-water and dried to a constant weight at 80 °C.

Infrared (FTIR) spectra were measured by an iS50 spectrometer (Thermo Nicolet Instruments Co., Madison, WI, USA) using the reflection method in the attenuated total reflection (ATR) mode (diamond crystal) in the spectral range of 400–4000 cm^{-1} with a resolution of 4 cm^{-1}, averaging 32 scans. All samples were scanned directly as observed (in powder form with variable grain size). The resulting spectra were processed using OMNIC 9 software.

The thermal stability of the AGW samples was determined by thermogravimetric (TG) analysis using a SETARAM-Setsys Evolution 18 thermal analyzer (SETARAM Instrumentation, Caluire-et-Cuire, France) with an Omnistar GSD 320 O3 mass spectrometer (Pfeiffer Vacuum Austria GmbH, Wien, Austria) for gas analysis (1–300 amu).

The microstructure of AGW and AGW-G composites was studied on a STEM Apreo S LoVac scanning electron microscope (ThermoFisher Scientific, Waltham, MA, USA) equipped with an ETD detector (Everhart-Thornley SED) in the high-vacuum mode. AGW fragments were adjusted to an aluminum stab using a carbon target, plastic conductive-carbon cement and platinum coated in Ar atmosphere on a Leica EM ACE600 sputter coater (Specion s.r.o., Prague, Czech Republic). The coating thickness was 16.01 nm. Geopolymer composite samples were polished using standard materialographic methods prior to mounting and coating, with the coating thickness being 10.85 nm. The SE (secondary-electron) mode with an acceleration voltage of 5.00 kV at magnifications of 200×–10,000× was used to observe detailed surface information in Standard and OptiPlan use.

Compressive strength was determined on a test press (Matest, Arcore, Italy) according to the EN-196-1:2016 standard [52] on 4 × 4 × 4 cm specimens. The resulting values are the average of three measurements.

3. Results and Discussion

3.1. The Characterization of Automotive Glass Waste

3.1.1. Sieve Analysis

The sieve analysis (Table 3) has shown the predominance of the 2–5 mm fraction for both types of automotive glass (31.15 wt.% for AGW1 and 34.87 wt.% for AGW2), followed by the 1.25–2.0 mm fraction. The initial assumption that the glass shard would contain a high proportion of fine particles has not been confirmed for either glass examined, as the particle contents below 0.063 mm are only 3.97 wt.% for AGW1 and 2.28 wt.% for AGW2. The photographs of AGW1 and AGW2 glass fractions are presented as Figures S3 and S4 in the Supplementary Files, respectively.

To assess the impact of individual fractions, specific fractions were singled out for analysis. Priority was given to fractions with the predominant content, namely those in the range of 2–5 mm and 1.25–2 mm. Subsequently, the 0.4–0.63 mm fraction, ranking third in content for AGW1, was included. Finally, the 0.16–0.25 mm fraction was selected to explore the effect of smaller particles despite its lower content. Fractions below 0.16 mm were not used for further experiments due to the workability of the geopolymer mixture.

Table 3. The sieve analyses of the crushed AGW1 and AGW2.

Fraction (mm)	AGW1 (wt.%)	AGW2 (wt.%)
>5	1.85	4.37
2–5	31.15	34.87
1.25–2	18.76	16.20
0.8–1.25	8.56	9.82
0.63–0.8	6.19	5.97
0.4–0.63	9.21	8.52
0.25–0.4	7.48	8.27
0.16–0.25	5.37	4.27
0.063–0.16	7.45	5.41
<0.063	3.97	2.28
Sum	99.99	99.98

3.1.2. X-ray Fluorescence Analysis

The results of the chemical (XRF) analyses of AGW samples (Table 4) are almost identical in the main glass-forming oxides (Na_2O, SiO_2, MgO, CaO, K_2O, etc.). The contents of Na_2O, MgO, CaO, K_2O, etc. are only marginally higher in AGW2 than in AGW1. In contrast, the content of SiO_2 is slightly higher in AGW1 than in AGW2. The samples differ in Fe_2O_3 content, which is higher in AGW1 than in AGW2. Another difference is evident for Cr_2O_3, whose value for AGW2 is almost twice as high as for AGW1. In the case of Bi, AGW2 contains much more of it than AGW1.

Table 4. The chemical analysis of AGW samples (the main oxides in wt.%).

Oxides	Na_2O	MgO	Al_2O_3	SiO_2	SO_3	K_2O	CaO	Cr_2O_3	Fe_2O_3	Bi	LOI	Residues
AGW1	10.21	3.14	0.30	71.28	0.32	0.12	13.15	0.06	1.24	0.0008	0.02	0.19
AGW2	10.74	4.12	0.50	69.57	0.26	0.47	13.52	0.15	0.18	0.22	0.14	0.13

3.1.3. X-ray Diffraction Analysis

The results of the XRD analyses of AGW1 and AGW2 (Figure 1) have confirmed the expected amorphous character of the glass with only a minor content of crystalline quartz (Q; PDF00-001-0649). The curves of both samples are very similar, with no visible phases coming from the safety film or the heating elements.

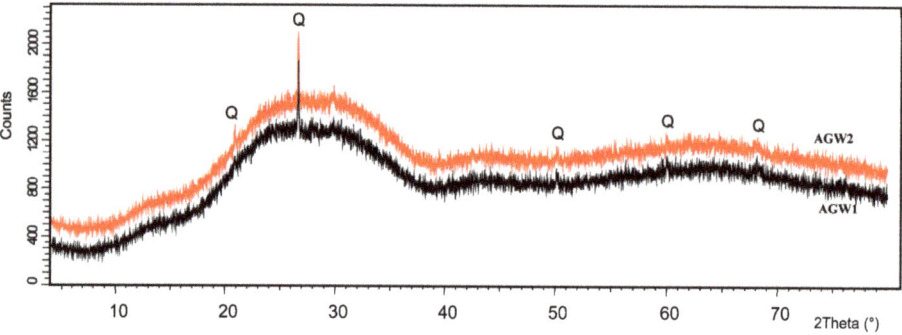

Figure 1. The XRD analyses of AGW1 and AGW2.

3.1.4. Thermogravimetric Analysis

AGW samples were further characterized by thermogravimetric analysis (Figure 2). Glass samples of 21–24 mg were heated in an inert helium atmosphere (at a flow rate of 20 mL·min^{-1}) to prevent oxidation of the samples and to avoid affecting their quality. The change in weight loss was monitored by heating the samples at a constant heating rate of

10 °C min^{-1} from 25 °C to 1000 °C, which is sufficient to determine the volatile content of the finely crushed glasses. The glass contains a safety layer, which is typically polyvinyl butyral (PVB), whose main degradation starts at temperatures in the range of 260–330 °C [47,49,53]. This decrease can be divided into two phases [54]. In the first phase (up to 200 °C), moisture evaporation occurs, and in the second phase (200–900 °C), organic decomposition occurs, which is consistent with our results. These mass losses are seen from the TG curves in Figure 2, where moisture evaporation occurs with the first temperature increase up to 150 °C and amounts to 0.16 wt.% (AGW1) and 0.21 wt.% (AGW2). The subsequent decrease in signal in the temperature range of 150–650 °C is related to the gradual release of organic matter (e.g., polycarbonate, polyethylene terephthalate, etc.). This resulted in the release of gaseous molecules CO, CO_2, CH_4 and H_2O from the hydroxyl groups bound to silica. The loss of organics was monitored up to a temperature of 1000 °C, which was sufficient to confirm that no undesirable substances were released.

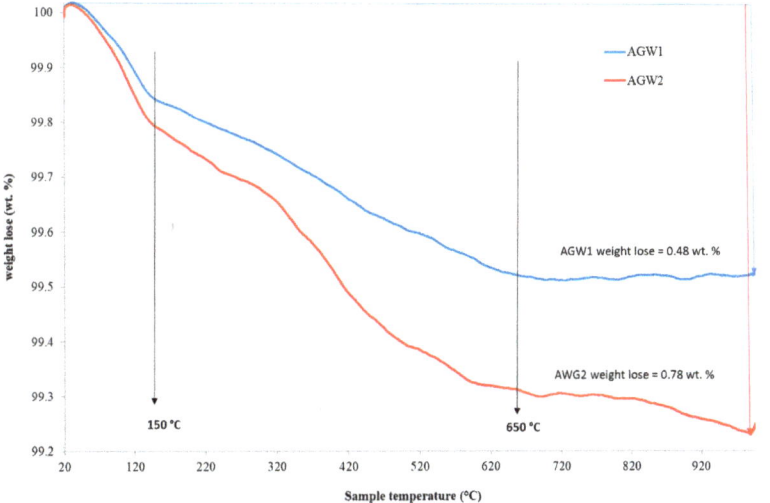

Figure 2. The TG curves of AGW1 and AGW2.

Sample AGW1 commenced melting at 680 °C, whereas sample AGW2 initiated melting at 650 °C. The minor decrease in the AWG2 curve above 650 °C suggests that the final temperature of 1000 °C was inadequate to fully decompose this glass, unlike the AGW1 sample, where there was no decrease in the curve, indicating adequacy. This contrast was evident in the overall weight loss of 0.5% for AGW1 and 0.8% for AGW2. All the resulting melts exhibited coloring.

3.1.5. Gas Chromatography/Mass Spectrometry Analysis

GC/MS was applied for the identification of organic substances in both types of AGW and selected fractions (0.16–0.25 mm, 0.4–0.63 mm, 1.25–2 mm and 2–5 mm). The extraction yields were low (0.01–0.04 wt.%), corresponding to the inorganic nature of the samples. The extracts exhibited discoloration of the solutions, which varied according to the type of AGW (Figure S5 of the Supplementary Files). In all AGW1 and AGW2 samples, triethylene glycol bis(2-ethylhexanoate) was identified (formula: $C_{22}H_{42}O_6$, MW: 402).

This compound serves as a plasticizer for various industrial materials, including polyvinyl butyral (PVB), which is widely used in glass lamination, especially in automotive applications [53]. In the case of windshields, the plasticizer triethylene glycol bis(2-ethylhexanoate) gives the film formability and flexibility and makes it possible to balance mechanical strength and elasticity, almost without affecting adhesion. The compound is insoluble, resistant to low temperatures and to ultraviolet radiation, has antistatic properties and is not listed as toxic or environmentally dangerous [55]. However, even if the substance is not labeled as hazardous, its release should be monitored and possibly restricted. This precaution is warranted due to documented instances of this substance being identified as a contact allergen [56]. Therefore, the influences of the AGW particle size on the release of triethylene glycol bis(2-ethylhexanoate) were investigated.

The concentrations of triethylene glycol bis(2-ethylhexanoate) in the AGW samples were quantitatively determined, ranging from 0.2 to 24.6 ng·kg^{-1} (Table 5).

Table 5. The results of the quantitative analysis of triethylene glycol bis(2-ethylhexanoate) in the fractions of AGW samples.

Fraction	AGW1 (ng·kg^{-1})	AGW2 (ng·kg^{-1})
0.16–0.25 mm	12.7 ± 0.2	24.6 ± 1.2
0.4–0.63 mm	8.5 ± 0.5	3.6 ± 0.3
1.25–2 mm	11.3 ± 0.8	13.7 ± 0.5
2–5 mm	13.3 ± 0.8	0.2 ± 0.1

The results show that for AGW1, the particle size has no significant effect on the presence of triethylene glycol bis(2-ethylhexanoate) in the samples. The measured values are very similar except for the 0.4–0.63 mm fraction, which is noticeably lower. These values range from 8.5 ng·kg^{-1} for the 0.4–0.63 mm fraction to 13.3 ng·kg^{-1} for the 2–5 mm fraction.

In contrast, a difference between the fractions has been found for AGW2. The contents of triethylene glycol bis(2-ethylhexanoate) in individual particle-size fractions varied. The highest concentration was found in the finest fraction, and the lowest concentration was determined in the coarsest fraction.

Surprisingly, the 0.4–0.63 mm fraction of both types of AGW showed the presence of a small amount of triethylene glycol bis(2-ethylhexanoate).

3.1.6. Scanning Electron Microscope Structure Study

Scanning electron microscopy (SEM) micrographs have been used to display the surfaces of AGW1 and AGW2 samples. Figure 3 shows the state of the ground glass before its exposure to the alkaline solution. Individual images are at the same magnification. It is evident from the pictures that the glass grit is sharp-edged after being crushed. In the larger fractions, it is possible to see traces of crushing as well as crushed-glass dust particles, especially in the case of the AGW2 1.25–2 mm fraction. These crushing residues (traces and dust particles) are more evident at 10,000× magnification (Figure S6 of the Supplementary Files).

Figure 3. An overview of different AGW fractions (200× magnification).

3.2. The Effect of the Alkaline Activator on AGW Materials

The effect of the alkaline activator on the waste glass has been determined for selected fractions of both AGW types (0.16–0.25 mm, 0.4–0.63 mm, 1.25–2 mm and 2–5 mm). The glass waste after exposure to the alkaline solution has been investigated by SEM analysis.

Figure 4 shows the AGW1 particles of different fractions after exposure to the alkaline solution at the same magnification (10,000×). The image reveals surface degradation in the form of oval, almost parallel grooves. The larger the original particles, the greater the degradation. Similar degradation traces have also been found in AGW2 (Figure 5). In comparison with AGW1, the incisions are almost parallel, but wherever they cross, the local damage is deeper. A comparison of the two images shows that AGW2 has more surface deterioration than AGW1, in both the extent and depth of damage.

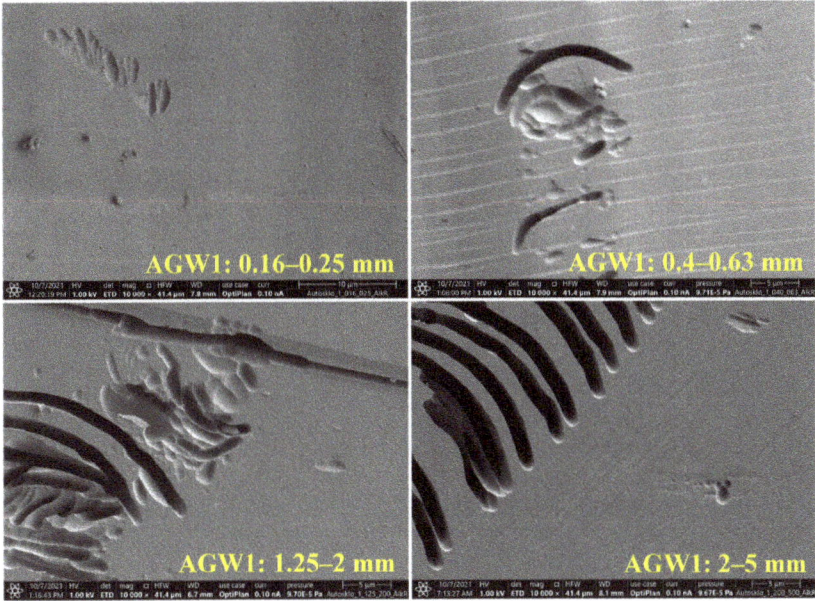

Figure 4. AGW1 particles after exposure to the alkaline solution (10,000× magnification).

The SEM study has confirmed the effect of the alkaline solution, with partial degradation occurring in the areas that were probably disrupted during crushing. However, it cannot be assumed that the same effect will be observed when the glass is embedded in the geopolymer matrix because no free-alkali activator will act on the glass after solidification (6–12 h). To confirm this hypothesis, geopolymer samples were prepared and subsequently analyzed.

3.3. The Characterization of Geopolymer Composites

Geopolymer composites (AGW-Gs) were prepared using the same fractions (0.16–0.25 mm, 0.4–0.63 mm, 1.25–2 mm and 2–5 mm) as those used to monitor the effect of the alkaline solution on AGW particles.

The geopolymer solids prepared (Figures S7 and S8 of the Supplementary Files) were studied mainly in terms of the geopolymer composition, bonds formed (FTIR spectroscopy), the microstructure, the effect of the geopolymer reaction on the glass-waste particles (SEM), and the effect of solidification on the release of organics (GC/MS) as a function of the fraction used. In addition, indicative compressive strength was determined.

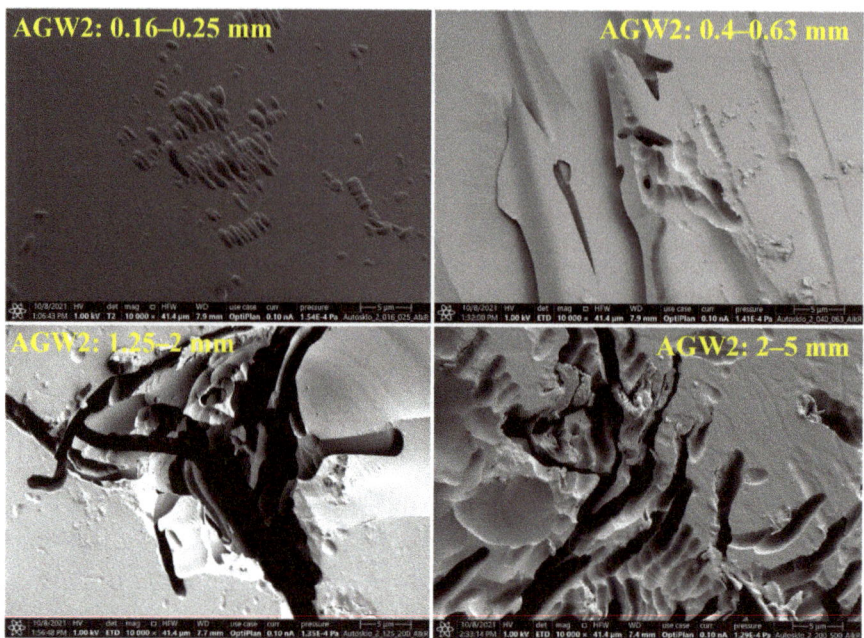

Figure 5. AGW2 particles after exposure to the alkaline solution (10,000× magnification).

3.3.1. X-ray Fluorescence Analysis

The chemical composition of geopolymer samples (Table 6) is very similar and corresponds to the composition of the raw materials from which the geopolymer was prepared (Tables 1 and 4). AGW1-G and AGW2-G samples contain a high percentage of SiO_2 (51.53 wt.% and 50.83 wt.%) and Al_2O_3 (19.88 wt.% and 20.68 wt.%), respectively. The increased concentration of Bi in AGW2-G correlates with the higher content identified in AGW2.

Table 6. The chemical analysis of AGW-G samples (the main oxides in wt.%).

Oxides	Na_2O	MgO	Al_2O_3	SiO_2	K_2O	CaO	Fe_2O_3	Bi	LOI	Residues
AGW1-G	2.04	0.53	19.88	51.53	11.73	3.71	0.96	0.02	8.37	1.23
AGW2-G	1.84	0.65	20.68	50.83	12.28	3.47	0.76	0.04	8.17	1.28

3.3.2. FTIR Analysis

The FTIR spectra of the geopolymer material with variable grain size and a comparison with the clay material and glass waste are provided in Figure 6 for the AGW1-G series and in Figure 7 for the AGW2-G series.

The FTIR spectra of both types of glass waste (AGW1 and AGW2) are identical and contain a broad, intense band in the spectral region of 1240–840 cm^{-1}, corresponding to the Si-O in asymmetric stretching vibrations (ν3) in the SiO_4 tetrahedra. Another band, at 450–470 cm^{-1}, is ascribed to the O-Si-O bending vibrations (ν4) in the SiO_4 tetrahedra [57]. The weak band in the region of 720–840 cm^{-1} is attributed to the Si–O–Si symmetric stretching vibrations of the bridging oxygens between the SiO_4 tetrahedra [58].

The metakaolin-based clay material contains bands attributed to the asymmetric stretching and bending vibrations of Si–O and the asymmetric stretching of Si–O–Al groups, visible at 1087 cm^{-1} and ~780 cm^{-1}. The band with more distinctive features at 799 and 777 cm^{-1} and another at 467 cm^{-1} may be attributed to quartz [57], which is present in L05 as a minor impurity. The spectral shoulder at ~560 cm^{-1} is attributable to silicates and/or aluminosilicates with a long-range structural order [59].

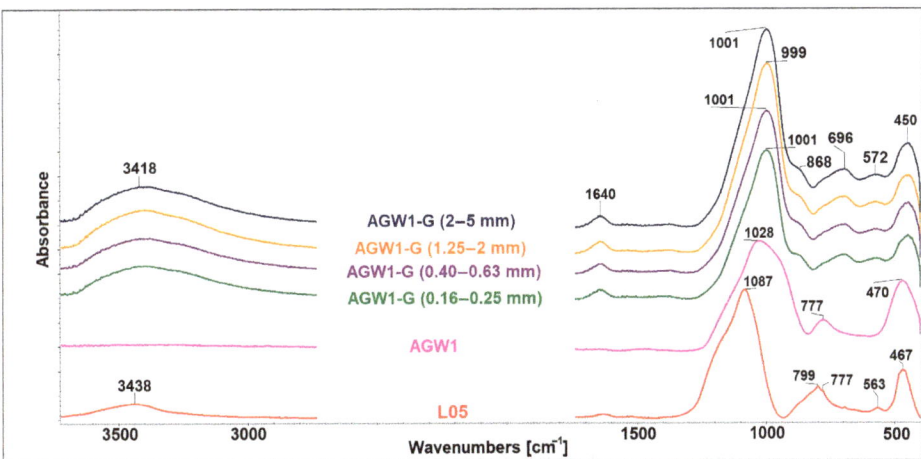

Figure 6. The ATR-FTIR spectra of the geopolymer materials (AGW1-G series with variable grain size) and a comparison with L05 and AGW1.

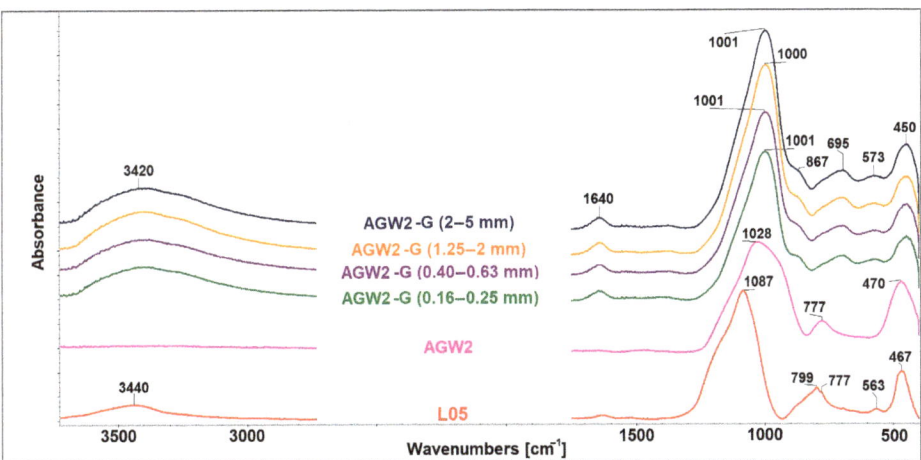

Figure 7. The ATR-FTIR spectra of the geopolymer materials (AGW2-G series with variable grain size) and a comparison with the L05 and AGW2.

The reaction of L05 with an alkaline activator leads to an increase in the substitution of the tetrahedral Al in the silicate network in the geopolymer matrix. This process is manifested by a spectral shift of the principal band, located at 1087 cm^{-1} in the L05 spectrum, to lower wave numbers (~1000 cm^{-1}) in the spectra of the geopolymer materials [60]—see Figures 6 and 7. Slight changes in wavenumber positions in the individual spectra of geopolymer materials (999–1001 cm^{-1}) are within the resolution limit of the method (up to 4 cm^{-1}). New weak bands positioned at around 866–867 cm^{-1} and 695–696 cm^{-1} can be ascribed to the bending vibration mode of Si-OH [61] and to the stretching and bending vibrations of Si-O-Al, respectively, providing another fingerprint for the formation of the geopolymer structure [62].

The spectra of L05 and geopolymer materials also contain a large peak at 3440 and 3420 cm^{-1} and a peak centered at 1640 cm^{-1}, which are associated with the stretching and bending modes of OH groups in water molecules due to the residual water and moisture content in the materials [62].

Spectroscopically, all final geopolymer materials (of both the AGW1-G and AGW2-G series) are identical. FTIR analysis has proven that the grain size of waste glass does not affect its final structural properties or phase changes.

3.3.3. Scanning Electron Microscope Structure Study

The SEM images of the geopolymer composites containing different fractions of AGW1 and AGW2 (AGW1-Gs and AGW2-Gs, respectively) are presented in Figure 8 and Figure 9, respectively. The glass waste is more compact and homogeneous than the geopolymer matrix.

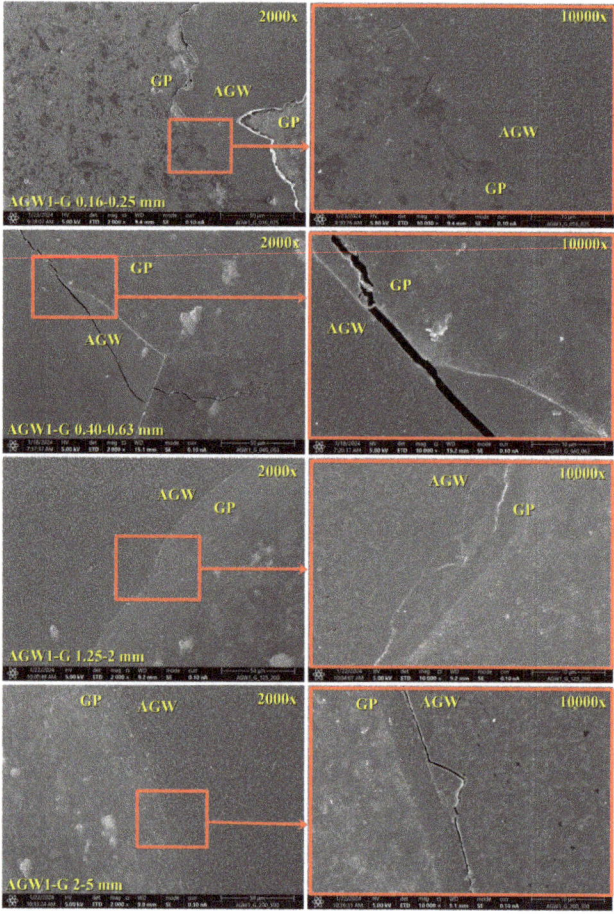

Figure 8. The SEM images of the geopolymer composite with AGW1 particles (AGW1-G): The left side shows the images at lower magnification (2000×), while the right side contains details of the images on the left at higher magnification (10,000×). The images depict the geopolymer matrix (GP) and particles of automotive glass waste (AGW).

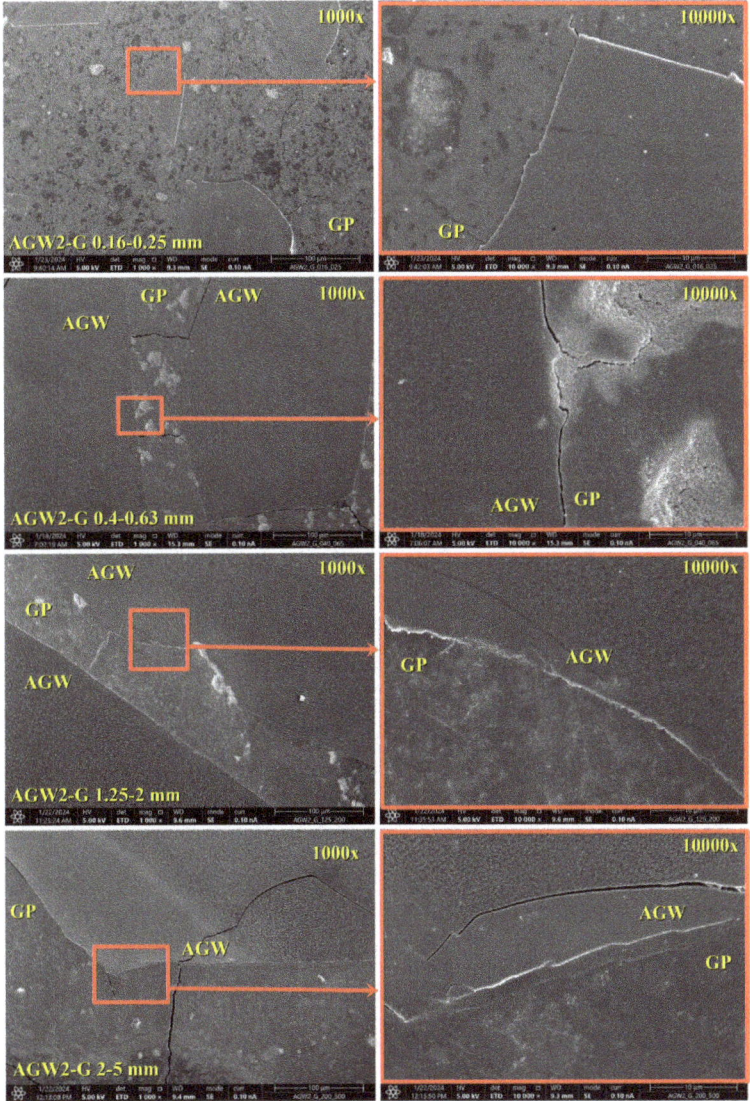

Figure 9. The SEM images of the geopolymer composite with AGW2 particles (AGW2-G): The left side shows the images at lower magnification (1000×), while the right side contains details of the images on the left at higher magnification (10,000×). The images depict the geopolymer matrix (GP) and particles of automotive glass waste (AGW).

In general, the geopolymer composites are very similar according to SEM investigations. No degradation of the glass is evident, as was the case of the glass after exposure to the alkaline solution (Figures 4 and 5). The composites are stable without significant signs of degradation. Only in the case of 0.40–0.63 mm AGW2, a slight surface distortion in the geopolymer is visible.

In most cases, the geopolymer/AGW interface shows good adhesion of the geopolymer to the AGW. In certain images within Figure 8 (0.40–0.63 mm and 2–5 mm AGW1-G), it is evident that microcracks have developed, likely either during the geopolymer maturation process or during the preparation of the polished blocks. The crack runs through the AGW particles, indicating a relatively strong bond between the geopolymer and the glass waste. A similar phenomenon also occurs in the second type of glass waste (Figure 9, 2–5 mm AGW2-G). Microcracks are further visible in the geopolymer matrix even in areas with the predominance of AGW particles, where the geopolymer is almost closed between them, for example 0.40–0.63 mm and 1.25–2 mm AGW1-G. In some composites, there is a visible interlayer, particularly noticeable in the larger particles of both AGWs (1.25–2 mm and 2–5 mm), implying potential ion transfer between the materials. Interlayer formation has also been observed at interfaces between geopolymers and glass particles or other materials [36,63,64].

3.3.4. Gas Chromatography/Mass Spectrometry Analysis

The extraction yields of both AGW-G geopolymer composites were low for all fractions (0.01–0.04 wt.%) and comparable with AGW extraction yields. The GC/MS analysis of the AGW-G extracts confirmed the presence of triethylene glycol bis(2-ethylhexanoate), with the determined concentrations ranging from 0.0 to 6.4 $ng \cdot kg^{-1}$.

The results (Table 7) demonstrate the presence of triethylene glycol bis(2-ethylhexanoate) in both types of geopolymer composites. Like in the case of AGW1, the effect of particle size is not evident for the AGW1-G composite either, with concentration values ranging from 0.7 $ng \cdot kg^{-1}$ for the 0.4–0.63 mm fraction to 4.1 $ng \cdot kg^{-1}$ for the 2–5 mm fraction. In AGW2-G composites, on the other hand, the effect of particle size is visible (with the values ranging from 0.0 $ng \cdot kg^{-1}$ for the 2–5 mm fraction to 6.4 $ng \cdot kg^{-1}$ for the 0.16–0.25 mm fraction), which is consistent with the trend of AGW2 results.

Table 7. The results of the quantitative analysis of triethylene glycol bis(2-ethylhexanoate) in the fractions of AGW-G samples.

Fraction	AGW1-G ($ng \cdot kg^{-1}$)	AGW2-G ($ng \cdot kg^{-1}$)
0.16–0.25 mm	3.1 ± 0.1	6.4 ± 0.2
0.4–0.63 mm	0.7 ± 0.1	0.6 ± 0.2
1.25–2 mm	1.9 ± 0.1	2.5 ± 0.4
2–5 mm	4.1 ± 0.2	0.0 ± 0.1

Nevertheless, it should be considered that the geopolymer composite contains only 35% of AGW. If we take only 35% of the values shown in Table 4 (AGW1—4.4 $ng \cdot kg^{-1}$, 3.0 $ng \cdot kg^{-1}$, 4.0 $ng \cdot kg^{-1}$ and 4.7 $ng \cdot kg^{-1}$ and AGW2—8.6 $ng \cdot kg^{-1}$, 1.3 $ng \cdot kg^{-1}$, 4.8 $ng \cdot kg^{-1}$ and 0.1 $ng \cdot kg^{-1}$ for the fractions of 0.16–0.25 mm, 0.4–0.63 mm, 1.25–2 mm and 2–5 mm, respectively), we obtain values comparable with the results in Table 7.

The evaluation shows that the incorporation of AGW particles into the geopolymer matrix reduces the release of triethylene glycol bis(2-ethylhexanoate) from both types of geopolymer composites, especially in the case of 0.4–0.63 mm particles.

3.3.5. Compressive Strength

For the estimation of mechanical properties, the compressive strength was determined only tentatively at 28 days for a ratio of 100 g of the geopolymer matrix and 40 g of the waste-glass fraction (see Sample Preparation). The results are shown in the graph (Figure 10). The pure geopolymer matrix serves as a blank sample (without aggregates).

The results show that geopolymer composites with the addition of AGW have sufficient mechanical properties. The 28-day compressive-strength values reach up to 81.3 MPa and 82.8 MPa for the finest fractions (0.16–0.25 mm) of AGW1 and AGW2, respectively. There is a noticeable trend for both types of glass waste where the compressive strength decreases with the increasing particle size of the glass waste. A similar decrease in compressive strengths has been observed by several authors, but in their case, AGW particles were applied as a partial replacement for sand aggregate in geopolymer [43,45]. It can be assumed that the filling for other fractions was not optimal, especially for the larger fractions, where sedimentation of the particles occurred. On the other hand, for the fraction of 0.16 to 0.25 mm whose content in the geopolymer matrix is almost optimal, the compressive strengths slightly exceed the strength of the geopolymer matrix (80.6 MPa), which is in agreement with the work of Srivastava et al. and Hajimohammadi et al. [42,46]. By optimizing the AGW content or by using a combination of several AGW fractions, possibly in combination with another type of filler, it would probably be possible to achieve higher values of mechanical properties.

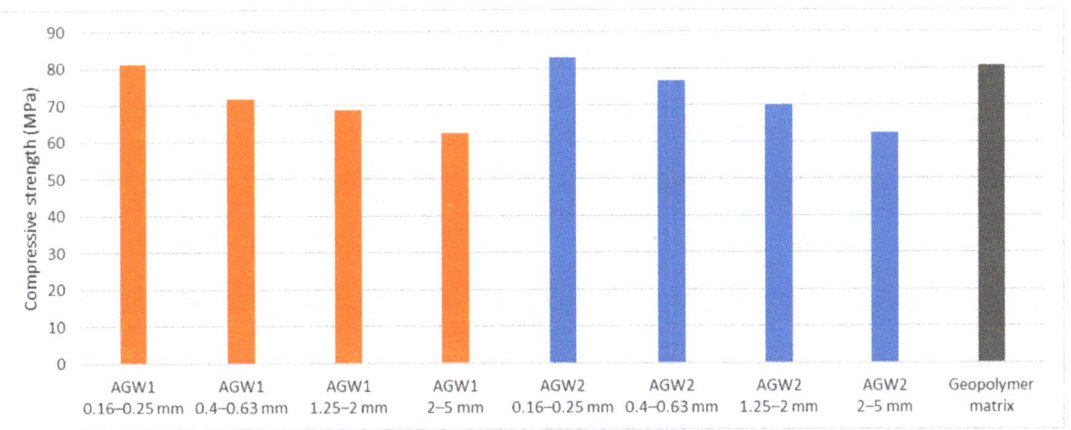

Figure 10. An overview of 28-day compressive-strength values for both AGW types and geopolymer matrix.

The main aim of this paper was to verify the possibility of using AGW as an alternative aggregate in a geopolymer matrix. Due to the composition of AGW, which includes organic safety films and heating elements, it was first necessary to characterize the AGW materials (AGW1 and AGW2) and determine the effect of an alkaline environment. Furthermore, the influence of the fractions used, both of the AGW itself and of the AGW incorporated in the geopolymer matrix, was also investigated.

Direct exposure to an alkaline solution partially degrades AGW particles. However, because of the setting time of the geopolymer matrix (from 2 to 12 h) [65], there is no long-term exposure of AGW particles to alkaline environments. Therefore, the deterioration of these particles is minor. It is possible to assume that this phenomenon is part of the glass particle dissolution process described by several authors [38–40,42] but due to the larger particle size, only partial degradation occurred.

AGW materials have been found to comprise mainly triethylene glycol bis(2-ethylhexanoate), originating from the contained safety films, which may be further released into the environment. GC/MS analysis has confirmed that the addition of AGW to the geopolymer matrix has lowered the release of triethylene glycol bis(2-ethylhexanoate).

The decline in the concentration of triethylene glycol bis(2-ethylhexanoate) in geopolymer composites may stem from the incorporation of this organic substance into the structure of the geopolymer because geopolymers are able to form hybrid bonds with organic substances (hybrid organic–geopolymer materials), as described by Reeb et al. [66]. Another possibility is the degradation within the alkaline environment of the geopolymer matrix [67]. Nevertheless, no byproducts of this degradation have been identified.

The effect of particle size is more evident in AGW2, where the concentration of triethylene glycol bis(2-ethylhexanoate) decreases with increasing particle size in both AGW2 and AGW2-G. The exception is the 0.4–0.63 mm fraction, where the concentrations are lower than would be expected for both types of glass waste.

The determination of the AGW1-G and AGW2-G mechanical properties showed relatively high 28-day compressive-strength values despite the non-optimal ratio of the geopolymer matrix to the aggregates for some fractions. In the case of larger particles, sedimentation occurred, which was evident in the geopolymer solids. It can also be assumed that finely ground AGW could be actively involved in the geopolymer reaction [42,68].

In view of the above, AGW materials can be considered potential alternative aggregates in metakaolin-based geopolymers, leading to a more efficient use of this non-recyclable material.

4. Conclusions

This paper has studied the possibility of using automotive glass waste as an alternative aggregate for geopolymer materials. The results have led to the following conclusions:

The effective utilization of AGW as an alternative aggregate in the production of geopolymer composites has resulted in solid, resilient materials that are insoluble in water.

FTIR analysis has demonstrated that the use of AGW materials does not affect the formation of geopolymer bonds.

It has been confirmed by GC/MS analysis that in the case of AGW1, the particle size used does not affect the release of triethylene glycol bis(2-ethylhexanoate) except for the 0.4–0.63 mm fraction, where its concentration was significantly lower than in the other fractions.

In the case of AGW2, which contains a heating element in addition to the safety film, the release of triethylene glycol bis(2-ethylhexanoate) has been found to be partially dependent on the size of the fractions used, both of the AGW itself and of the AGW incorporated in the geopolymer matrix.

The results have shown that the incorporation of AGW into the geopolymer matrix slightly reduces the release of triethylene glycol bis(2-ethylhexanoate).

Preliminary mechanical-property tests have demonstrated the potential of AGW-G composites. However, further research will be required to optimize the content ratios and fractions of the AGW materials used.

It can thus be assumed that AGW materials can be utilized as alternative aggregates. The use of this waste material (AGW) in geopolymer materials saves primary raw materials, thereby reducing the amount of landfill waste, which is in accordance with the EU environmental strategy.

Supplementary Materials: The following supporting information can be downloaded at: https://www.mdpi.com/article/10.3390/app14083439/s1, Figure S1: Automotive glass waste used (A and B—general and detailed view on windscreen with protective film, C and D—general and detailed view on windscreen with protective film and heat elements); Figure S2: Windscreen cutting (A—general view, B—detailed view); Figure S3: The photographs of AGW1 glass fractions; Figure S4: The photographs of AGW2 glass fractions; Figure S5: Photographs of the extracts prepared for GC analysis; Figure S6. An overview of different AGW fractions (10,000× magnification); Figure S7. Photographs of the AGW1-G solids; Figure S8. Photographs of the AGW2-G solids.

Author Contributions: Conceptualization: I.P.; methodology: I.P., M.H., M.Š., M.Ž. and O.B.; formal analysis: I.P., M.H., M.Š., M.Ž. and O.B.; investigation: I.P., M.H., M.Š., M.Ž. and O.B.; data curation: I.P.; writing—original draft preparation: I.P.; writing—review and editing: I.P., M.H., M.Š., M.Ž. and O.B.; supervision, I.P. All authors have read and agreed to the published version of the manuscript.

Funding: This work has been carried out thanks to the support of the long-term project for the conceptual development of the research organization No. 67985891 and the Strategy AV21, activity of the Czech Academy of Sciences, the research program VP23—City as a Laboratory of Change: Historical Heritage and Place for Safe and Quality Life.

Data Availability Statement: The raw data supporting the conclusions of this article will be made available by the authors on request.

Conflicts of Interest: The authors declare no conflicts of interest.

References

1. Eurostat Statistics Explained. Recycling—Secondary Material Price Indicator. [Online]. Available online: http://ec.europa.eu/eurostat/statistics-explained/index.php/Recycling_%E2%80%93_secondary_material_price_indicator (accessed on 12 January 2024).
2. Blengini, G.A.; Busto, M.; Fantoni, M.; Fino, D. Eco-Efficient Waste Glass Recycling: Integrated Waste Management and Green Product Development through LCA. *Waste Manag.* **2012**, *32*, 1000–1008. [CrossRef] [PubMed]
3. Akinwumi, I.I.; Booth, C.A.; Ojuri, O.O.; Ogbiye, A.S.; Coker, A.O. Containment of Pollution from Urban Waste Disposal Sites. In *Urban Pollution*; Wiley: Hoboken, NJ, USA, 2018; pp. 223–234.
4. Pant, D.; Singh, P. Pollution Due to Hazardous Glass Waste. *Environ. Sci. Pollut. Res.* **2014**, *21*, 2414–2436. [CrossRef] [PubMed]
5. Butler, J.H.; Hooper, P.D. Glass Waste. In *Waste: A Handbook for Management*; Academic Press: Cambridge, MA, USA, 2019; pp. 307–322. [CrossRef]
6. Musgraves, J.D.; Hu, J.; Calvez, L. (Eds.) *Springer Handbook of Glass*; Springer: Cham, Switzerland, 2019.
7. Wu, Y.; Lu, B.; Bai, T.; Wang, H.; Du, F.; Zhang, Y.; Cai, L.; Jiang, C.; Wang, W. Geopolymer, Green Alkali Activated Cementitious Material: Synthesis, Applications and Challenges. *Constr. Build. Mater.* **2019**, *224*, 930–949. [CrossRef]
8. Singh, N.B.; Kumar, M.; Rai, S. Geopolymer Cement and Concrete: Properties. *Mater. Today Proc.* **2019**, *29*, 743–748. [CrossRef]
9. Elahi, M.M.A.; Hossain, M.M.; Karim, M.R.; Zain, M.F.M.; Shearer, C. A Review on Alkali-Activated Binders: Materials Composition and Fresh Properties of Concrete. *Constr. Build. Mater.* **2020**, *260*, 119788. [CrossRef]
10. Amran, Y.H.M.; Alyousef, R.; Alabduljabbar, H.; El-Zeadani, M. Clean Production and Properties of Geopolymer Concrete; A Review. *J. Clean. Prod.* **2020**, *251*, 119679. [CrossRef]
11. Ren, B.; Zhao, Y.; Bai, H.; Kang, S.; Zhang, T.; Song, S. Eco-Friendly Geopolymer Prepared from Solid Wastes: A Critical Review. *Chemosphere* **2021**, *267*, 128900. [CrossRef] [PubMed]
12. Chowdhury, S.; Mohapatra, S.; Gaur, A.; Dwivedi, G.; Soni, A. Study of Various Properties of Geopolymer Concrete—A Review. *Mater. Today Proc.* **2021**, *46*, 5687–5695. [CrossRef]
13. Perná, I.; Novotná, M.; Hanzlíček, T.; Šupová, M.; Řimnáčová, D. Metakaolin-Based Geopolymer Formation and Properties: The Influence of the Maturation Period and Environment (Air, Demineralized and Sea Water). *J. Ind. Eng. Chem.* **2024**, *134*, 415–424. [CrossRef]
14. Reddy, D.V.; Edouard, J.-B.; Sobhan, K. Durability of Fly Ash-Based Geopolymer Structural Concrete in the Marine Environment. *J. Mater. Civ. Eng.* **2013**, *25*, 781–787. [CrossRef]
15. Rashad, A.M. Insulating and Fire-Resistant Behaviour of Metakaolin and Fly Ash Geopolymer Mortars. *Proc. Inst. Civ. Eng.-Constr. Mater.* **2019**, *172*, 37–44. [CrossRef]
16. Bakharev, T. Resistance of Geopolymer Materials to Acid Attack. *Cem. Concr. Res.* **2005**, *35*, 658–670. [CrossRef]
17. Lee, N.K.; Lee, H.K. Influence of the Slag Content on the Chloride and Sulfuric Acid Resistances of Alkali-Activated Fly Ash/Slag Paste. *Cem. Concr. Compos.* **2016**, *72*, 168–179. [CrossRef]
18. Hager, I.; Sitarz, M.; Mróz, K. Fly-Ash Based Geopolymer Mortar for High-Temperature Application—Effect of Slag Addition. *J. Clean. Prod.* **2021**, *316*, 128168. [CrossRef]
19. Davidovits, J. *Geopolymer Chemistry and Applications*, 5th ed.; Institute Geopolymer: Saint-Quentin, France, 2020.
20. Liew, Y.M.; Heah, C.Y.; Mohd Mustafa, A.B.; Kamarudin, H. Structure and Properties of Clay-Based Geopolymer Cements: A Review. *Prog. Mater. Sci.* **2016**, *83*, 595–629. [CrossRef]
21. Rashad, A.M. Alkali-Activated Metakaolin: A Short Guide for Civil Engineer—An Overview. *Constr. Build. Mater.* **2013**, *41*, 751–765. [CrossRef]
22. John, S.K.; Nadir, Y.; Girija, K. Effect of Source Materials, Additives on the Mechanical Properties and Durability of Fly Ash and Fly Ash-Slag Geopolymer Mortar: A Review. *Constr. Build. Mater.* **2021**, *280*, 122443. [CrossRef]
23. Amer, I.; Kohail, M.; El-Feky, M.S.; Rashad, A.; Khalaf, M.A. A Review on Alkali-Activated Slag Concrete. *Ain Shams Eng. J.* **2021**, *12*, 1475–1499. [CrossRef]
24. Perná, I.; Hanzlíček, T. The Solidification of Aluminum Production Waste in Geopolymer Matrix. *J. Clean. Prod.* **2014**, *84*, 657–662. [CrossRef]
25. Perná, I.; Šupová, M.; Hanzlíček, T.; Špaldoňová, A. The Synthesis and Characterization of Geopolymers Based on Metakaolin and High LOI Straw Ash. *Constr. Build. Mater.* **2019**, *228*, 116765. [CrossRef]
26. Ng, C.; Johnson Alengaram, U.; Sing Wong, L.; Hung Mo, K.; Zamin Jumaat, M.; Ramesh, S. A Review on Microstructural Study and Compressive Strength of Geopolymer Mortar, Paste and Concrete. *Constr. Build. Mater.* **2018**, *280*, 550–576. [CrossRef]
27. Duxson, P.; Provis, J.L.; Lukey, G.C.; Mallicoat, S.W.; Kriven, W.M.; van Deventer, J.S. Understanding the Relationship between Geopolymer Composition, Microstructure and Mechanical Properties. *Physicochem. Eng. Asp.* **2005**, *269*, 47–58. [CrossRef]
28. Fu, Q.; Xu, W.; Zhao, X.; Bu, M.; Yuan, Q.; Niu, D. The Microstructure and Durability of Fly Ash-Based Geopolymer Concrete: A Review. *Ceram. Int.* **2021**, *47*, 29550–29566. [CrossRef]
29. Parathi, S.; Nagarajan, P.; Pallikkara, S.A. Ecofriendly Geopolymer Concrete: A Comprehensive Review. *Clean Technol. Environ. Policy* **2021**, *23*, 1701–1713. [CrossRef]
30. Bai, T.; Song, Z.; Wang, H.; Wu, Y.; Huang, W. Performance Evaluation of Metakaolin Geopolymer Modified by Different Solid Wastes. *J. Clean. Prod.* **2019**, *226*, 114–121. [CrossRef]
31. Novotná, M.; Perná, I.; Hanzlíček, T. Review of Possible Fillers and Additives for Geopolymer Materials. *Waste Forum* **2020**, *2*, 78–89.

32. Tian, Q.; Bai, Y.; Pan, Y.; Chen, C.; Yao, S.; Sasaki, K.; Zhang, H. Application of Geopolymer in Stabilization/Solidification of Hazardous Pollutants: A Review. *Molecules* **2022**, *27*, 4570. [CrossRef] [PubMed]
33. Rasaki, S.A.; Bingxue, Z.; Guarecuco, R.; Thomas, T.; Minghui, Y. Geopolymer for Use in Heavy Metals Adsorption, and Advanced Oxidative Processes: A Critical Review. *J. Clean. Prod.* **2019**, *213*, 42–58. [CrossRef]
34. Khater, H.M.; El Naggar, A. Combination between Organic Polymer and Geopolymer for Production of Eco-Friendly Metakaolin Composite. *J. Aust. Ceram. Soc.* **2020**, *56*, 599–608. [CrossRef]
35. Siddika, A.; Hajimohammadi, A.; Al Mamun, M.A.; Alyousef, R.; Ferdous, W. Waste Glass in Cement and Geopolymer Concretes: A Review on Durability and Challenges. *Polymers* **2021**, *13*, 2071. [CrossRef]
36. Rios, L.M.H.; Hoyos Triviño, A.F.; Villaquirán-Caicedo, M.A.; Mejía De Gutiérrez, R. Effect of the Use of Waste Glass (as Precursor, and Alkali Activator) in the Manufacture of Geopolymer Rendering Mortars and Architectural Tiles. *Constr. Build. Mater.* **2023**, *363*, 129760. [CrossRef]
37. Fouad Alnahhal, M.; Kim, T.; Hajimohammadi, A. Waste-Derived Activators for Alkali-Activated Materials: A Review. *Cem. Concr. Compos.* **2021**, *118*, 103980. [CrossRef]
38. Torres-Carrasco, M.; Puertas, F. Waste Glass as a Precursor in Alkaline Activation: Chemical Process and Hydration Products. *Constr. Build. Mater.* **2017**, *139*, 342–354. [CrossRef]
39. Tchakouté, H.K.; Rüscher, C.H.; Kong, S.; Kamseu, E.; Leonelli, C. Geopolymer Binders from Metakaolin Using Sodium Waterglass from Waste Glass and Rice Husk Ash as Alternative Activators: A Comparative Study. *Constr. Build. Mater.* **2016**, *114*, 276–289. [CrossRef]
40. Puertas, F.; Torres-Carrasco, M. Use of Glass Waste as an Activator in the Preparation of Alkali-Activated Slag. Mechanical Strength and Paste Characterisation. *Cem. Concr. Res.* **2013**, *57*, 95–104. [CrossRef]
41. El-Naggar, M.R.; El-Dessouky, M.I. Re-Use of Waste Glass in Improving Properties of Metakaolin-Based Geopolymers: Mechanical and Microstructure Examinations. *Constr. Build. Mater.* **2016**, *132*, 543–555. [CrossRef]
42. Hajimohammadi, A.; Ngo, T.; Kashani, A. Glass Waste versus Sand as Aggregates: The Characteristics of the Evolving Geopolymer Binders. *J. Clean. Prod.* **2018**, *193*, 593–603. [CrossRef]
43. Tahwia, A.M.; Heniegal, A.M.; Abdellatief, M.; Tayeh, B.A.; Elrahman, M.A. Properties of Ultra-High Performance Geopolymer Concrete Incorporating Recycled Waste Glass. *Case Stud. Constr. Mater.* **2022**, *17*, e01393. [CrossRef]
44. Chindaprasirt, P.; Lao-un, J.; Zaetang, Y.; Wongkvanklom, A.; Phoo-ngernkham, T.; Wongsa, A.; Sata, V. Thermal Insulating and Fire Resistance Performances of Geopolymer Mortar Containing Auto Glass Waste as Fine Aggregate. *J. Build. Eng.* **2022**, *60*, 105178. [CrossRef]
45. Kuri, J.C.; Hosan, A.; Uddin, F.; Shaikh, A.; Biswas, W.K. The Effect of Recycled Waste Glass as a Coarse Aggregate on the Properties of Portland Cement Concrete and Geopolymer Concrete. *Buildings* **2023**, *13*, 586. [CrossRef]
46. Srivastava, V.; Gautam, S.P.; Agarwal, V.C.; Mehta, P.K. Glass Wastes as Coarse Aggregate in Concrete. *J. Environ. Nanotechnol.* **2014**, *3*, 2319–5541. [CrossRef]
47. Swain, B.; Ryang Park, J.; Yoon Shin, D.; Park, K.S.; Hwan Hong, M.; Gi Lee, C. Recycling of Waste Automotive Laminated Glass and Valorization of Polyvinyl Butyral through Mechanochemical Separation. *Environ. Res.* **2015**, *142*, 615–623. [CrossRef] [PubMed]
48. Farzana, R.; Rajarao, R.; Sahajwalla, V. Characteristics of Waste Automotive Glasses as Silica Resource in Ferrosilicon Synthesis. *Waste Manag. Res.* **2015**, *34*, 113–121. [CrossRef]
49. Farzana, R.; Rajarao, R.; Sahajwalla, V. Synthesis of Ferrosilicon Alloy Using Waste Glass and Plastic. *Mater. Lett.* **2014**, *116*, 101–103. [CrossRef]
50. Farzana, R.; Sahajwalla, V. Recycling Automotive Waste Glass and Plastic—An Innovative Approach. In Proceedings of the 8th Pacific Rim International Congress on Advanced Materials and Processing 2013, PRICM 8, Waikoloa, HI, USA, 4–9 August 2013; Volume 3, pp. 2267–2276.
51. Munhoz, A.H.; Faldini, S.B.; de Miranda, L.F.; Masson, T.J.; Maeda, C.Y.; Zandonadi, A.R. Recycling of Automotive Laminated Waste Glass in Ceramic. *Mater. Sci. Forum* **2014**, *798–799*, 588–593. [CrossRef]
52. EN 196-1; Methods of Testing Cement–Part 1: Determination of Strength. European Committee for Standardization: Brusel, Belgium, 2016.
53. Dhaliwal, A.K.; Hay, J.N. The Characterization of Polyvinyl Butyral by Thermal Analysis. *Thermochim. Acta* **2002**, *391*, 245–255. [CrossRef]
54. Siddika, A.; Hajimohammadi, A.; Sahajwalla, V. Stabilisation of Pores in Glass Foam by Using a Modified Curing-Sintering Process: Sustainable Recycling of Automotive Vehicles' Waste Glass. *Resour. Conserv. Recycl.* **2022**, *179*, 106145. [CrossRef]
55. Li, X.; Zhang, Q.; Li, H.; Gao, X. A Novel Process for the Production of Triethylene Glycol Di-2-Ethylhexoate by Reactive Distillation Using a Sulfated Zirconia Catalyst. *Ind. Eng. Chem. Res.* **2020**, *59*, 9242–9253. [CrossRef]
56. Andersen, K.E.; Vestergaard, M.E.; Christensen, L.P. Triethylene Glycol Bis(2-Ethylhexanoate)—A New Contact Allergen Identified in a Spectacle Frame. *Contact Dermat.* **2014**, *70*, 112–116. [CrossRef]
57. Torres-Carrasco, M.; Palomo, J.G.; Puertas, F. Sodium Silicate Solutions from Dissolution of Glasswastes. Statistical Analysis. *Mater. Constr.* **2014**, *64*, e014. [CrossRef]
58. ElBatal, H.A.; Hassaan, M.Y.; Fanny, M.A.; Ibrahim, M.M. Optical and FT Infrared Absorption Spectra of Soda Lime Silicate Glasses Containing Nano Fe_2O_3 and Effects of Gamma Irradiation. *Silicon* **2017**, *9*, 511–517. [CrossRef]

59. Sitarz, M.; Mozgawa, W.; Handke, M. Vibrational Spectra of Complex Ring Silicate Anions-Method of Recognition. *J. Mol. Struct.* **1997**, *404*, 193–197. [CrossRef]
60. Rees, C.A.; Provis, J.L.; Lukey, G.C.; Van Deventer, J.S.J. Attenuated Total Reflectance Fourier Transform Infrared Analysis of Fly Ash Geopolymer Gel Aging. *Langmuir* **2007**, *23*, 8170–8179. [CrossRef] [PubMed]
61. Kouamo Tchakouté, H.; Henning Rüscher, C.; Hinsch, M.; Noël, J.; Djobo, Y.; Kamseu, E.; Leonelli, C. Utilization of Sodium Waterglass from Sugar Cane Bagasse Ash as a New Alternative Hardener for Producing Metakaolin-Based Geopolymer Cement. *Geochemistry* **2017**, *77*, 257–266. [CrossRef]
62. Zheng, J.; Li, X.; Bai, C.; Zheng, K.; Wang, X.; Sun, G.; Zheng, T.; Zhang, X.; Colombo, P. Rapid Fabrication of Porous Metakaolin-Based Geopolymer via Microwave Foaming. *Appl. Clay Sci.* **2024**, *249*, 107238. [CrossRef]
63. Perná, I.; Šupová, M.; Hanzlíček, T. The Characterization of the Ca-K Geopolymer/Solidified Fluid Fly-Ash Interlayer. *Ceram.-Silik.* **2017**, *61*, 26–33. [CrossRef]
64. Perná, I.; Hanzlíček, T.; Žaloudková, M. Microscopic Study of the Concrete/Geopolymer Coating Interface. *Ceram.-Silik.* **2020**, *64*, 68–74. [CrossRef]
65. Perná, I.; Hanzlíček, T. The Setting Time of a Clay-Slag Geopolymer Matrix: The Influence of Blast-Furnace-Slag Addition and the Mixing Method. *J. Clean. Prod.* **2016**, *112*, 1150–1155. [CrossRef]
66. Reeb, C.; Pierlot, C.; Davy, C.; Lambertin, D. Incorporation of Organic Liquids into Geopolymer Materials—A Review of Processing, Properties and Applications. *Ceram. Int.* **2021**, *47*, 7369–7385. [CrossRef]
67. Balazs, D.J.; Triandafillu, K.; Wood, P.; Chevolot, Y.; Van Delden, C.; Harms, H.; Hollenstein, C.; Mathieu, H.J. Inhibition of Bacterial Adhesion on PVC Endotracheal Tubes by RF-Oxygen Glow Discharge, Sodium Hydroxide and Silver Nitrate Treatments. *Biomaterials* **2004**, *25*, 2139–2151. [CrossRef]
68. Toniolo, N.; Boccaccini, A.R. Fly Ash-Based Geopolymers Containing Added Silicate Waste. A Review. *Ceram. Int.* **2017**, *43*, 14545–14551. [CrossRef]

Disclaimer/Publisher's Note: The statements, opinions and data contained in all publications are solely those of the individual author(s) and contributor(s) and not of MDPI and/or the editor(s). MDPI and/or the editor(s) disclaim responsibility for any injury to people or property resulting from any ideas, methods, instructions or products referred to in the content.

Article

Recycled Eco-Concretes Containing Fine and/or Coarse Concrete Aggregates. Mechanical Performance

Pablo Plaza [1,*], Isabel Fuencisla Sáez del Bosque [1], Javier Sánchez [2] and César Medina [2,*]

[1] School of Engineering, University of Extremadura, UEx-CSIC Partnering Unit, Institute for Sustainable Regional Development (INTERRA), 10003 Cáceres, Spain; isaezdelu@unex.es

[2] Eduardo Torroja Institute for Construction Science, Spanish National Research Council (CSIC), 28033 Madrid, Spain; javier.sanchez@csic.es

* Correspondence: pablopc@unex.es (P.P.); cmedinam@unex.es (C.M.)

Abstract: This study analysed the effect of substituting different percentages of natural aggregate with recycled aggregate from concrete crushing, using a coarse fraction as well as a fine fraction. Natural and recycled materials were classified in order to analyse the mechanical performance and impermeability of these eco-concretes in the fresh state as well as in the hardened state. A statistical analysis also determined whether the performance loss was significant from a statistical point of view, finding strength decreases of less than 13% in compressive strength and losses of less than 20% in flexural strength. An increasing trend was found in permeability as the percentage of recycled aggregate in the mix increased.

Keywords: mechanical performance; coarse and fine recycled aggregates; permeability; fresh-state concrete properties

Citation: Plaza, P.; Sáez del Bosque, I.F.; Sánchez, J.; Medina, C. Recycled Eco-Concretes Containing Fine and/or Coarse Concrete Aggregates. Mechanical Performance. *Appl. Sci.* **2024**, *14*, 3995. https://doi.org/10.3390/app14103995

Academic Editors: Mouhamadou Amar and Nor Edine Abriak

Received: 27 March 2024
Revised: 1 May 2024
Accepted: 5 May 2024
Published: 8 May 2024

Copyright: © 2024 by the authors. Licensee MDPI, Basel, Switzerland. This article is an open access article distributed under the terms and conditions of the Creative Commons Attribution (CC BY) license (https://creativecommons.org/licenses/by/4.0/).

1. Introduction

The construction sector has a high demand for natural resources and is one of the activities that generates the largest amount of waste during all phases of the construction process (construction, maintenance and demolition). Throughout the EU specifically, the amount of waste generated in the construction sector has been increasing in recent years. According to Eurostat [1], a total of 6.81 Gt of waste was generated in the period from 2004–2020 in the construction sector in the EU-27 as a whole, increasing its magnitude year by year, rising from 29.66% of total waste in 2004 to 37.11% in 2020.

The construction sector also has high CO_2 emissions in the extraction, raw material manufacturing and transport processes and high energy consumption. It is estimated that the construction sector is globally responsible for 33% of annual CO_2 emissions [2] and 40% of global energy consumption [3]. The high energy consumption and emissions are due to manufacturing concrete components, especially cement, which accounts for 73% of total emissions in the sector [4], and transporting these components.

In this context, the use of recycled materials can be a great advantage in logistical terms, since the distance of transporting raw materials is reduced by having construction and demolition waste territory that can be converted into recycled aggregates available throughout the [5], especially in areas where natural aggregate is scarce and/or impossible to extract [6]. The impact on the environment is also significantly reduced by reducing the exploitation of natural resources in quarries as well as waste deposits in landfills [7]. As a whole, the use of coarse recycled aggregate can reduce greenhouse gas emissions by up to 65% [8], a percentage that could increase even further if fine aggregate is added, which would be advantageous, since sand consumption worldwide is increasing year by year, with an estimate of 47.5 billion tonnes by 2023 [9] and up to 60 billion tonnes by 2030 [10].

However, the main problem in the use of recycled aggregates for manufacturing concrete is limitations at the regulatory level, which restrict the use of recycled aggregates, in most cases enabling substitution percentages of less than 60% for the coarse fraction of recycled aggregate from concrete crushing. Table 1 shows the amount of recycled concrete aggregate permitted within the different international regulations, indicating the permitted granulometric fraction (coarse or fine), maximum substitution percentage and maximum strength class.

Table 1. Regulatory framework for the use of recycled aggregates in concrete manufacture.

Country	Aggregate Type	Fraction	Max. Substitution (%)	Concrete Type	Strength Class
Australia AS 1141.62/HB 155:2002 [11]	RCA (Class 1A)	Coarse	30	Structural	C40/50
China GB/T-25177 [12]	RCA—Type I	Coarse	100	Structural	No limit
	RCA—Type II	Coarse	30	Structural	C40/50
	RCA—Type III	Coarse	30	Structural	C25/30
	RCA—Type I	Fine	100	Structural	C40/50
	RCA—Type II	Fine	30	Structural	C25/30
	RCA—Type III	Fine	30	Non structural	-
Korea KS-F-2573 [13]	RCA	Coarse	30	Structural	27 MPa
	RCA	Coarse + Fine	30	Non structural	21 MPa
Hong Kong CS-3:2013/HKBD 2009/WBTC-No. 12 [14]	RCA	Coarse	20	Structural	C25/30–C35/45
	RCA	Coarse	100	Non structural	
Japan JIS-5021 [15]/JIS-5022 [16]/ JIS-5023 [17]	RCA—HQ	Coarse	100	Structural	C45/55
		Fine	100		
	RCA—MQ	Coarse	100	Structural	C35/45
		Fine	100		
	RCA—LQ	Coarse	No limit	Non structural	-
		Fine			
Belgium PTV 406-2003 [18]/NBN B 15-001 [19]	RCA—Type A	Coarse	50, 30, 20	Structural	C30/37
Germany DIN 4226-101, DAfStb [20]	RCA—Type A	Coarse	45, 35, 25	Structural	C30/37
Italy NTC-2008 [21]	RCA 1	Coarse	30	Structural	C30/37
			60		C25/30
	RCA 2		15		C45/55
Denmark DS 2426/DCA No. 34 [22]	RCA 1	Coarse	100	Structural	C40/50
	RCA 2	Coarse and fine	100		
Netherlands NEN-5905 [23]	RCA	Coarse	20	Structural	C55/67
Portugal LNEC-E471 [24]	RCA 1	Coarse	25	Structural	C40/50
Switzerland MB-2030 [25]	RCA 1	Coarse	100	Structural	No limit
	RCA 2		100		

Table 1. Cont.

Country	Aggregate Type	Fraction	Max. Substitution (%)	Concrete Type	Strength Class
United Kingdom BS 8500-2 [26]	RCA	Coarse/Fine	20	Structural	C40/50
France NF P 18-545 [27]	RCA 1	Coarse	60, 30, 20	Structural	No limit
	RCA 2		40, 15		
Spain Structural Code [28]	RCA	Coarse	20	Structural	C40/50
			100	Non structural	-
EN 206 [29]	RCA	Coarse	50, 30	Structural	No limit
RILEM	RCA	Coarse	100	Structural	C50/60
Brazil NBR 15116 [30]	RCA	Coarse/Fine	100	Non structural	-

There are numerous works that have studied the use of recycled aggregates from concrete crushing in structural concrete manufacture. Jayasuriya [31] statistically analysed a large database with experimental results of concretes that included a coarse fraction of recycled concrete aggregates, reaching the following conclusions: (i) the optimal substitution percentage is below 20%, where the best properties in concrete are obtained; (ii) although strength losses occur as the substitution percentage and the use of homogeneous aggregates increases (as in the case of total substitution), it provides better results as the entire fraction has the same properties; (iii) the strength of concrete with recycled aggregates is affected by an increase in the effective water/cement ratio; and (iv) fracture behaviour is unpredictable due to the notable differences in rigidity within the concrete matrix, on which point further study is required.

Several authors also conclude that recycled concrete aggregates can also be used regardless of the fraction and substitution percentage if optimal quality of the new concrete is achieved. Etxeberría [32] maintains that recycled coarse aggregate can be used for concrete in medium–low-strength concrete (20–45 MPa) even if strength variations of up to 25% are recorded for total substitution of natural coarse aggregate by concrete aggregate. Limiting strength makes it possible to avoid an increase in the amount of cement, which would be counter-productive from an economic and environmental point of view. Along the same lines, McNeil [33] maintains that although there are differences in behaviour, the requirements specified in structural standards are met in real structures, which means that recycled aggregates can be used in structural concrete even if the quality is slightly lower. If the fine fraction is also substituted with recycled concrete sand, the strength reduction is often greater, since the fine fraction has greater influence on strength than the coarse fraction. Tang [34] analysed the properties of concrete with 100% recycled aggregates, observing losses of up to 28.6% for total substitution of both fractions. Regarding other properties, the elastic modulus and tensile strength have similar behaviour, with losses of between 20% and 33% depending on the amount of recycled aggregate and the fraction chosen. Along the same lines, Kenai [35] observed decreases in all the mechanical properties of concrete using recycled aggregates until reaching total substitution. The decrease in compressive strength, of approximately 50% in concrete completely composed of recycled aggregates, presents a considerably greater water demand, mainly because the recycled aggregate has greater absorption. This behaviour can be explained by analysing the micro-structure of the concrete. Several authors [36,37] maintain that the performance loss of concrete with recycled aggregate is due to lower-strength ITZ between the mortar layer adhering to the recycled aggregate and the mortar that is formed while mixing the concrete. The cracks and pores present in this layer of old mortar make it the weakest point of the new concrete, which also has a greater demand for water due to the increased absorption of this layer of mortar [38].

The variability of results mentioned above lies in the difference in the quality of the recycled aggregates depending on the concrete of origin. Kumar [39] observed that the use of aggregates from medium–low-strength (30 MPa) but high-quality concrete did not affect the strength for substitution percentages of up to 20%. On the contrary, substituting the coarse and fine fractions, separately as well as together, obtained slightly higher strengths with small adjustments in the dosage, obtaining a high-performance concrete (HPC). Other authors have also observed recycled concrete that is more resistant than the reference mixes in the long term, using a coarse fraction to completely substitute the natural aggregate [40] as well as both fractions, substituting the fine fraction in lower percentages [41], defining the optimal percentage below 60% [42].

This work attempts to deepen the study of the physical/mechanical behaviour of concrete that simultaneously includes coarse and fine aggregate from concrete crushing. Specifically, this work analyses the effect of partially (25, 50 and 75%) or totally (100%) substituting the coarse fraction of the natural aggregate with recycled aggregate and at the same time partially substituting (10, 20 and 50%) the fine fraction of the aggregate in the design of class C30/37 structural concretes (characteristic strength of 30 N/mm^2). The properties in the fresh state (density, consistency and entrained air) and hardened state (compressive and flexural strength and water penetration under pressure) of the concrete have thus been studied. These results have been analysed from a statistical point of view using two techniques: in the case of density, a linear regression model was studied based on the total percentage of substituted aggregate and in the rest of the properties, an analysis of the variance (ANOVA) with two factors and interaction, which analyses the relative effects and interferences of substituting the coarse and fine fractions separately as well as simultaneously, indicating which changes in the properties studied are significant from a statistical point of view.

2. Materials and Methods

2.1. Materials

The natural aggregate used to manufacture the concrete comes from crushing greywacke (Figure 1). It has an irregular shape and marked edges and is supplied in three granulometric fractions: 0/6 mm (finely crushed stone, CS-F), 6/12 mm (medially crushed stone, CS-M) and 12/20 mm (coarsely crushed stone, NG-C). Regarding its chemical composition, it is a siliceous aggregate with around 60% SiO_2, as well as other oxides in a smaller proportion (Al_2O_3, Fe_2O_3, MgO and Na_2O). From the mineralogical point of view it is characterised by having quartz, feldspars and phyllosilicates.

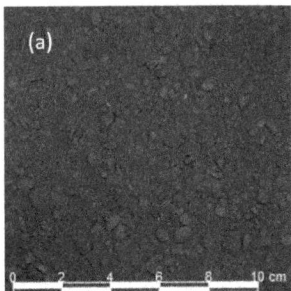

Figure 1. (**a**) Finely crushed stone (CS-F); (**b**) medially crushed stone (CS-M); (**c**) coarsely crushed stone (CS-C).

Also, the recycled aggregate comes exclusively from crushing concrete (Figure 2). As with natural aggregates, it is supplied in three granulometric fractions: 0/6 mm (recycled concrete sand, RCF), 6/12 mm (recycled crushed gravel, RCG) and 12/20 mm (recycled crushed concrete, RCC).

Figure 2. (**a**) Recycled Concrete Sand (RCS); (**b**) Recycled Concrete Gravel (RCG) and (**c**) Recycled Crushed Croncrete (RCC).

All aggregates were supplied by ARAPLASA, an aggregate recycling plant located in Plasencia, north of the province of Cáceres (Spain).

The Portland cement used was a CEM I 42.5 R supplied by the Lafarge Holcim plant located in Villaluenga de la Sagra, in the Spanish province of Toledo. This cement meets all the requirements of the EN-197 standard [43].

Finally, the super-plasticising additive FUCHS BRYTEN NF supplied by FUCHS Lubricantes S.A.U. (Bacerlona, Spain) was used, which consists of a modified water-based polycarboxylate. This brown additive is free from chlorides and has a density of 1.1 g/cm^3, pH = 8.0 and 20% solids content.

2.2. Aggregate Characterisation

Figure 3 shows the composition of gravel and fine gravel recycled from concrete, respectively. It shows that regardless of the recycled coarse fraction, the content of concrete or mortar (Rc) and unbound aggregates (Ru) is ≥95% by weight. Based on this result and according to the classification proposed by the Structural Code (CodE), they can be classified as aggregates from concrete crushing (Rc+Ru ≥ 95%).

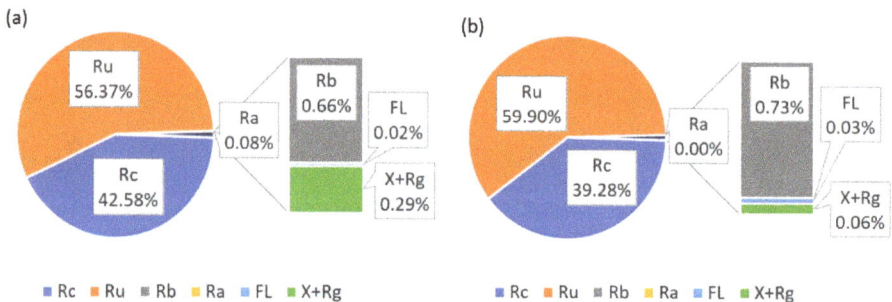

Figure 3. Classification for the constituents of coarse recycled aggregate. (**a**) RCG, (**b**) RCC (EN 933-11) [44].

Table 2 shows the physical, chemical and mechanical properties of the aggregates used in the formulation of concrete, as well as the EN 12620 standard requirements [45].

Table 2. Physical, chemical and mechanical properties of the aggregates.

Property [Standard]	CS-F	RCF	CS-M	CS-C	RCG	RCC	EN 12620
Density (Mg/m^3) [46]	2.82	2.79	2.78	2.77	2.72	2.73	-
Sorptivity (wt%) [46]	1.18	4.42	0.88	0.78	5.40	3.63	<5
Fine equivalent (wt%) [47]	73	61	-	-	-	-	>70 *
LA coefficient (wt%) [48]	-	-	16	18	27	27	≤40
Flakiness index (wt%) [49]	-	-	20.36	24.79	16.08	20.85	<35
Water-soluble chlorides (wt%) [50]			<0.01				<0.05
Acid soluble sulphates (wt%) [50]			<0.002				<0.80
Total sulphates (wt%) [50]			<0.001				<1

* Aggregates intended for concrete elements exposed to exposure class X0 or XC. Note: CS-F: natural sand, RCF: recycled sand, CS-M: natural gravel—medium, CS-C: natural gravel—coarse, RCG: recycled gravel—medium, RCC: recycled gravel—coarse.

In terms of density, the values of the recycled aggregates are lower than those of natural aggregates, with a decrease of 1.44%, 2.16% and 1.06% for the coarse aggregate (RCC), the medium (RCG) and the fine (RCF), respectively. This decrease is mainly due to the adhered mortar layer present in this type of recycled aggregate, which is less dense and more porous than natural aggregate. The values obtained are similar to those observed by Andreu [51] and Gao [52], who observed values of 5.21% and 4.50%, respectively, for recycled concrete aggregates.

Water absorption of recycled aggregates is between 3.5 and 6 times greater than that of natural aggregates depending on the fraction, coarse or fine, analysed. In all cases however, the provisions of the Structural Code are complied with, which limits the absorption value to 7% for recycled aggregates and 4.5% for natural aggregates. Additionally, the granular skeleton has a water absorption of less than 5% by weight, the maximum limit imposed by the EN 12620 standard. The values obtained are similar to those reported by other authors [51,53,54], who observed values of 3.74% and 5.91%, respectively, for all the recycled concrete aggregate fractions studied.

Regarding the values obtained from the Los Angeles (LA) coefficient, the recycled coarse fractions have a value slightly higher proportion (27% by weight) than that corresponding to the natural fraction (16–18%) due especially to the lower resistance to fragmentation of the adhered mortar, which is more friable than natural aggregates. However, the recorded values are below the 40% by weight (LA40) and 50% by weight (LA50) required for recycled concrete crushing aggregates by the CodE and recycled concrete aggregates type A and B pursuant to the EN 206 standard, respectively. The values obtained are also similar [53], although Andreu [51] observed very low values (10–16%) analysing recycled aggregates from high-strength concrete.

Regarding chemical properties, it should be noted that all aggregates comply with the Structural Code limitations regarding the content of organic matter, water-soluble chlorides and sulphates, including total as well as acid-soluble.

Regarding geometric properties, the flakiness index of recycled aggregates is lower than that of natural aggregates, due to its more rounded shape and less sharp edges (see Figure 2) associated with the construction and demolition waste crushing process and the presence once again of adhered mortar. These values are below the 35% required in the CodE and are similar to those found by other authors [51,53]. Additionally, regarding the quality of the fines, the sand equivalent of the recycled aggregate (RCF) is lower than the minimum required by the CodE for the 0/4 fraction used in concrete elements exposed to exposure classes X0 and XC.

Figure 4 shows the granulometric distribution of the aggregates used, as well as the upper, lower and fine limit content (<0.063 mm) established in the CodE for the fine fraction, observing that all granulometric fractions 0/6, 6/12 and 12/22, regardless of their nature (natural or recycled), have a similar granulometric distribution.

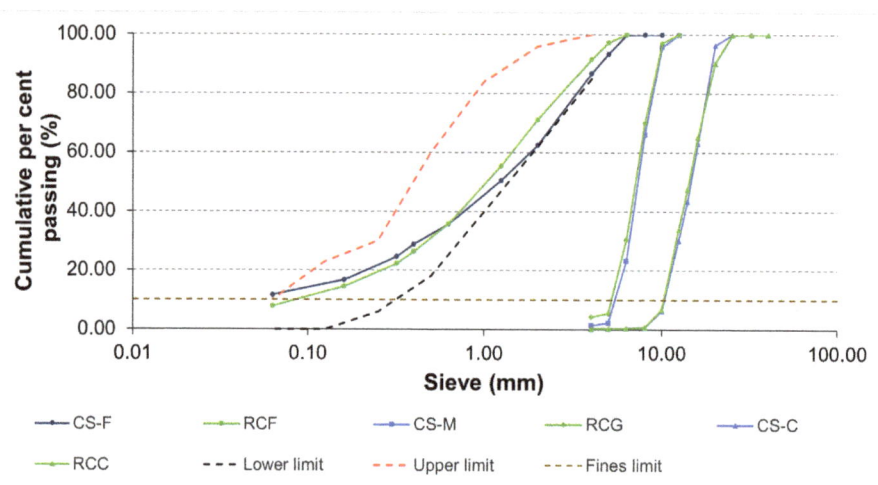

Figure 4. Aggregate particle size distribution.

Finally, the fine content (<0.063 mm) of the natural aggregate (CS-F) is 11.59%, slightly over the limit established for concrete with non-limestone aggregates subject to general exposure classes X0 or XC and not subject to any specific XA, XF or XM exposure class. In the case of recycled sand (RCF), the fine content is 7.78%.

2.3. Mix Design

Table 3 shows the composition of the 20 formulated mixes: (i) 1 reference mix (HP) with 100% natural aggregate; (ii) 4 mixes with recycled coarse aggregate in different percentages (25%, 50%, 75% and 100%) and 0% recycled sand (HR-25, HR-50, HR-75 and HR-100); (iii) 5 mixes with recycled coarse aggregate in different percentages (0%, 25%, 50%, 75% and 100%) and 10% recycled sand (HR-0+10, HR-25+10, HR-50+10, HR-75+10 and HR-100+10); (iv) 5 mixes with recycled coarse aggregate in different percentages (0%, 25%, 50%, 75% and 100%) and 20% recycled sand (HR-0+20, HR-25+20, HR-50+20, HR-75+20 and HR-100+20); and (v) 5 mixes with recycled coarse aggregate in different percentages (0%, 25%, 50%, 75% and 100%) and 50% recycled sand (HR-0+50, HR-25+50, HR-50+50, HR-75+50 and HR-100+50).

The starting data required to design the mixes according to the DOE British Method [55] were: (i) characteristic design strength (f_{ck}) of 30 MPa (C30/37); (ii) effective water-to-cement ratio (w/c) of 0.45; and (iii) 20 mm maximum size of the coarse aggregate. It is also considered that the aggregates are dry, and 70% of the water absorption of the recycled aggregates has been added to the theoretical water content resulting from the dosing process, thus guaranteeing that all mixes have the same amount of water available for cement hydration regardless of the aggregate mix. For manufacturing all the mixes, a super-plasticising additive (6.20 kg/m^3) was also added with an amount of 1.55% by weight of cement.

Finally, all mixes meet the minimum dosage requirements (maximum water/cement ratio and minimum cement content) specified in article 43.2.1 of the CodE for use as structural concrete for exposure classes XC1/XC2 and XC3/XC4 with maximum water/cement ratio of 0.60 and 0.55, respectively. Regarding the minimum cement content, a value of 275 kg/m^3 and 300 kg/m^3 of cement is established for classes XC1/XC2 and XC3/XC4, respectively.

Table 3. Mix batching.

Mix	Components (kg/m³)							
	NS	RS	NG-M	NG-C	RG-M	RG-C	Cement	Water
HP	732.36	0.00	382.96	766.69	0.00	0.00	400.00	193.03
HR-0+10	655.65	70.59	380.95	762.65	0.00	0.00	400.00	195.20
HR-0+20	581.26	140.81	379.94	760.63	0.00	0.00	400.00	197.40
HR-0+50	360.4	349.21	376.92	754.58	0.00	0.00	400.00	203.93
HR-25	724.65	0.00	284.20	568.96	92.01	186.72	400.00	197.37
HR-25+10	648.72	109.24	282.69	565.94	91.52	185.72	400.00	202.36
HR-25+20	576.64	218.48	282.69	565.94	91.52	185.72	400.00	206.34
HR-25+50	358.47	347.35	281.18	562.91	184.73	281.18	400.00	209.38
HR-50	716.94	0.00	187.45	375.27	182.06	369.46	400.00	203.86
HR-50+10	643.51	108.37	186.95	374.26	181.57	368.47	400.00	207.72
HR-50+20	572.01	216.73	186.95	374.26	181.57	368.47	400.00	211.67
HR-50+50	354.62	343.61	185.43	371.24	180.10	365.49	400.00	214.55
HR-75	711.16	0.00	92.97	186.12	270.88	549.72	400.00	209.18
HR-75+10	638.31	68.72	92.72	185.62	270.15	548.23	400.00	211.29
HR-75+20	567.38	137.44	92.72	185.62	270.15	548.23	400.00	213.49
HR-75+50	349.80	338.94	91.46	183.10	266.48	540.79	400.00	219.48
HR-100	701.52	0.00	0.00	0.00	356.29	723.03	400.00	214.20
HR-100+10	627.90	67.60	0.00	0.00	356.29	723.03	400.00	216.15
HR-100+20	556.59	134.83	0.00	0.00	353.35	717.07	400.00	218.19
HR-100+50	345.94	335.21	0.00	0.00	351.39	713.10	400.00	224.40

2.4. Experimental

Prior to the design of the different mixes, the physical, mechanical and chemical properties (see point 2.2) of the aggregates used in this research were analysed. Next, the concrete mixes previously described were theoretically formulated (see Section 2.3) and adjusted on a laboratory scale. The properties of the manufactured concrete (see Table 4) were then studied in their fresh state (density, entrained air and consistency) and in their hardened state (density, compressive and flexural strength and water penetration under pressure) (see Table 4).

Table 4. Concrete properties studied.

Property	Standard	Sample Size (cm)	NS/M	Testing Age (Days)
Fresh state				
Density	EN 12350-6 [56]			
Entrained air	EN 12350-7 [57]	Evaluated during the manufacturing process		
Consistency	EN 12350-2 [58]			
Hardened state				
Density	EN 12390-7 [59]	15 × 15 × 15	3	28
Compressive strength	EN 12390-3 [60]	15 × 15 × 15	9	7, 28, 90
Flexural strength	EN 12390-5 [61]	10 × 10 × 40	3	28
Water penetration under pressure	EN 12390-8 [62]	Ø15 × 30	3	28

Note: NS/M: number of samples/mix.

Figure 5 shows the procedure followed for mixing the designed mixes, manufacturing and subsequent curing of the samples. The first phase of the mixing process consists of loading and homogenising the materials. To do so, the granular skeleton consisting of the aggregates was first placed into the mixer and then mixed for 30 s. The cement was then added, mixing for another 60 s. The second phase consists of the mixing itself, which begins by diluting the additive in 10% of the mixing water. This mix is added to the mixer for 45 s, after which 70% of the mixing water is added. Finally, the remaining 20% is added, and everything is mixed for 240 s. Finally, the third phase consists of filling the moulds and compacting them by chopping with a bar as established by the EN 12390-1 standard [63].

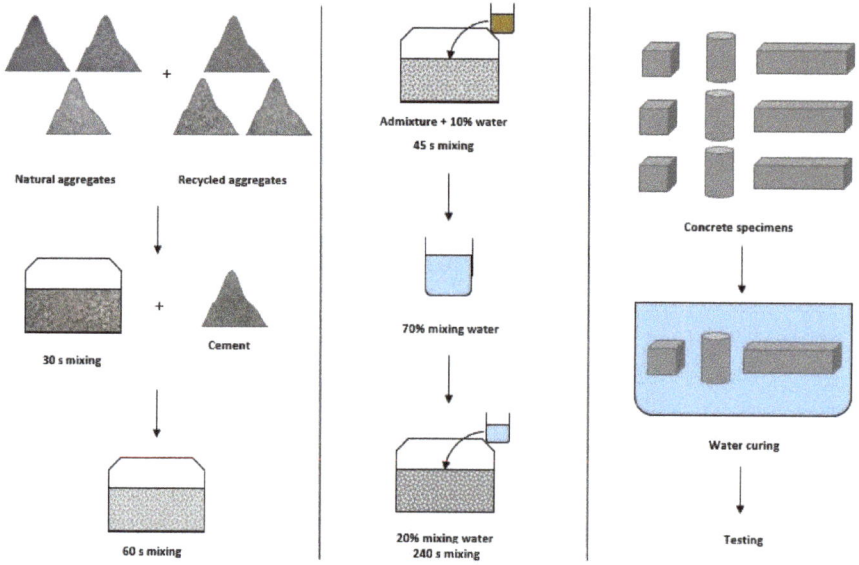

Figure 5. Mixing process.

2.5. Statistical Analysis

To evaluate the influence of replacing natural aggregate with recycled aggregate on concrete properties, two different techniques were used, depending on the property studied. First, to study the mechanical properties of the concrete, an analysis of variance (ANOVA) was carried out for each property and age studied, carrying out a total of four analyses. The statistical software "R", version 4.0.5, was used to make the calculations.

The proposed model (Equation (1)) to carry out the analyses corresponds has two factors (% substitution of coarse aggregate and % substitution of fine aggregate) and interactions.

$$Y_{ijk} = \mu_{11} + \alpha_i + \beta_j + \alpha\beta_{ij} + \varepsilon_{ijk}, i = 1, \ldots, 4; j = 2; k = 1, \ldots, 3 \qquad (1)$$

This model obtains a value for the response variable Y_{ijk} (strength studied in each case) by adding different values: (i) μ_{11} is the average value corresponding to the reference mix (HP) in each case; (ii) α_i quantifies the relative effect corresponding to the first factor (% coarse aggregate substitution); (iii) β_j quantifies the relative effect corresponding to the second factor (% fine aggregate substitution); (iv) $\alpha\beta_{ij}$ is the relative effect due to the interaction that occurs when simultaneously substituting both fractions; and (v) ε_{ijk} indicates the perturbation of the model.

To check whether the analyses carried out are valid, the homoscedasticity and normality assumptions must first be checked using the Bartlett and Shapiro–Wilk tests, respectively. Table 5 shows the *p*-values of both tests. As can be seen, both assumptions are met in all analyses, so the analyses are therefore considered valid.

Table 5. Homoscedasticity and normality tests.

Contrast Type	Compressive			Flexural	Water under Pressure
	7 Days	28 Days	90 Days		
Bartlett	0.456	0.803	0.381	0.605	0.075
Shapiro–Wilk	0.237	0.627	0.556	0.231	0.532

Once the initial assumptions were tested, the model calculates the model coefficients, indicating whether or not these values are significant. In other words, for what percentages of substitution is strength variation relevant or not from a statistical point of view (p-value < 0.05).

In the density study (apparent as well as fresh-state), expressions were also observed that relate the total percentage of substituted aggregate (%) with the density (D) studied in each case. In both cases, the model proposed for the optimal adjustment was obtained using linear regression (Equation (2)), in both cases obtaining very good correlation coefficients ($R^2 > 0.8$):

$$D = a\% + b \qquad (2)$$

This model returns a density value through an affine function with parameters including the y-intercept (b), which corresponds to the reference value for the mix without recycled aggregate (RC) and the gradient (a) that adjusts the line to the data obtained through a least-squares fitting.

3. Results

3.1. Fresh-State Properties

Figure 6 shows the fresh-state density as a function of the percentage of total substitution of natural aggregate by recycled aggregate, observing that the density decreases as the recycled aggregate content increases, recording losses of less than 5% in all cases. This decrease is mainly due to the lower density of the recycled aggregates, as well as the greater amount of entrained air (see Table 6) in the mixes with a greater amount of recycled aggregate. This trend was observed by other authors [64,65] incorporating recycled concrete aggregates in all fractions, with decreases of around 3% for total substitution of coarse aggregate.

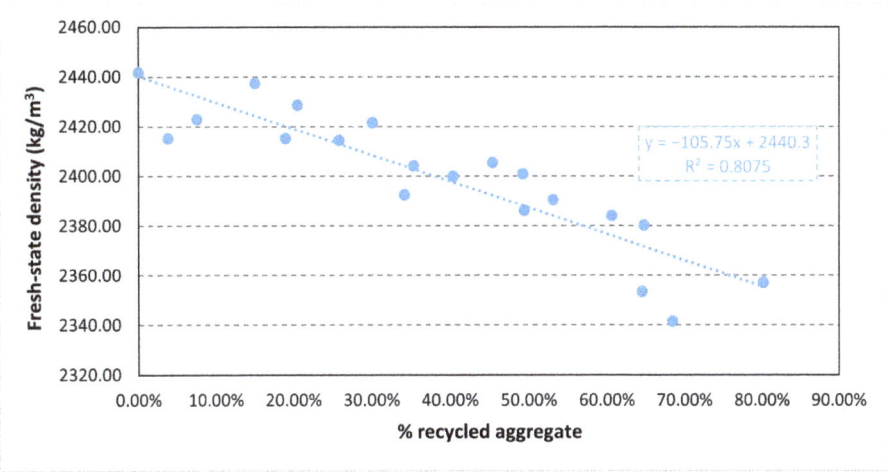

Figure 6. Fresh-state density.

Table 6. Settlement and entrained air.

Mix	Slump (mm)	Entrained Air (Vol. %)
HP	82.8	1.63
HR-0+10	65.0	1.58
HR-0+20	75.0	1.60
HR-0+50	80.0	1.68
HR-25	78.0	1.66
HR-25+10	77.5	1.78
HR-25+20	77.1	1.74
HR-25+50	87.0	1.80
HR-50	89.0	1.66
HR-50+10	82.0	1.66
HR-50+20	60.0	1.82
HR-50+50	77.5	1.74
HR-75	66.3	1.73
HR-75+10	90.0	1.60
HR-75+20	75.0	1.73
HR-75+50	92.5	1.68
HR-100	72.0	1.82
HR-100F10	90.0	1.90
HR-100F20	90.0	1.90
HR-100F50	80.0	1.90

From a statistical point of view, the proposed expression shows a clear linear relationship ($R^2 > 0.8$) between the substitution percentage and the density, enabling the density to be calculated based on the amount of aggregate substituted.

Table 6 shows the slump and entrained air values observed for the different mixes, first showing that all the mixes have a soft consistency (50–90 mm) pursuant to article 33.5 of the CodE, which corresponds to a type S2 settlement according to EN 206. These results confirm that all samples have the same workability regardless of the amount of aggregate substituted, because the amount of super-plasticiser additive as well as the effective water/cement ratio are maintained.

Finally, the entrained air content varies slightly for the different mixes, obtaining values between 1.58% and 1.90%. In this case, the lower density and greater porosity of the recycled aggregates result in a slight increase in air content [66]. Simsek [65] observed a similar behaviour, with slight variations but an increasing trend as the substitution percentage increases, for the coarse fraction as well as for the fine fraction.

3.2. Hardened-State Properties

3.2.1. Bulk Density

Figure 7 shows the bulk density data as a function of the amount of recycled aggregate of the different mixes, as well as the adjustment made and the proposed mathematical expression. In this case, the proposed expression also shows the linear relationship ($R^2 > 0.8$) between the density and the percentage of recycled aggregate.

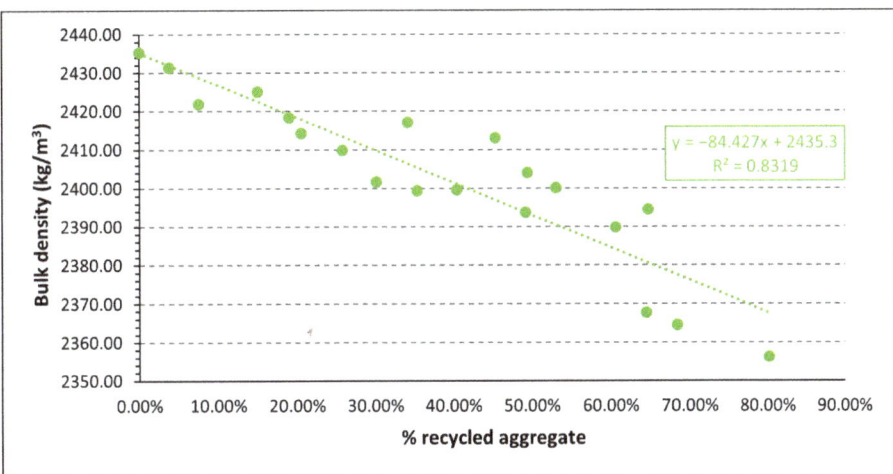

Figure 7. Apparent density.

The density behaviour is very similar to that previously described for the fresh-state density, although with slightly lower values. The range of density loss (0.16–3.25%) is lower than that obtained in the fresh state, although the data trend is very similar. Tuyan [67] maintains that this decrease in apparent density is directly related to the presence of macropores in the mortar adhered to recycled concrete aggregates, with a direct relationship between the amount of aggregate substituted and the decrease in density.

3.2.2. Compressive Strength

Figure 8 shows the results of compressive strength at age 7, 28 and 90 days in a 150 × 150 × 150 mm cubic sample, observing that all the mixes exceed the corresponding design strength with a C30/37 concrete (f_{ck} = 30 N/mm^2), which indicates that all mixes, regardless of the content and recycled fraction used, could be used from a point of view of this property in manufacturing structural concrete.

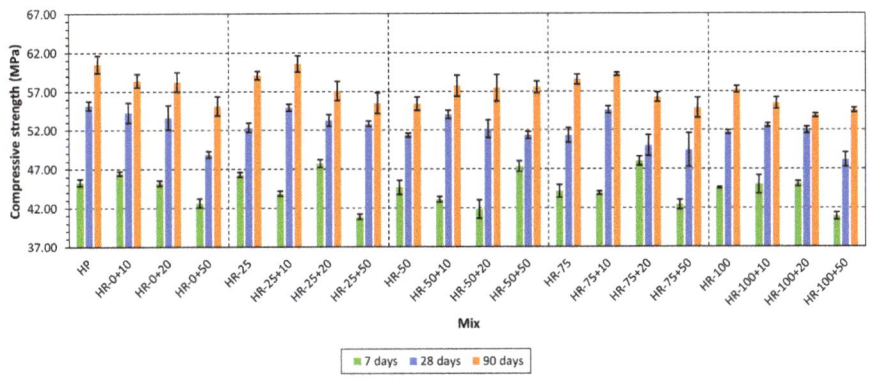

Figure 8. Simple compressive strength.

In the case of mixes that include only the coarse fraction of the aggregate, there are small strength losses, which increase slightly as the age increases. At 7 days, the HR-25 mix even presents a higher strength than the reference mix (+2.22%). However, strength decreases in the rest of the mixes between 1.42% and 2.5%. At 28 and 90 days, the losses are very similar in all mixes, with decreases between 5.20% and 7.04% at 28 days and between 2.39% and 8.45% at 90 days. This behaviour is similar to that found by Chang [68], who observed a 7-day strength increase of 1.35% for a 25% substitution of coarse concrete aggregate. At 28 days, losses of 3.86% and 7% were observed for substitution percentages of 25% and 75%, respectively. Pedro [69] observed losses of between 3.2% and 7.6% for concrete with a target f_{ck} of 45 MPa, substituting 25% to 100% of the coarse fraction with recycled concrete aggregates.

Regarding the fine fraction, the strength loss is similar to that resulting from substituting the coarse fraction for percentages up to 20%. At 7 days, the behaviour is very similar to that described above, even recording a slight increase (2.61%) for the HR-0+10 mix. At 28 and 90 days, the losses range between 1.67% and 5.83%. In the case of the HR-0+50 mix, the strength losses are significantly higher, with losses between 5.83% and 11.46%. This behaviour agrees with the results observed by other authors. Mohammed [70] recorded a decrease in strength of approximately 14% for a 50% substitution of the fine fraction with concrete aggregate, associating these losses with the significant increase in the absorption of the recycled aggregate. For lower substitution percentages, the results observed by Zega [71] are very similar to those found in this work, with losses around 2% substituting 20% of the fine fraction with concrete sand. At 90 days, the difference with respect to the reference concrete is reduced to approximately 1.6%. In this case, a reduction in the effective water/cement ratio leads to an improvement in the interface (ITZ), which results in behaviour very similar to the reference concrete, even improving other durable properties.

Finally, the simultaneous addition of both fractions mitigates the effect of strength loss, obtaining concretes with strength very similar to the previous ones but with a higher recycled aggregate content. The effect of simultaneous addition can be observed with the data from the statistical analysis, whose parameters are shown in Table 7.

From a statistical point of view, the effect of adding coarse aggregate is generally negative, as reflected in the data shown in Figure 9. However, the behaviour at 7 days has non-significant factors that reveal that the trend is not as clear as in the other ages. In the case of fine aggregate, the factors follow the same trend, which corroborates the idea that the addition of any fraction of the aggregate results in a decrease in strength.

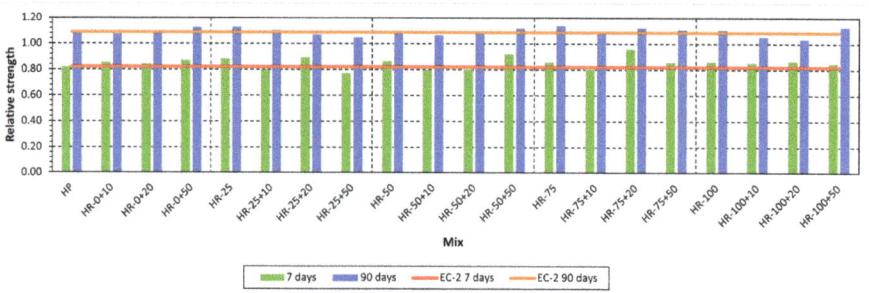

Figure 9. Relative strength.

Additionally, the interaction factors reveal positive values for 28 and 90 days in all cases, which indicates that the strength loss is mitigated when both fractions are added simultaneously, which also contributes to obtaining concretes with similar strength to those that have only one fraction type (coarse or fine), although increasing the percentage of recycled aggregate in the mixes.

Table 7. Estimation of the parameters of the different models.

Parameter	Compressive			Flexural	Water under Pressure
	7 Days	28 Days	90 Days		
μ_{11}	45.27	55.20	60.53	7.12	4.45
α_2	1.00	−2.87	NS	NS	NS
α_3	NS	−3.84	−5.12	−0.42	5.98
α_4	−1.13	−3.89	−1.97	−0.49	3.38
α_5	NS	−3.49	−3.28	−0.66	NS
β_2	1.18	NS	−2.12	NS	−2.51
β_3	NS	−1.56	−2.28	NS	NS
β_4	−2.64	−6.33	−5.39	−0.25	NS
$\alpha\beta_{22}$	−3.60	3.53	3.62	No interaction	3.54
$\alpha\beta_{32}$	−2.72	3.57	4.44		−3.28
$\alpha\beta_{42}$	−1.44	4.24	2.78		NS
$\alpha\beta_{52}$	NS	NS	NS		NS
$\alpha\beta_{32}$	1.51	2.47	NS		NS
$\alpha\beta_{33}$	−2.71	2.36	4.32		NS
$\alpha\beta_{43}$	3.96	NS	NS		NS
$\alpha\beta_{53}$	NS	NS	NS		NS
$\alpha\beta_{23}$	−2.80	6.81	NS		NS
$\alpha\beta_{33}$	5.34	6.30	7.55		−4.83
$\alpha\beta_{43}$	NS	5.64	NS		NS
$\alpha\beta_{53}$	NS	2.79	2.72		NS

Note: NS: not significant.

Regarding the evolution of strength over time, Figure 9 shows the relative strength values of the mixes with respect to the reference value at 28 days, as well as the relative strength estimated in the Eurocode-2 (EC-2) at 7 and 90 days. It shows that all the mixes present a similar evolution in strength gain, in most cases reaching greater strength than expected at 7 and at 90 days. This result shows that the addition of the recycled coarse and/or fine concrete fraction does not influence the cement hydration process. It also indicates that the behaviour recorded at 7 days, in which greater strength is observed than expected by the EC, is concordant with the results of Surendar [72], which evaluated the behaviour of mixes with between 10% and 75% substitution of the coarse fraction with concrete washed aggregate.

Finally, it indicates that all the mixes present the same failure mode regardless of the age studied. This failure mode is classified in the EN 12390-3 standard as satisfactory (Figure 10).

Figure 10. Failure mode in HR-100+50 mix.

3.2.3. Flexural Strength

Figure 11 shows the flexural strength at 28 days of the formulated concrete, as well as the estimated strength according to the expression included ($f_{ct,m,fl} = 1.6 - h/1000)f_{ct,m}$) in point 3.1.8 of the CodE. Additionally, the flexural strength has been estimated from the compressive strength, using the expression included in article A19.3.1.8, combined with the expression for average tensile strength and taking 90% of the compressive strength obtained in the cubic samples as f_{ck}. The resulting expression is shown in the Equation (3)

$$f_{ct,m,fl} = 0.405 \cdot f_{cm}^{2/3} \tag{3}$$

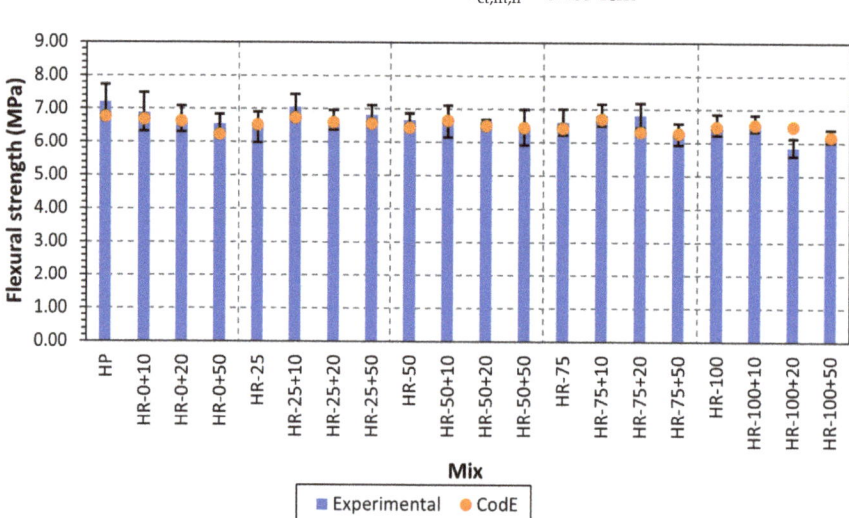

Figure 11. Flexural strength.

In the mixes that only substitute the coarse fraction of the aggregate, lower strengths were observed, with losses of between 7.41% and 10.73%, with very similar values regardless of the substitution percentage. In the case of fine aggregate, the decrease in strength is greater as the amount of recycled aggregate increases, with losses of between 4.37% and

9.01%. In the mixes that incorporate both fractions, the values are similar except for mixes with 50% fine aggregate, which generally present a higher strength loss, although all values remain below 20%. These values are similar to those recorded by other authors. Saini [73] and Yaba [74] found losses of 4.1% and 10% for 50% and 100% substitution of the coarse fraction, respectively. Regarding the fine fraction, Mohammed [70] observed a loss of 14% by substituting 50% of the fine aggregate with recycled aggregate.

Mohammed [70] and Xiao [75] state that the reduction in strength occurs due to the presence of micro-cracks in the adhered mortar in the recycled fractions, as well as the intrinsic properties (thickness and micro elastic) of the ITZs old mortar/new mortar and old aggregate/old mortar, whose values are worse than those shown by new aggregate/new mortar.

It should also be noted that in all mixes, regardless of their granular skeleton, the failure mechanism consisted of a single large crack that was initiated in the flexural span and with a normal orientation to the tensile stresses generated due to flexure [76].

From a statistical point of view (Table 7), the effect of adding the coarse aggregate is negative, with a non-significant value for the HR-25 mix. Regarding the mixes that incorporate only the fine fraction, there is only one significant factor for the HR-0+50 mix, so there is no clear trend. Likewise, there is no interaction, so the combination of the coarse and fine fractions has no significant effect on the model studied.

3.2.4. Water Penetration under Pressure

Table 8 shows the average and maximum depth of water penetration under pressure, showing that all the mixes, regardless of the percentage and recycled fraction added, have average and maximum depth values below the limits established in the article. 43.3.2 of the CodE ($P_{med} \leq 20$ mm and $P_{max} \leq 30$ mm) for exposure classes (XS3 and XA3). Therefore, all formulated concretes have a sufficiently impermeable structure against water penetration.

Table 8. Average and maximum depths of water penetration under pressure.

Mix	Average Depth (mm)	Maximum Depth (mm)
HP	4.45	14.54
HR-0+50	5.35	9.96
HR-25	5.25	9.09
HR-25+10	6.28	16.06
HR-25+20	5.50	12.58
HR-25+50	5.81	11.43
HR-50	9.51	14.64
HR-50+10	4.64	9.48
HR-50+20	8.27	17.94
HR-50+50	6.49	11.44
HR-75	7.83	15.02
HR-75+10	5.32	10.17
HR-75+20	7.53	14.78
HR-75+50	6.91	12.50
HR-100	6.21	12.97
HR-100+50	6.80	13.87

It firstly indicates that for the mixes that incorporate only the coarse fraction of the recycled aggregate, the average penetration values are higher, presenting high variability in the data obtained, with increases from 17.99% to 113.63%. However, the maximum penetration values are very similar or even lower in some cases. The fine fraction only has the HR-0+50 mix, which has an average depth 20.15% greater than the reference mix.

In general terms, the variability of the measurements does not allow a clear trend to be established from a statistical point of view, although a slight upward trend can be observed as the recycled aggregate content increases. Analysing the model parameters (Table 7), the substitution of the coarse fraction produces a positive relative effect (increase in the average depth) regardless of the substitution percentage. However, the effect is the opposite for low substitution percentages of the fine fraction (10%). Regarding interferences, there is no clear trend from a statistical point of view, with few significant values.

In general terms, the literature reveals that the depth values increase as the amount of recycled aggregate increases, both coarse and fine [77], although some authors do not consider it significant [78]. Zega [71] observed that the penetration values are slightly higher if the fine fraction of recycled aggregates is used, although the behaviour is very similar for substitution percentages less than 30%. Kapoor [79] recorded depth increases of 30% for total substitution of the coarse fraction with concrete aggregate, as well as increases of 18% when both fractions were combined (100% of the coarse fraction and 50% of the fine). Velardo [80] also observed similar values for average penetration (~8 mm) and somewhat higher values in the case of maximum penetration (~18 mm) using mixed aggregates.

4. Conclusions

The conclusions obtained in this work are presented below:
- Recycled aggregates have greater absorption, as well as lower LA coefficient, density and flakiness index than natural aggregates.
- The coarse recycled fractions (gravel and gravel) and fine (sand) comply with the mechanical, physical and chemical requirements set forth in the current regulations on aggregates for concrete.
- The workability of the concrete is not affected by the addition of the recycled fractions (coarse and/or fine), all of which show a soft consistency.
- The density of the concrete with recycled aggregate is lower than that of the reference concrete in all cases, in the fresh state as well as in the hardened state. The density decreases as the proportion of recycled aggregate in the mix increases, registering density variations of less than 5% in all cases.
- The entrained air content increases slightly as the amount of recycled aggregate increases, although remaining within the usual values for conventional reinforced concrete, not exceeding 1.9% in the mixes with the highest recycled aggregate content.
- The compressive strength of concrete with recycled aggregate is lower than that of the reference mix, with losses of less than 13% in all cases. The greatest losses are recorded in mixes that include a higher percentage (50%) of fine recycled aggregate.
- The flexural behaviour is similar to that recorded in compressive, slightly increasing the maximum loss percentage to 19%. Losses are generally greater in mixes that include a high percentage of recycled aggregate, coarse as well as fine.
- All mixes are therefore suitable for use in class C30/37 structural concrete.
- The expression included in the structural code for estimating the flexural strength is correct, showing values with differences of less than 10% compared to the experimental values for all mixes.
- The penetration depths of water under pressure present great variability, with increases of up to 100%. However, the provisions of the regulations are complied with in all cases.

The conclusions indicate that eco-concretes can be used to manufacture structural concrete analyzing mechanical performance. However, it is necessary to complement the tests carried out by analysing the durability of all mixes, as well as on full-scale structural pieces. Furthermore, in order to properly study the environmental benefit, a Life Cycle Analysis (LCA) would be necessary to estimate the environmental benefit taking into account the variables concerning waste treatment or transportation.

Author Contributions: Conceptualization, P.P., I.F.S.d.B., J.S. and C.M.; methodology, P.P. and I.F.S.d.B.; software, P.P.; validation, P.P.; formal analysis, P.P.; investigation, P.P. and I.F.S.d.B.; writing—original draft preparation, P.P.; writing—review and editing, P.P., I.F.S.d.B., J.S. and C.M.; supervision, I.F.S.d.B., J.S. and C.M.; project administration, J.S. and C.M.; funding acquisition, C.M. All authors have read and agreed to the published version of the manuscript.

Funding: This research was funded by the Spanish Ministry for Science and Innovation under project PDC2022-133285-C21 funded by MCIN/AEI/10.13039/501100011033 and, by the 'European Union NextGenerationEU/PRTR', by Spanish Ministry for Science, Innovation and Universities under project PID2022-136244OB-I00 funded MICIU/AEI/10.13039/501100011033 and by "FEDER/UE" and the IB 20131 research project financed by the Consejería de Economía, Ciencia y Agenda Digital de la Junta de Extremadura and by the European Union Regional Development Fund (ERDF). Author Pablo Plaza benefitted from Spanish Ministry of Education, Culture and Sport pre-doctoral grant FPU19/06704.

Institutional Review Board Statement: Not applicable.

Informed Consent Statement: Not applicable.

Data Availability Statement: The datasets presented in this article are not readily available because the data are part of an ongoing study. Requests to access the datasets should be directed to cmedinam@unex.es.

Conflicts of Interest: The authors declare no conflicts of interest.

List of Abbreviations

CDW	Construction and demolition waste
RCA	Recycled concrete aggregate
CS-F	Natural sand
CS-M	Natural gravel—medium
CS-C	Natural gravel—coarse
RCF	Recycled sand
RCG	Recycled gravel—medium
RCC	Recycled gravel—coarse

References

1. Comission, E. Waste Statistics. 2020. Available online: https://ec.europa.eu/eurostat/statistics-explained/index.php?title=Waste_statistics (accessed on 4 May 2024).
2. Ürge-Vorsatz, D.; Novikova, A. Potentials and costs of carbon dioxide mitigation in the world's buildings. *Energy Policy* **2008**, *36*, 642–661. [CrossRef]
3. Peng, C. Calculation of a building's life cycle carbon emissions based on Ecotect and building information modeling. *J. Clean. Prod.* **2016**, *112*, 453–465. [CrossRef]
4. Zhang, Z.; Wang, B. Research on the life-cycle CO2 emission of China's construction sector. *Energy Build.* **2016**, *112*, 244–255. [CrossRef]
5. Zhao, M.-Z.; Wang, Y.-Y.; Lehman, D.E.; Geng, Y.; Roeder, C.W. Response and modeling of axially-loaded concrete-filled steel columns with recycled coarse and fine aggregate. *Eng. Struct.* **2021**, *234*, 111733. [CrossRef]
6. Katerusha, D. Investigation of the optimal price for recycled aggregate concrete—An experimental approach. *J. Clean. Prod.* **2022**, *365*, 132857. [CrossRef]
7. Algourdin, N.; Pliya, P.; Beaucour, A.L.; Noumowé, A.; di Coste, D. Effect of fine and coarse recycled aggregates on high-temperature behaviour and residual properties of concrete. *Constr. Build. Mater.* **2022**, *341*, 127847. [CrossRef]
8. Wang, B.; Yan, L.; Fu, Q.; Kasal, B. A Comprehensive Review on Recycled Aggregate and Recycled Aggregate Concrete. *Resour. Conserv. Recycl.* **2021**, *171*, 105565. [CrossRef]
9. Group, F. *Global Construction Aggregates—Demand and Sales Forecasts, Market Share, Market Size, Market Leaders*; Freedonia Group: Cleveland, OH, USA, 2016.
10. Da, S.; Le Billon, P. Sand mining: Stopping the grind of unregulated supply chains. *Extr. Ind. Soc.* **2022**, *10*, 101070. [CrossRef]
11. *AS 1141.62/HB 155:2002*; Guide to the Use of Recycled Concrete and Masonry Materials. Commonwealth Scientific and Industrial Research Organisation: Canberra, Australia, 2002.
12. *GB/T-25177*; Recycled Coarse Aggregate for Concrete. China Academy of Building Research: Beijing, China, 2010.
13. *KS-F-2573*; Recycled Aggregates for Concrete. Korean Standards and Certification: Seoul, Republic of Korea, 2023.

14. CS-3:2013/HKBD 2009/WBTC-No. 12; Aggregates for Concrete. Civil Engineering and Development Department: Hong Kong, China, 2016.
15. JIS A 5021; Recycled Aggregates for Concrete—Class H. Japanese Industrial Standard: Tokyo, Japan, 2018.
16. JIS A 5022; Recycled Aggregate Concrete—Class M. Japanese Industrial Standard: Tokyo, Japan, 2018.
17. >JIS A 5023; Recycled Concrete Using Recycled Aggregate—Class L. Japanese Industrial Standard: Tokyo, Japan, 2022.
18. PTV 406-2003; Recycled Granulates. Aggregates of Concrete Debris, Mixed Debris, Masonry Debris and Asphalt Debris. Impartial Control Body for Construction Products: Brussels, Belgium, 2003.
19. NBN B 15-001; Concrete—Specification, Performance, Production and Conformity—National Supplement to NBN EN 206. Belgian Standards: Brussels, Belgium, 2012.
20. DIN 4226-101; DAfStb; Aggregates for Concrete and Mortar—Part 100: Recycled Aggregates. German Institute for Standardization: Berlin, Germany, 2017.
21. NTC-2008; Technical Standard for Constructions. Ministry of Infrastructures and Transport: Rome, Italy, 2008.
22. DS 2426/DCA No. 34; Concrete—Materials—Rules for Application of EN 206-1. Dansk Standard: Copenhage, Denmark, 2011.
23. NEN-5905; Aggregates for Concrete. Royal Netherlands Standardization Institute: Delft, The Netherlands, 2005.
24. E 471-2009; Guide for the Use of Coarse Recycled Aggregates in Concrete. National Laboratory of Civil Engineering: Lisbon, Portugal, 2009.
25. MB 2030; Recycling Concrete. Swiss Association of Engineers and Architects: Zürich, Switzerland, 2010.
26. BS 8500-2; Concrete. Complementary British Standard to BS EN 206. Specification for Constituent Materials and Concrete. British Standards: London, UK, 2019.
27. NF P18-545; Aggregates—Elements of Definition, Conformity and Coding. French Standardization Association: Paris, France, 2021.
28. Structural Code; Ministry of Transport Mobility and Urban Agenda: Madrid, Spain, 2021; p. 1789.
29. EN 206:2013+A2:2021/1M:2022; Concrete. Specification, Performance, Production and Conformity. Standardization, E.C.f: Brussels, Belgium, 2022; p. 142.
30. NBR 15116; Recycled Aggregates for Uses in Mortar and Concrete—Requirements and Test Methods. Brazilian Association of Technical Standards: Rio de Janeiro, Brazil, 2021; p. 16.
31. Jayasuriya, A.; Shibata, E.S.; Chen, T.; Adams, M.P. Development and statistical database analysis of hardened concrete properties made with recycled concrete aggregates. *Resour. Conserv. Recycl.* **2021**, *164*, 105121. [CrossRef]
32. Etxeberria, M.; Vazquez, E.; Mari, A.; Barra, M. Influence of amount of recycled coarse aggregates and production process on properties of recycled aggregate concrete. *Cem. Concr. Res.* **2007**, *37*, 735–742. [CrossRef]
33. McNeil, K.; Kang, T.H.K. Recycled Concrete Aggregates: A Review. *Int. J. Concr. Struct. Mater.* **2013**, *7*, 61–69. [CrossRef]
34. Tang, Y.; Xiao, J.; Zhang, H.; Duan, Z.; Xia, B. Mechanical properties and uniaxial compressive stress-strain behavior of fully recycled aggregate concrete. *Constr. Build. Mater.* **2022**, *323*, 126546. [CrossRef]
35. Kenai, S.; Debieb, F.; Azzouz, L. Mechanical properties and durability of concrete made with coarse and fine recycled concrete aggregates. In *Challenges of Concrete Construction: Volume 5, Sustainable Concrete Construction*; Thomas Telford Publishing: London, UK, 2002; pp. 383–392.
36. Zheng, Y.; Zhang, Y.; Zhang, P. Methods for improving the durability of recycled aggregate concrete: A review. *J. Mater. Res. Technol.* **2021**, *15*, 6367–6386. [CrossRef]
37. Liang, C.; Cai, Z.; Wu, H.; Xiao, J.; Zhang, Y.; Ma, Z. Chloride transport and induced steel corrosion in recycled aggregate concrete: A review. *Constr. Build. Mater.* **2021**, *282*, 122547. [CrossRef]
38. Poon, C.S.; Shui, Z.H.; Lam, L.; Fok, H.; Kou, S.C. Influence of moisture states of natural and recycled aggregates on the slump and compressive strength of concrete. *Cem. Concr. Res.* **2004**, *34*, 31–36. [CrossRef]
39. Vinay Kumar, B.M.; Ananthan, H.; Balaji, K.V.A. Experimental studies on utilization of recycled coarse and fine aggregates in high performance concrete mixes. *Alex. Eng. J.* **2018**, *57*, 1749–1759. [CrossRef]
40. Kou, S.-C.; Poon, C.-S.; Etxeberria, M. Influence of recycled aggregates on long term mechanical properties and pore size distribution of concrete. *Cem. Concr. Compos.* **2011**, *33*, 286–291. [CrossRef]
41. Singh, R.; Nayak, D.; Pandey, A.; Kumar, R.; Kumar, V. Effects of recycled fine aggregates on properties of concrete containing natural or recycled coarse aggregates: A comparative study. *J. Build. Eng.* **2022**, *45*, 103442. [CrossRef]
42. Kaarthik, M.; Maruthachalam, D. A sustainable approach of characteristic strength of concrete using recycled fine aggregate. *Mater. Today Proc.* **2021**, *45*, 6377–6380. [CrossRef]
43. EN 197-1:2001; Cement, Part 1: Composition, Specifications and Conformity Criteria for Common Cements. European Committee for Standardization: Brussels, Belgium, 2001; p. 40.
44. EN 933-11:2009/AC:2009; Tests for Geometrical Properties of Aggregates. Part 11: Classification Test for the Constituents of Coarse Recycled Aggregate. Ministry of Transport Mobility and Urban Agenda: Madrid, Spain, 2009; p. 18.
45. EN 12620:2003+A1:2009; Aggregates for Concrete. European Committee for Standardization: Brussels, Belgium, 2009; p. 60.
46. EN 1097-6:2014; Test for Mechanical and Physical Properties of Aggregates. Part 6: Determination of Particle Density and Water Absorption. European Committee for Standardization: Brussels, Belgium, 2014; p. 54.
47. EN 933-8:2011+A1:2015/1M:2016; Test for Geometrical Properties of Aggregates. Part 8: Assesment of Fines. Sand Equivalent Test. European Committee for Standardization: Brussels, Belgium, 2016; p. 26.

48. *EN 1097-2:2021*; Test for Mechanical and Physical Properties of Aggregates. Part 2: Methods for the Determination of Resistance to Fragmentation. European Committee for Standardization: Brussels, Belgium, 2021; p. 50.
49. *EN 933-3:2012*; Test for Geometrical Properties of Aggregates. Part 3: Determination of Particle Shape. Flakiness Index. European Committee for Standardization: Brussels, Belgium, 2012; p. 14.
50. *EN 1744-1:2010+A1:2013*; Tests for Chemical Properties of Aggregates. Part 1: Chemical Analysis. European Committee for Standardization: Brussels, Belgium, 2013; p. 64.
51. Andreu, G.; Miren, E. Experimental analysis of properties of high performance recycled aggregate concrete. *Constr. Build. Mater.* **2014**, *52*, 227–235. [CrossRef]
52. Gao, D.; Zhu, W.; Fang, D.; Tang, J.; Zhu, H. Shear behavior analysis and capacity prediction for the steel fiber reinforced concrete beam with recycled fine aggregate and recycled coarse aggregate. *Structures* **2022**, *37*, 44–55. [CrossRef]
53. Bairagi, N.K.; Ravande, K.; Pareek, V.K. Behavior of concrete with different proportions of natural and recycled aggregates. *Resour. Conserv. Recycl.* **1993**, *9*, 109–126. [CrossRef]
54. Tran, D.L.; Mouret, M.; Cassagnabère, F.; Phung, Q.T. Effects of intrinsic granular porosity and mineral admixtures on durability and transport properties of recycled aggregate concretes. *Mater. Today Commun.* **2022**, *33*, 104709. [CrossRef]
55. Teychenné, D.C.; Franklin, R.E.; Erntroy, H.C. *Design of Normal Concrete Mixes*; IHS BRE Press: Watford, UK, 2010.
56. *EN 12350-6:2020*; Testing Fresh Concrete. Part 6: Density. European Committee for Standardization: Brussels, Belgium, 2020; p. 15.
57. *EN 1350-7:2020+AC:2022*; Testing Fresh Concrete. Part 7: Air Content. Pressure Methods. European Committee for Standardization: Brussels, Belgium, 2022; p. 33.
58. *EN 12350-2:2020*; Testing Fresh Concrete. Part 2: Slump Test. European Committee for Standardization: Brussels, Belgium, 2020; p. 13.
59. *EN 12390-7:2020/AC:2020*; Testing Hardened Concrete. Part 7: Density of Hardened Concrete. European Committee for Standardization: Brussels, Belgium, 2020; p. 19.
60. *EN 12390-3:2020*; Testing Hardened Concrete. Part 3: Compressive Strength of Test Specimens. European Committee for Standardization: Brussels, Belgium, 2020; p. 24.
61. *EN 12390-5:2020*; Testing Hardened Concrete. Part 5: Flexural Strength of Test Specimens. European Committee for Standardization: Brussels, Belgium, 2020; p. 14.
62. *EN 12390-8:2009/1M:2011*; Testing Hardened Concrete. Part 8: Depth of Penetration of Water under Pressure. European Committee for Standardization: Brussels, Belgium, 2011; p. 16.
63. *EN 12390-1*; Testing Hardened Concrete. Part 1: Shape, Dimensions and Other Requirements for Specimens and Moulds. European Commttee for Standardization: Brussels, Belgium, 2022; p. 16.
64. Malešev, M.; Radonjanin, V.; Marinković, S. Recycled concrete as aggregate for structural concrete production. *Sustainability* **2010**, *2*, 1204–1225. [CrossRef]
65. Şimşek, O.; Pourghadri Sefidehkhan, H.; Gökçe, H.S. Performance of fly ash-blended Portland cement concrete developed by using fine or coarse recycled concrete aggregate. *Constr. Build. Mater.* **2022**, *357*, 129431. [CrossRef]
66. Kurda, R.; de Brito, J.; Silvestre, J.D. Combined influence of recycled concrete aggregates and high contents of fly ash on concrete properties. *Constr. Build. Mater.* **2017**, *157*, 554–572. [CrossRef]
67. Tuyan, M.; Mardani-Aghabaglou, A.; Ramyar, K. Freeze–thaw resistance, mechanical and transport properties of self-consolidating concrete incorporating coarse recycled concrete aggregate. *Mater. Des.* **2014**, *53*, 983–991. [CrossRef]
68. Chang, Y.-C.; Wang, Y.-Y.; Zhang, H.; Chen, J.; Geng, Y. Different influence of replacement ratio of recycled aggregate on uniaxial stress-strain relationship for recycled concrete with different concrete strengths. *Structures* **2022**, *42*, 284–308. [CrossRef]
69. Pedro, D.; de Brito, J.; Evangelista, L. Influence of the use of recycled concrete aggregates from different sources on structural concrete. *Constr. Build. Mater.* **2014**, *71*, 141–151. [CrossRef]
70. Mohammed, S.I.; Najim, K.B. Mechanical strength, flexural behavior and fracture energy of Recycled Concrete Aggregate self-compacting concrete. *Structures* **2020**, *23*, 34–43. [CrossRef]
71. Zega, C.J.; Di Maio, Á.A. Use of recycled fine aggregate in concretes with durable requirements. *Waste Manag.* **2011**, *31*, 2336–2340. [CrossRef] [PubMed]
72. Surendar, M.; Beulah Gnana Ananthi, G.; Sharaniya, M.; Deepak, M.S.; Soundarya, T.V. Mechanical properties of concrete with recycled aggregate and M—sand. *Mater. Today Proc.* **2021**, *44*, 1723–1730. [CrossRef]
73. Saini, B.S.; Singh, S.P. Flexural fatigue strength prediction of self compacting concrete made with recycled concrete aggregates and blended cements. *Constr. Build. Mater.* **2020**, *264*, 120233. [CrossRef]
74. Yaba, H.K.; Naji, H.S.; Younis, K.H.; Ibrahim, T.K. Compressive and flexural strengths of recycled aggregate concrete: Effect of different contents of metakaolin. *Mater. Today Proc.* **2021**, *45*, 4719–4723. [CrossRef]
75. Xiao, J.; Tang, Y.; Chen, H.; Zhang, H.; Xia, B. Effects of recycled aggregate combinations and recycled powder contents on fracture behavior of fully recycled aggregate concrete. *J. Clean. Prod.* **2022**, *366*, 132895. [CrossRef]
76. Kumar, R.; Gurram, S.C.B.; Minocha, A.K. Influence of recycled fine aggregate on microstructure and hardened properties of concrete. *Mag. Concr. Res.* **2017**, *69*, 1288–1295. [CrossRef]
77. Thomas, C.; Setien, J.; Polanco, J.A. Structural recycled aggregate concrete made with precast wastes. *Constr. Build. Mater.* **2016**, *114*, 536–546. [CrossRef]

78. Faella, C.; Lima, C.; Martinelli, E.; Pepe, M.; Realfonzo, R. Mechanical and durability performance of sustainable structural concretes: An experimental study. *Cem. Concr. Compos.* **2016**, *71*, 85–96. [CrossRef]
79. Kapoor, K.; Singh, S.P.; Singh, B. Water Permeation Properties of Self Compacting Concrete Made with Coarse and Fine Recycled Concrete Aggregates. *Int. J. Civ. Eng.* **2018**, *16*, 47–56. [CrossRef]
80. Velardo, P.; Sáez del Bosque, I.F.; Sánchez de Rojas, M.I.; De Belie, N.; Medina, C. Durability of concrete bearing polymer-treated mixed recycled aggregate. *Constr. Build. Mater.* **2022**, *315*, 125781. [CrossRef]

Disclaimer/Publisher's Note: The statements, opinions and data contained in all publications are solely those of the individual author(s) and contributor(s) and not of MDPI and/or the editor(s). MDPI and/or the editor(s) disclaim responsibility for any injury to people or property resulting from any ideas, methods, instructions or products referred to in the content.

Recycling 3D Printed Concrete Waste for Normal Strength Concrete Production

Girum Mindaye Mengistu and Rita Nemes *

Department of Construction Materials and Technologies, Budapest University of Technology and Economics, 1111 Budapest, Hungary; gmengistu@edu.bme.hu
* Correspondence: nemes.rita@emk.bme.hu

Abstract: As the use of 3D-printed concrete becomes more prevalent, the need for effective recycling methods becomes paramount. This study addresses this concern by exploring the repurposing of 3D-printed concrete waste as an aggregate in normal-strength concrete for C30/37 and C40/50 classes, covering both fine and coarse aggregates in its particle size distribution. The extent of recycled aggregate (RA) replacement was determined through sieve analysis. A two-stage investigation assessed the compressive strength performance of the concrete specimens. The initial stage produced reference specimens with no replacement, representing conventional concrete. In the second stage, variable specimens incorporated 50% and 67% recycled aggregate (RA) from 3D-printed concrete waste. Results revealed that in C40/50, both the 50% and 67% replacements consistently exhibited a higher strength than 0%. In C30/37, the 50% replacement displayed decreased strength compared to the 0% and 67%, while the 67% replacement consistently showed superior strength. Adjusting the water content impacted strength; at 67%, slight variations occurred, while at 50%, extra water led to a significant decrease. An overarching discovery is that the efficacy of the 67% replacement level holds regardless of the concrete strength class.

Keywords: recycled 3D-printed concrete aggregate; particle size distribution; mix design; compressive strength; normal strength concrete; interfacial transition zone

1. Introduction

In recent years, the construction industry has witnessed a transformative shift, with the advent of 3D printing technology revolutionizing traditional building methods. One of the most groundbreaking developments in this domain is 3D-printed concrete, a cutting-edge construction technique that has the potential to reshape the way we build structures. 3D-printed concrete, also known as the 3D printing of concrete or concrete 3D printing, is an innovative construction technology that uses computer-controlled robotic systems or large-scale 3D printers to deposit layers of concrete material in a precise and predetermined manner [1,2]. 3D-printed concrete (3DPC) finds applications across residential, commercial, and infrastructure projects, ranging from walls and buildings to bridges and smaller structures, all while continuing to evolve and shape the future of construction.

However, as the adoption of 3DPC grows, the need for sustainable and environmentally-responsible practices becomes increasingly critical. This environmental consideration is expected to gain further prominence in the future, particularly as 3D-printed concrete is applied in the construction of temporary buildings. The swift production pace inherent to this technology necessitates the prompt encounter and management of demolition waste in such projects. This dual emphasis on sustainability and efficiency positions 3D-printed concrete as a promising solution not only for current construction needs but also for the evolving demands of future structures. The construction industry's increasing inclination towards environmentally-conscious and resource-efficient solutions underscores the transformative potential of 3D-printed concrete in shaping the future of construction methodologies [3,4].

The incorporation of recycled coarse aggregate (RCA) into concrete mixes presents a promising avenue for sustainable construction practices. Various studies have delved into the impact of RCA replacement ratios on both the mechanical properties and durability of concrete structures. Research indicates that up to 30% RCA replacement has no significant impact on mechanical properties [5]. At 50% RCA replacement, there is typically a 10–15% reduction, and at 100% replacement, it can reduce mechanical properties by 30–40% [6–9]. Over a 2-year period, Abed et al. [10] investigated the compressive strength of concrete, incorporating 0%, 25%, and 50% of RCA alongside supplementary cementitious materials, revealing increased strength in recycled aggregate (RA) mixes over natural aggregate (NA) mixes and suggesting enhanced long-term performance attributed to the continuous hydration of attached mortar on the RCA, even during freeze-thaw cycles. Gonzalez and Moriconi [11] recommend that in regions prone to severe seismic activity, it is sufficient to replace up to 30% of conventional NA with RA to maintain acceptable structural performance. Kou and Poon [12] assessed the mechanical properties of RA produced from parent concrete (PC) with 28-day compressive strengths ranging from 30 to 100 MPa. After 28 days, concrete derived from PC with 80 and 100 MPa compressive strengths exhibited strengths exceeding 65 MPa, slightly surpassing those of natural aggregate concrete. Over a 90-day period, both mixtures achieved compressive strengths of 75 MPa. The highest compressive strength reported by Tu et al. [13] was 42 MPa after 91 days at a water-cement (w/c) ratio of 0.32.

In addressing the potential adverse effects on strength and durability in structural applications due to RCA, various guidelines have been established to provide specific recommendations. One such guideline, HB155-2002 [14], proposes measures to control the use of RCA in different concrete grades. For instance, it suggests limiting the replacement of conventional coarse aggregate with Class 1A RCA to 30% in Grade 1 structural concrete. In contrast, Grade 2 non-structural concrete is permitted to have a 100% replacement with RCA. Adding to this, BV-MI 01: 2005(H) [15] introduces recommendations that involve varying levels of RCA replacement, ranging from 30% to 100%, based on class strength and the geometrical properties of aggregates. However, it advises not to use RCA replacement for C50/60 and above class strength. It is noteworthy that different countries have adopted varying approaches to regulate the percentage of RCA replacement. For example, Brazil, China, Spain, and the UK have set a lower limit, restricting RCA replacement to 20% for structural concrete [16]. The diversification in these guidelines reflects regional considerations, experiences, and research findings, emphasizing the need for a nuanced approach in addressing the challenges associated with incorporating recycled materials in concrete production.

In conventional concrete, the use of aggregates with a diverse range of particle sizes is a common practice aimed at enhancing packing density and overall strength. However, in the realm of 3D-printed concrete, a departure from this conventional approach is observed. Instead, a more deliberate emphasis is placed on achieving a controlled and uniform particle size distribution of the fine aggregate [17]. This strategic adjustment is designed to facilitate a consistent flow through the printer nozzle, thereby ensuring the smooth extrusion of the material during the 3D printing process.

Hence, it is essential to determine the optimal replacement level for the recycled aggregate (RA) sourced from 3D-printed concrete waste to uphold the mechanical properties of the concrete. Additionally, a mechanism needs to be developed for incorporating the fine particle size RA into the mix. The decision to include fine recycled aggregate is influenced by the fact that the parental material (i.e., 3D-printed concrete) consists solely of fine aggregate, introducing an additional aspect for investigation.

This study aims to investigate the feasibility of recycling 3D-printed concrete waste as an aggregate in normal strength concrete, focusing on compressive strengths for the C30/37 and C40/50 MPa classes, covering both fine and coarse aggregates in particle size distribution.

Despite the increasing interest in 3D-printed concrete, there is a notable gap in comprehensive research on utilizing recycled aggregate (RA) from this material in conventional concrete production. This study seeks to bridge the divide between cutting-edge construction techniques and sustainable resource management. By repurposing waste materials from 3D-printed concrete processes, the objective is to showcase the potential for establishing a closed-loop system within the construction industry.

2. Materials and Methods

2.1. Materials

The material employed in the fabrication of 3D-printed concrete is the premixed white powder known as Sikacrete®-752 3D, combined solely with water. Sikacrete®-752 3D incorporates a fine aggregate with a maximum size of 2 mm, along with Portland cement and additives [18]. The specific type and proportions of these constituents are proprietary to the Sika company (Biatorbágy, Hungary). When mixing this powder with water, the company prescribes a water-to-powder ratio (w/p) ranging from 14% to 16%, based on the weight of the powder [18].

The 3D-printed concrete waste, obtained after Kovács [19] conducted a comprehensive investigation into the material properties of Sikacrete®-752 using both conventional and 3D printing casting methods for her MSc thesis work, is outlined in Table 1. The table includes relevant information on casting types, the material's water-to-powder ratio (w/p), and printing directions, with the symbol N/A indicating 'Not Applicable'. Following this study, the 3D-printed concrete waste was processed by crushing it into fine and coarse aggregates using a jaw crusher in one step.

Table 1. Physical and mechanical properties of waste concrete.

Type of Casting	w/p (%)	Direction of Printing	Density (kg/m^3)	Tensile Strength (MPa)		Compressive Strength (MPa)
				Flexural	Splitting	
Conventional	14.0	N/A	2084	9.4	3.9	63.8
	14.5	N/A	2084	8.8	3.2	63.1
	15.0	N/A	2066	8.7	4.2	61.1
3D Printing	14.0	horizontal	2116	6.9	3.3	53.8
	14.5	horizontal	2140	5.4	2.8	53.4
		vertical	2140	2.5	3.2	44.2
	15.0	horizontal	2119	3.6	3.5	53.0
		vertical	2119	2.4	3.1	48.1

In the recycling approach, all crushed particle sizes, including both fine and coarse particles, were used as recycled aggregates. This approach towards recycled aggregate usage emphasizes a comprehensive and sustainable recycling strategy. By incorporating the full spectrum of particle sizes, the aim is to optimize the potential for reusing 3D-printed concrete waste in a manner that enhances the overall material sustainability and minimizes environmental impact.

In this experimental study, a two-stage investigation was conducted to evaluate the performance of concrete specimens in terms of compressive strength using a 150 mm cube mold [20]. The initial stage concentrated on producing reference specimens, representing conventional concrete with no replacement. The materials used in this phase included CEM II/AS 42.5 N [21], river sand, and quartz gravel, aiming to establish a baseline for standard or reference mixes. These reference specimens, following conventional practices, serve as a reference point for subsequent comparisons. Having completed the first stage, the transition to the second phase emphasized the incorporation of recycled aggregate (RA) sourced from 3D-printed concrete. Visual representations of both quartz gravel and recycled aggregates from the 3D printed concrete are presented in Figure 1.

Figure 1. Aggregates: (**a**) Quartz gravel; (**b**) RA of 3D-printed concrete waste.

The specimens in this stage, still utilizing a 150 mm cube mold, showcase the potential of recycled materials in concrete production.

2.2. Particle Size Distribution

The meticulous attention to adhering to established particle distribution grading curves underscores the importance of maintaining compatibility with industry standards, particularly in the pursuit of sustainable and performance-enhancing concrete formulations. To determine the extent of recycled aggregate (RA) replacement, a precise sieve analysis was performed on the RA sourced from 3D-printed concrete and the river sand in accordance with the specifications outlined in EN 933-1:2012 [22]. This standardized procedure ensures the accurate assessment and comparison of particle size distribution for both types of aggregates. The resulting grading curves for each aggregate are visually represented in Figures 2 and 3, providing a clear depiction of the size distribution characteristics.

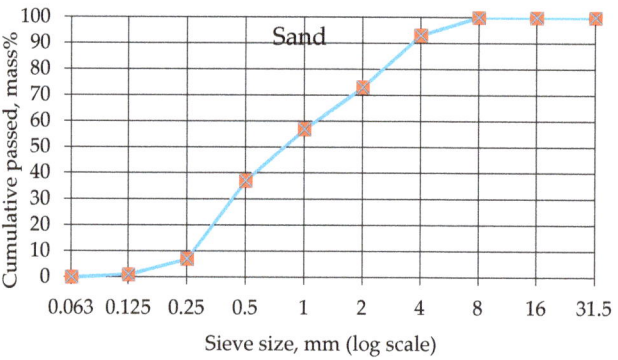

Figure 2. Grading curve of river sand.

Notably, river sand predominantly functions as a fine aggregate; however, a more detailed examination, illustrated in Figure 2, reveals the presence of 7% coarse aggregate with a particle size of 4mm by mass. This observation requires careful consideration during the mix design phase when proportioning the various ingredients for the overall construction material. By addressing these intricacies in the mix design, the goal is to achieve a composition that aligns with the specified requirements and ensures the desired performance characteristics of the resulting construction material.

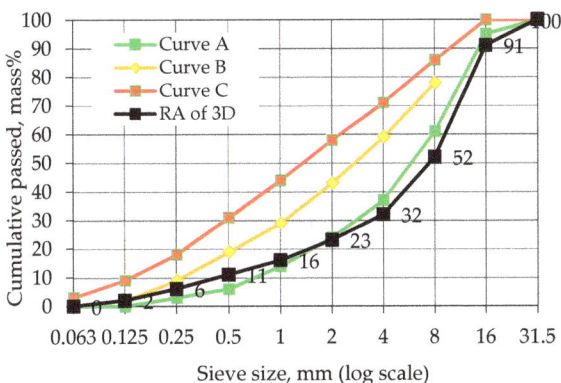

Figure 3. Grading curve of crushed 3D-printed concrete.

In Figure 3, the data reveals that 68% of the recycled aggregate, based on mass, constitutes a coarse aggregate. Especially, its grading curve deviates from the specified boundaries of curves A and C, contravening the stipulations outlined in MSZ 4798-1 [23] regarding grading distribution.

To rectify this discrepancy, a crucial step involves blending the recycled aggregate with sand in carefully determined proportions. The key objective is to establish a specific grading for the combined fine and coarse aggregates, ensuring an optimal distribution that maximizes the aggregate content while minimizing void spaces. This approach serves to reduce reliance on paste (water and cementitious material) in concrete formulations, consequently enhancing dimensional stability and overall durability.

Adhering to this guiding principle, the integration of recycled aggregates with sand has been executed with a specific approach, taking into account both the maximum level of replacement required to achieve a Class I and Class II aggregate size distribution [15,23]. Illustrated in Figures 4 and 5 are the replacement levels that successfully meet these criteria, represented by recycled aggregate to sand mass ratios of 2:1 (67% replacement) and 1:1 (50% replacement). This visual representation serves as a clear demonstration of the effectiveness of these specific ratios in achieving the desired aggregate classifications.

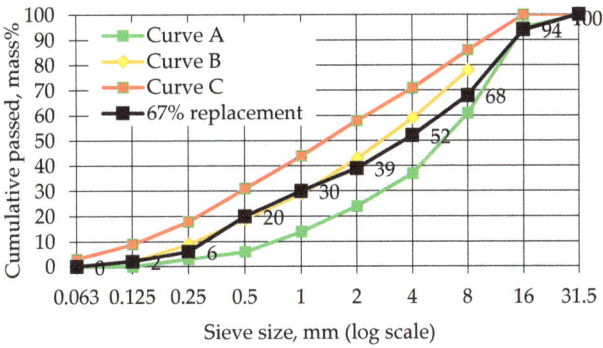

Figure 4. Grading curve for 67% replacement ratio (2:1).

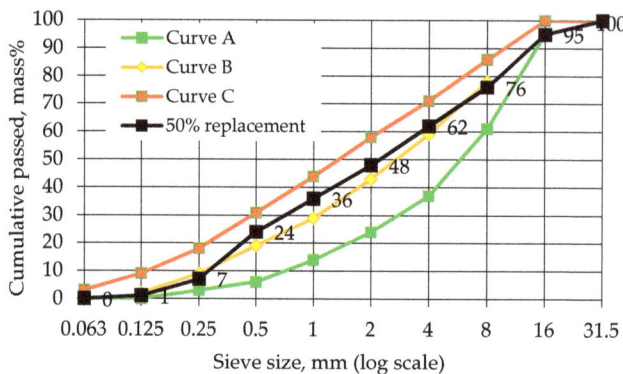

Figure 5. Grading curve for 50% replacement ratio (1:1).

By strategically incorporating these replacement levels, the aim is to strike a balance between sustainable practices and meeting the specified standards for aggregate distribution. The selection of these replacement levels is informed by careful consideration of the desired concrete properties and the need to optimize the use of recycled materials.

The concrete selected for this study belongs to the strength classes of C30/37 and C40/50, both of which fall within the category of normal strength concrete, as they are under the C50/60 threshold [24]. In summary, Figure 6 illustrates the grading curve of the aggregate, representing the combined fine and coarse aggregate particle size distribution, for both strength classes of concrete when there is 0% (i.e., no replacement level), 50% and 67% replacements.

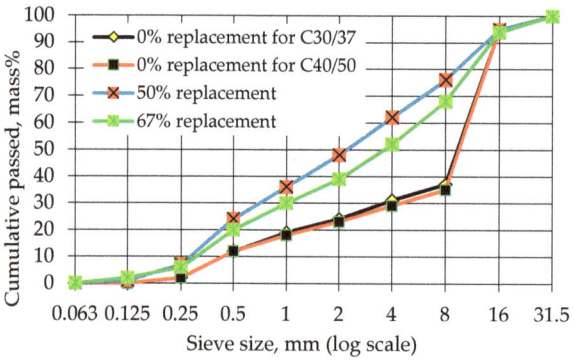

Figure 6. Grading curve of aggregate.

However, when we examine the scenarios with 50% and 67% replacement levels, irrespective of the concrete's strength class, a notable observation emerges. In these instances, the grading curve of the aggregate remains identical, as depicted in Figure 6. This indicates a consistent pattern in particle size distribution when a substantial portion of the aggregate is replaced, emphasizing a crucial aspect of the study's findings.

2.3. Mix Design

Table 2 provides a comprehensive breakdown of mix designations and the respective percentages of different materials incorporated into the mix. In parallel, Table 3 offers a detailed presentation of the mix proportions measured in kg/m^3, accompanied by the fresh concrete density for various mixtures. The focal points of this study encompass three main parameters: (i) the type of aggregate (natural or recycled), with varying proportions, (ii) the variation of water amount with constant cement content, and (iii) the strength class of the concrete, represented by C30/37 and C40/50.

Table 2. Mix designations and proportions of different ingredients.

Mix	Mix Designation	w/c	Superplasticizer (1% of Binder)	Aggregate (%) NA	RA
Mix 1	C40-RA0	0.41	-	100	-
Mix 2	C40-RA50	0.41	1	50	50
Mix 3	C40-RA50w+	0.51	-	50	50
Mix 4	C40-RA67	0.41	1	33	67
Mix 5	C40-RA67w−	0.38	1	33	67
Mix 6	C30-RA0	0.51	-	100	-
Mix 7	C30-RA50	0.51	1	50	50
Mix 8	C30-RA50w+	0.70	-	50	50
Mix 9	C30-RA67	0.51	1	33	67
Mix 10	C30-RA67w−	0.41	1	33	67

Table 3. Mix proportions and fresh density in kg/m^3.

Mix Designation	Cement	River Sand	River Gravel	RA [1]	Water	Superplasticizer (1% of Binder)	Fresh Density
C40-RA0	459	542	1206	-	188	-	2388
C40-RA50	459	874	-	874	188	4.59	2236
C40-RA50w+	459	874	-	874	234	-	2191
C40-RA67	459	583	-	1165	188	4.59	2207
C40-RA67w−	459	583	-	1165	174	4.59	2196
C30-RA0	343	621	1261	-	175	-	2382
C30-RA50	343	941	-	941	175	3.43	2178
C30-RA50w+	343	941	-	941	240	-	2156
C30-RA67	343	627	-	1255	175	3.43	2184
C30-RA67w−	343	627	-	1255	141	3.43	2147

[1] All crushed particle sizes, encompassing both fine and coarse particles.

To systematically explore the impact of these parameters, ten concrete mixtures, each comprising four specimens, were carefully prepared. These mixtures encompass different aggregate replacement percentages and water amounts, tailored for the selected concrete strength classes. The mix designations outlined in Tables 2 and 3 utilize specific labels: 'C' for concrete, 'NA' for natural aggregates, and 'RA' for recycled aggregates. The numerical values following 'C' and 'RA' denote the cylindrical characteristic compressive strength in MPa at the age of 28 days and the percentage of recycled aggregate by weight in the total aggregate, respectively. The symbols 'w+' and 'w−' represent the utilization of additional water without superplasticizer and the use of superplasticizer with a reduced water amount compared to the reference mix to achieve the desired workability in the recycled concrete mix. This systematic nomenclature provides a clear and organized framework for understanding the composition and characteristics of each concrete mixture in the study.

The primary source of natural coarse aggregate (NCA) is quartz gravel, with a minor fraction sourced from sand—this accounts for a maximum of 3%. River sand is employed as the natural fine aggregate. Recycled coarse and fine aggregates (RCA and RFA) are derived from crushed 3D-printed concrete waste.

The water-cement (w/c) ratio is a pivotal factor with a direct impact on the properties and workability of concrete mixes. In light of this, we introduce different water-cement ratios (w/c) beyond those of the reference mixes. In the absence of a superplasticizer, determining the highest permissible w/c ratio becomes crucial, aligning with the natural workability of recycled concrete. Conversely, the introduction of a superplasticizer into the mix presents another dimension. Superplasticizers, as chemical admixtures, enhance the workability of concrete without significantly altering the w/c ratio. In this scenario, selecting the lowest possible w/c ratio becomes imperative to maintain the desired workability with the assistance of the plasticizer. This allows for greater control over the mix's consistency and flow, even at a lower w/c ratio, offering advantages in achieving the desired properties in the final concrete while conserving water.

In both scenarios, the main emphasis is on adjusting the water content, while keeping the other mixing ingredients constant. This is done to evaluate the necessity of pre-saturation of the recycled aggregates (RA) with water before concrete mixing, given that 3D-printed concrete waste has a water absorption capability of 4.5% in its saturated state and a particle density of 2116 kg/m^3. This precautionary measure aims to prevent water absorption by the RA during the mixing process, aligning with recommendations from various researchers and guidelines [14,15]. This exploration also contributes to an enhanced understanding of how moisture conditions influence the performance of recycled aggregates in concrete mixes.

Four specimens of 150 mm cube size were cast using steel molds for each concrete mix. Subsequently, these cubes underwent a carefully devised mixed curing process. During the initial seven days, the cubes were immersed in water, providing a controlled environment for the crucial early stages of hydration. Following this submerged period, the specimens were exposed to the open air for the subsequent 21 days. This mixed curing approach is strategically designed to optimize the concrete's development of strength and durability. The initial submersion in water serves to create favorable conditions for the hydration process, crucial for the formation of a robust concrete structure. The subsequent exposure to open air replicates real-world atmospheric conditions, contributing to the concrete's long-term curing and enhancing its overall performance.

2.4. Testing Methods

The testing method employed is the standard compression test, conducted in accordance with EN 12390-3:2001 [20], to determine the compressive strength of cubic specimens. The results of the test are then checked for adherence to the criteria specified in EN 206:2013 [25]. The compressive strength testing utilizes a standardized compression testing machine, specifically the FORMTEST ALPHA 3-3000. Within this controlled testing environment, the specimens undergo a carefully regulated testing process, exposed to a consistent rate of loading set at 11.5 kN/s. As per EN 1992-1-1:2004 [26] standard, the cylindrical mean compressive strength value ($f_{cm,cyl}$) in MPa is determined by adding 8 to its cylindrical characteristic compressive strength value ($f_{ck,cyl}$). It is important to note that this relationship specifically applies to cylindrical samples measuring 150 mm in diameter, 300 mm in height, aged for 28 days, and cured underwater throughout the entire duration (wet cured) [27]. In practical concrete technology, the conformity of compressive strength is commonly assessed using 150 mm-sized cubes, tested at the age of 28 days, and subjected to mixed curing. For such scenarios, the required minimum mean cubic compressive strength values ($f_{cm,cube,H}$) for concrete classes up to C45/55 are provided in the BV-MI 01:2005 (H) [15], specifically for specimens aged 28 days and subjected to mixed curing as per MSZ 4798-1:2004 [23]. To meet the conformity criteria of EN 206:2013 [23] for compressive strength, this value should be less than or equal to the measured average value of its compressive strength ($f_{cm,cube,test,H}$) in accordance with BV-MI 01:2005 (H) [15].

In concrete, the interfacial transition zones (ITZs) serve as a bridge between the mortar matrix and the coarse aggregate particles. Even when the individual components are of high stiffness, the stiffness of the concrete may be low because of breaks in these bridges (i.e., voids and micro-cracks in the interfacial transition zone), which do not permit stress transfer [28]. In recycled aggregate concrete, there are more interfacial transition zones than normal aggregate concrete [29]. Therefore, the interfacial transition zone should be considered when the strength of the concrete is evaluated. Taking this into account, scanning electron microscopy (SEM) testing is conducted to assess the microstructure of the concrete at ITZ. The significance of the interfacial bond between the aggregate and the cementitious matrix has been underscored by various researchers [29–31]. The ITZ is commonly regarded as the weakest region in concrete, impacting both mechanical properties and durability. Understanding the stress deformation behavior of concrete is crucial, particularly in this zone, as it differs from that of its individual components; namely, hydrated cement paste and aggregate. This test examines the SEM images of the specimens at the interfacial transition zone (ITZ) to assess how the aggregate types influence the nature of aggregate-cement paste interfacial bonding.

3. Results and Discussion

3.1. Compressive Strength

3.1.1. Influence of Replacement Level

Following the testing of specimens at 28 days to determine both their individual and mean compressive strength, the equations specified in EN 206:2013 [25] are applied. These calculations lead to the determination of the measured average compressive strength value of the specimens ($f_{cm,cube,test,H}$), a crucial parameter used for classifying the concrete strength class according to BV-MI 01:2005 (H) [15]. The specific classifications are detailed in Table 4, providing a clear reference for the concrete's strength characteristics based on industry standards. This systematic approach ensures that the concrete production process is aligned with the established guidelines and results in accurate strength class classifications. These mixes are utilized to investigate the impact of replacement levels on compressive strength.

Table 4. Concrete strength with variable RA replacement.

Mix Designation	Density (kg/m^3)	$f_{cm,cube,test,H}$ (MPa)	Strength Class [15]
C40-RA0	2377	69	C40/50
C40-RA50	2220	75	C45/55
C40-RA67	2174	75	C45/55
C30-RA0	2363	57	C30/37
C30-RA50	2155	49	C30/37
C30-RA67	2165	58	C30/37

A comparison is made between the measured average compressive strength value of specimens ($f_{cm,cube,test,H}$) and the corresponding minimum required compressive strength value ($f_{cm,cube,H}$) for both C40/50 and C30/37 at the age of 28 days, which are specified as 67 MPa and 49 MPa respectively [15]. The results clearly indicate that the measured compressive strength average value for each concrete mix surpasses the mean compressive strength requirement value stipulated for the concrete strength class. This observation holds true for instances of mixed curing.

In other words, the measured mean compressive strength values consistently meet or exceed the specified requirement values for the respective concrete strength classes. This alignment underscores the robustness of the concrete mixes, confirming their adherence to the established standards [20–27] and emphasizing the reliability of the mean compressive strength results under the conditions of the study.

Figures 7 and 8 provide a comprehensive depiction of the variations in compressive strength and density with respect to the replacement levels for the mix series, respectively. The inclusion of error bars, based on standard deviation, enhances the precision and insightfulness of the presented data.

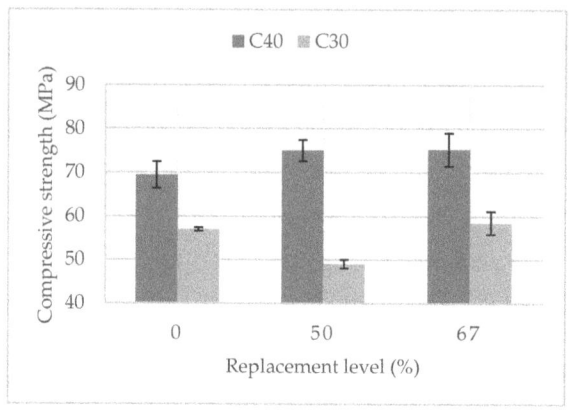

Figure 7. Measured compressive strength of C40 and C30 series.

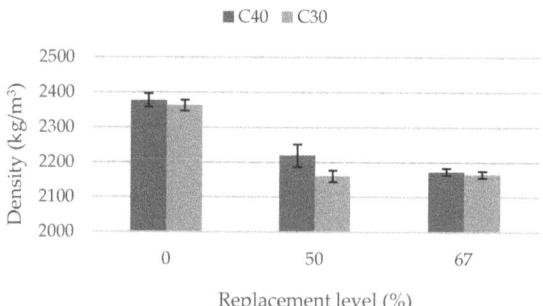

Figure 8. Air dry density of C40 and C30 series.

Focusing on the C40 series, Figure 7 reveals a noteworthy observation: the compressive strength at both 50% and 67% replacement levels is identical, and notably higher than that at the no replacement level (0%). Simultaneously, Figure 8 illustrates a consistent trend of decreasing density with an increasing replacement level.

In the context of the C30 series, Figures 7 and 8 show a significant trend where the 50% replacement level exhibits lower compressive strength and density compared to both the 0% (no replacement) and 67% replacement levels. This indicates that a 50% replacement negatively impacts the compressive strength and density of the concrete mix. One potential explanation for this is the high percentage of fine aggregate and poor compaction during casting.

The significant finding is that the 67% replacement level emerges as particularly effective. Remarkably, it not only aligns with the 0% (no replacement) scenario but even surpasses the compressive strength exhibited by the reference specimens. The primary reasons for low strength of the reference specimens are attributed to the smooth surface of the quartz gravel and the poor grading of the natural aggregate, as depicted in Figures 1 and 6, respectively.

Following the assessment of the conformity criteria for compressive strength and the analysis of the impact of replacement levels, the focus shifts to comparing the compressive ratio between the variable and reference specimens. In the majority of cases, the variable specimens with 50% and 60% replacement levels exhibit compressive strengths that are at least on par with, if not superior to, the reference specimens. A particularly noteworthy observation emerges: at a 67% replacement level, regardless of the concrete strength class, there is a consistent improvement in compressive strength. This implies that the 67%

replacement level is optimal for enhancing compressive strength, irrespective of the initial concrete strength class, highlighting a potentially advantageous mix design strategy.

3.1.2. Influence of Water Content

To demonstrate the impact of water content variation in the mixing process, we opted for the mix specified in Table 5. This particular mix features diverse water content levels while maintaining the same recycled aggregate (RA), replacement level, and concrete strength class. For the purpose of this comparison, C40-RA50, C40-RA67, C30-RA50, and C30-RA67 have been selected as reference mixes. These reference mixes share a similar water-cement ratio (w/c), with the mixes having 0% replacement of RA within their respective strength class [see, Table 2].

Table 5. Influence of water amount in the C40 and C30 mix series.

Mix Designation	Density (kg/m^3)	$f_{cm,cube,test,H}$ (MPa)	Strength Class [15]
C40-RA50	2220	75	C45/55
C40-RA50w+	2174	60	C35/45
C40-RA67	2174	75	C45/55
C40-RA67w−	2196	74	C40/50
C30-RA50	2155	49	C30/37
C30-RA50w+	2128	37	C20/25
C30-RA67	2165	58	C30/37
C30-RA67w−	2147	59	C30/37

Initially, we compared the measured mean compressive strength against the required minimum mean compressive strength for the respect strength class concrete, as illustrated in Figure 9.

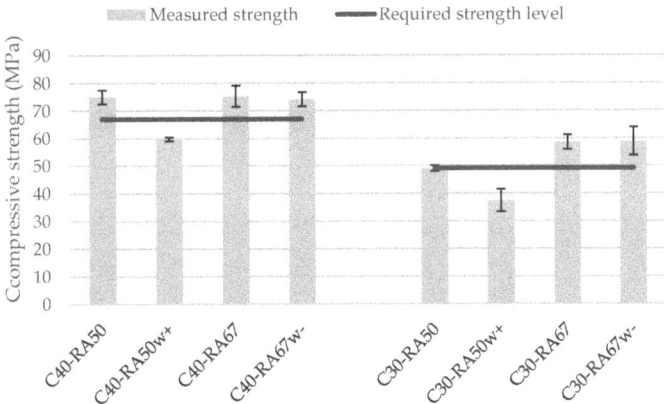

Figure 9. Influence of water content on compressive strength.

A prominent observation from Figure 9 indicates that the utilization of higher amounts of water, particularly for the 50% replacement level, fails to meet the required minimum compressive strength for the targeted strength class. Consequently, this leads to a downgrade in the strength class, as detailed in Table 5. Conversely, in the case of a lower amount of water used for the 67% replacement, although it satisfies the conformity of the mean compressive strength, the slight difference in the measured compressive strength results in a change in the concrete strength class, as indicated in Table 5.

Therefore, altering the water content in the mix results in a decrease in compressive strength compared to the reference mix. Additionally, the pre-saturation of the recycled aggregate in 3D-printed concrete waste is deemed unnecessary before mixing.

3.2. Aggregate-Cement Paste Interfacial Bonding (Interfacial Transition Zone)

After conducting the compressive test on both normal concrete and recycled concrete specimens, the shape of the developed cracks in the failed specimens was examined using SEM, as depicted in Figure 10. In the case of normal concrete, the specimens failed with the development of cracks in the interfacial transition zone (ITZ) between the natural coarse aggregate (quartz gravel) and the cement mortar. Furthermore, after the compressive test on normal concrete, the quartz gravel easily separated from the mortar due to the failure in the interfacial transition zone between the quartz gravel and mortar. Despite the bond cracks in the interfacial transition zone being less wide than those in recycled concrete, the smooth surface of the quartz gravel led to a lack of mechanical interlock between the quartz gravel and mortar. On the other hand, the specimens of recycled concrete failed with the development of cracks in the interfacial transition zone between the recycled coarse aggregate and the cement mortar, as well as the failure of the aggregate itself.

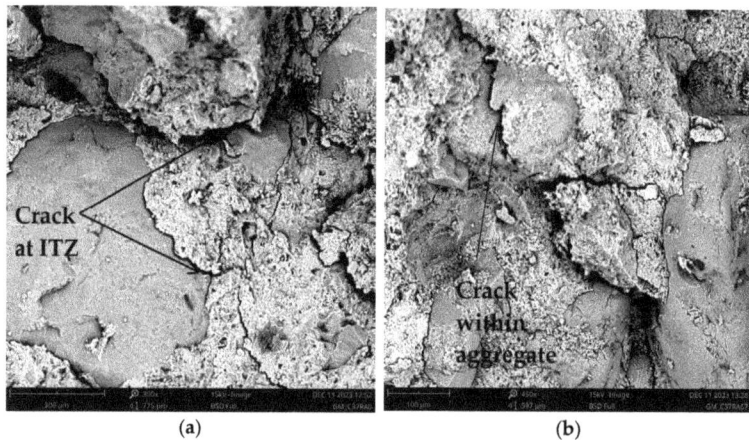

Figure 10. SEM images at ITZ. (**a**) Normal concrete; (**b**) Recycled concrete.

4. Conclusions

In conclusion, empirical testing of the mean compressive strength confirms the robustness of the concrete mixes, aligning with BV-MI 01: 2005(H) standards for respective strength classes. In C40/50 concrete, both the 50% and 67% replacement levels consistently exhibit higher compressive strength than the 0% replacement level. Transitioning to C30/37, a notable trend emerges: the 50% replacement level shows decreased strength compared to both the 0% and 67% replacement levels, while the 67% replacement level consistently displays superior compressive strength. In this range of replacement, the use of crushed concrete waste from 3D-printed concrete (3D RCA) is more similar to that of the RCAs from normal concrete. Additionally, adjusting the water content impacts the compressive strength in concrete mixes. At a 67% replacement, slight variations occur, while at 50%, additional water leads to a significant strength decrease.

SEM analysis of normal and recycled concrete specimens post-compressive test revealed distinctive cracking patterns. Normal concrete showed cracks in the interfacial transition zone (ITZ) between the natural coarse aggregate and the cement mortar, resulting in easy separation. Recycled concrete exhibited cracks in the ITZ between the recycled coarse aggregate and the cement mortar, along with aggregate failure. These findings highlight divergent failure mechanisms and underscore the importance of crack analysis for concrete performance enhancement.

An overarching discovery is that the efficacy of the 67% replacement level holds regardless of the concrete strength class. Remarkably, it not only outperforms the no-replacement scenario but even exceeds the compressive strength of reference specimens. This underscores the strategic advantage of a 67% replacement level in optimizing compressive strength across diverse concrete classes.

Author Contributions: Conceptualization, R.N.; methodology, G.M.M. and R.N.; software, G.M.M.; validation, R.N.; formal analysis, G.M.M.; investigation, G.M.M. and R.N.; resources, G.M.M. and R.N.; data curation, R.N.; writing—original draft preparation, G.M.M.; writing—review and editing, R.N.; visualization, G.M.M.; supervision, R.N. All authors have read and agreed to the published version of the manuscript.

Funding: The authors of the present manuscript are grateful to the Hungarian Scientific Research Fund (OTKA K 146724 project) for support.

Institutional Review Board Statement: Not applicable.

Informed Consent Statement: Not applicable.

Data Availability Statement: Data are contained within the article.

Acknowledgments: The authors of this manuscript would like to thank Duna-Dráva Cement Ltd., (Vác, Hungary) for the cement provided for the experiments.

Conflicts of Interest: The authors declare no conflict of interest.

References

1. Jay, S.; Ali, N.; Behzad, N. *3D Concrete Printing Technology*, 1st ed.; Butterworth-Heinemann: Kidlington, UK, 2019; pp. 1–10.
2. Zhang, J.; Wang, J.; Dong, S.; Yu, X.; Han, B. A review of the current progress and application of 3D printed concrete. *Compos. Part A Appl. Sci. Manuf.* **2019**, *125*, 105533. [CrossRef]
3. Suhendro, B. Toward green concrete for better sustainable environment. *Procedia Eng.* **2014**, *95*, 305–320. [CrossRef]
4. Agustí-Juan, I.; Habert, G. Environmental design guidelines for digital fabrication. *J. Clean. Prod.* **2017**, *142*, 2780–2791. [CrossRef]
5. Kofi, A. State of the art review of current literature and development studies on recycled aggregate concrete. *Bull. Eng.* **2018**, *11*, 121–134.
6. Xu, J.J.; Chen, Z.P.; Ozbakkaloglu, T.; Zhao, X.Y.; Demartino, C. A critical assessment of the compressive behavior of reinforced recycled aggregate concrete columns. *Eng. Struct.* **2018**, *161*, 161–175. [CrossRef]
7. Kisku, N.; Joshi, H.; Ansari, M.; Panda, S.K.; Sanket, N.; Sekhar, C.D. A critical review and assessment for usage of recycled aggregate as sustainable construction material. *Constr. Build. Mater.* **2017**, *131*, 721–740. [CrossRef]
8. Tam, V.-W.-Y.; Gao, X.-F.; Tam, C.M. Comparing performance of modified two-stage mixing approach for producing recycled aggregate concrete. *Mag. Concr. Res.* **2006**, *58*, 477–484. [CrossRef]
9. Guo, J.; Chen, Q.; Chen, W.; Cai, J. Tests and numerical studies on strain-rate effect on compressive strength of recycled aggregate concrete. *J. Mater. Civ. Eng.* **2019**, *31*, 04019281. [CrossRef]
10. Abed, M.A.; Tayeh, B.A.; Abu Bakar, B.H.; Nemes, R. Two-year non-destructive evaluation of eco-efficient concrete at ambient temperature and after freeze-thaw cycles. *Sustainability* **2021**, *13*, 10605. [CrossRef]
11. Corinaldesi, V.; Moriconi, G. Influence of mineral additions on the performance of 100% recycled aggregate concrete. *Constr. Build. Mater.* **2009**, *23*, 2869–2876. [CrossRef]
12. Kou, S.C.; Poon, C.S. Effect of the quality of parent concrete on the properties of high performance recycled aggregate concrete. *Constr. Build. Mater.* **2015**, *77*, 501–508. [CrossRef]
13. Tu, T.Y.; Chen, Y.Y.; Hwang, C.L. Properties of HPC with recycled aggregates. *Cem. Concr. Res.* **2006**, *36*, 943–950. [CrossRef]
14. *HB155: 2002*; Guide to the Use of Recycled Concrete and Masonry Materials. Standards Australia: Sydney, Australia, 2002.
15. *BV-MI 01: 2005(H)*; Technical Guideline for the Production and Utilization of Concrete out of Recycled Aggregates. Budapest, Hungary, 2005.
16. Tam, V.W.Y. Comparing the implementation of concrete recycling in the Australian and Japanese construction industries. *J. Clean. Prod.* **2009**, *17*, 688–702. [CrossRef]
17. Le, T.T.; Austin, S.A.; Lim, S.; Buswell, R.A.; Law, R. Mix design and fresh properties for high-performance printing concrete. *Mater. Struct.* **2012**, *45*, 1221–1232. [CrossRef]
18. Product Data Sheet. Sikacrete®-752 3D—1 Part Micro-Concrete for 3D Printing. Available online: https://gcc.sika.com/dms/getdocument.get/3e3a8234-36f9-4a74-bae8-9cf5b3eb2a1c/sikacrete_-752_3d.pdf (accessed on 1 December 2023).

19. Kovács, I. 3D Printed Concrete–Material Structure Properties. Master's Thesis, BME, Faculty of Civil Engineering, Budapest, Hungary, 2023.
20. *EN 12390-3:2001*; Testing Hardened Concrete—Part 3: Compressive Strength of Test Specimens. European Committee for Standardization:: Brussels, Belgium, 2001.
21. *EN 197-1:2000*; Cement—Part 1: Composition, Specifications and Conformity Criteria for Common Cements. European Committee for Standardization: Brussels, Belgium, 2000.
22. *EN 933-1:2012*; Tests for Geometrical Properties of Aggregates. European Committee for Standardization: Brussels, Belgium, 2012.
23. *MSZ 4798-1:2004*; Concrete Part 1: Specification, Performance Production, Conformity, and Rules of Application of MSZ EN 206-1. Budapest, Hungary, 2004.
24. *Fib Model Code for Concrete Structures*; Ernst &Sohn: Berlin, Germany, 2010; p. 75.
25. *EN 206:2013*; Concrete—Specification, Performance, Production, and Conformity. European Committee for Standardization: Brussels, Belgium, 2013.
26. *EN 1992-1-1:2004*; Design of Concrete Structures—Part 1-1: General Rules and Rules for Buildings. European Committee for Standardization: Brussels, Belgium, 2004.
27. *EN 12390-2:2019*; Testing Hardened Concrete—Part 2: Making and Curing Specimens for Strength Tests. European Committee for Standardization: Brussels, Belgium, 2019.
28. Povindar, K.M.; Paulo, J.M. *Concrete: Microstructure, Properties and Materials*, 3rd ed.; McGraw-Hill: New York, NY, USA, 2005; pp. 42–47.
29. Lee, G.C.; Choi, H.B. Study on interfacial transition zone properties of recycled aggregate by micro-hardness test. *Constr. Build. Mater.* **2013**, *40*, 455–460. [CrossRef]
30. Poon, C.S.; Shui, Z.H.; Lam, L. Effect of microstructure of ITZ on compressive strength of concrete prepared with recycled aggregates. *Constr. Build. Mater.* **2004**, *18*, 461–468. [CrossRef]
31. Tasong, W.A.; Lynsdale, C.J.; Cripps, J.C. Aggregate-cement paste interface. II: Influence of aggregate physical properties. *Cem. Concr. Res.* **1998**, *28*, 1453–1465. [CrossRef]

Disclaimer/Publisher's Note: The statements, opinions and data contained in all publications are solely those of the individual author(s) and contributor(s) and not of MDPI and/or the editor(s). MDPI and/or the editor(s) disclaim responsibility for any injury to people or property resulting from any ideas, methods, instructions or products referred to in the content.

Article

Prediction Models for Mechanical Properties of Cement-Bound Aggregate with Waste Rubber

Matija Zvonarić [1], Mirta Benšić [2], Ivana Barišić [1,*] and Tihomir Dokšanović [1]

[1] Faculty of Civil Engineering and Architecture Osijek, Josip Juraj Strossmayer University of Osijek, 31000 Osijek, Croatia; mzvonaric@gfos.hr (M.Z.); tdoksanovic@gfos.hr (T.D.)
[2] Department of Mathematics, Josip Juraj Strossmayer University of Osijek, 31000 Osijek, Croatia; mirta@mathos.hr
* Correspondence: ivana@gfos.hr

Abstract: The high stiffness of cement-bound aggregate (CBA) is recognized as its main drawback. The stiffness is described by the modulus of elasticity, which is difficult to determine precisely in CBA. Incorporating rubber in these mixtures reduces their stiffness, but mathematical models of the influence of rubber on the mechanical characteristics have not previously been defined. The scope of this research was to define a prediction model for the compressive strength (f_c), dynamic modulus of elasticity (E_{dyn}) and static modulus of elasticity (E_{st}) based on the measured ultrasonic pulse velocity as a non-destructive test method. The difference between these two moduli is based on the measurement method. Within this research, the cement and waste rubber content were varied, and the mechanical properties were determined for three curing periods. The E_{dyn} was measured using the ultrasonic pulse velocity (UPV), while the E_{st} was determined using three-dimensional digital image correlation (3D DIC). The influence of the amount of cement and rubber and the curing period on the UPV was determined. The development of prediction models for estimating the f_c and E_{st} of CBA modified with waste rubber based on the non-destructive test results is highlighted as the most significant contribution of this work. The curing period was statistically significant for the prediction of the E_{st}, which points to the development of CBA elastic properties through different stages during the cement-hydration process. By contrast, the curing period was not statistically significant when estimating the f_c, resulting in a simplified, practical and usable prediction model.

Keywords: prediction models; cement-bound aggregate; waste rubber; compressive strength; modulus of elasticity; ultrasonic pulse velocity; non-destructive testing

1. Introduction

In semi-flexible pavements, cement-bound aggregate (CBA) is used as a bearing layer. This layer provides improved bearing capacity and freeze—thaw resistance while presenting an even surface for installing asphalt layers. Despite all the benefits, these materials are prone to cracking due to cement hydration and the expansion of these cracks under the influence of repeated traffic loads [1]. Recently, waste rubber has been used in this material to release internal stresses and reduce the occurrence of cracks. The quality of this material is primarily described by its compressive strength. The compressive strength of CBA is usually tested after 7 and 28 days [2], but it is often measured after 90, 180 and 360 curing days for CBA modified with materials possessing postponed pozzolanic activity. The strength is significantly lower than that of conventional concrete, and satisfactory 7-day compressive strength ranges from 2.1 to 2.8 MPa [3,4]. Determining the compressive strength implies measuring the breaking force of a sample exposed to a uniaxial compressive load. The destruction of a sample is an acceptable way of testing laboratory-prepared samples when a sufficient number of samples can be produced. However, when evaluating the material incorporated in a pavement, there is a limited number of cored specimens. Preserving the sample for many test procedures is very useful in this case. Additionally, conducting a field evaluation of the inbuilt bearing layer using non-destructive testing is preferable.

The other important characteristic of cement-bound aggregate is its elasticity modulus (E). The dynamic (E_{dyn}) and static (E_{st}) moduli of elasticity can be measured within cement-based materials [5]. The E_{st} is determined from the linear relationship of the stresses and strains during the compression strength test. An obstacle in determining the static modulus of elasticity is the rough surface of this material [6], because the procedure entails the precise measurement of microscopic vertical displacements of points on the sample derivatives during the change in compression force. In addition to the difficulty of ensuring precise measurements, this is also a destructive method. At the same time, the static modulus of elasticity is significantly lower than the dynamic modulus [6].

On the other hand, the dynamic modulus of elasticity is usually measured using the ultrasonic pulse velocity. This is a non-destructive method, usually used in concrete testing, after which the sample is ready for further testing, and it can be applied to various materials. Ultrasonic pulses are pulses with frequencies over 20 Hz. There are several conventional ultrasonic testing methods, such as the pulse-echo ultrasonic, pitch-catch ultrasonic, immersion-based ultrasonic, air-coupled ultrasonic, oblique incidence, phase array ultrasonic and laser-ultrasonics and non-contact laser-ultrasonic techniques [7]. The choice of method depends on the tested material, the size of the specimen and external conditions. As mentioned above, cement hydration is a time-dependent process, so the passage of time significantly affects the development of the material stiffness. Guotang et al. [8] explain three typical stages of UPV development during the first 55 h of cement-stabilized aggregate microstructure formation. In the first stage, the UPV is stable at low values, followed by the second stage, where, due to cement hydration, the UPV rapidly increases. In the third stage, the UPV gradually becomes stable due to a rigid and stable matrix.

These methods apply to all materials used in road construction, starting from the stabilized soil through bearing layers to asphalt materials. Raavi and Tripura [9] developed prediction models for compressive and indirect tensile strength estimation of unstabilized and stabilized rammed earth based on UPV measurement. The authors also encourage using UPV measurement as an effective strength-estimation method. Furthermore, in [10], the authors emphasize that to develop prediction models, it is necessary to increase the number of UPV measurements in each direction (x, y, z) to four to increase the precision of the results. In addition, the importance of not carrying out measurements at the same point on the sample is emphasized. A prediction model for shear modulus estimation was developed in [11] by applying this method to cement-stabilized clays. The UPV proved helpful in multifunctional analysis, which predicts the compressive strength and rebound value [11]. Furthermore, this non-destructive method achieved reliable results in evaluating cement-bound aggregate. Barišić et al. [12] observed a strong relationship between the UPV and the compressive- and indirect-tensile-strength values and emphasized polynomial and exponential laws as the most appropriate to describe the relationship between strength and UPV. They also defined a range of UPVs in which CBA achieves satisfactory characteristics, which is helpful when making decisions during an examination. This paper develops models for steel-slag-stabilized mixtures for three different curing ages: 7, 28 and 90 days. Liu et al. observed a difference between compression and tension modulus and developed a power function decay model for these two parameters [13]. According to Mandal et al. [14], the UPV can be used to estimate the mechanical properties of most cement-stabilized materials, except cement-stabilized clay, which behaves differently from other stabilized materials. This paper presents strong correlations in the developed models between the flexural strength and the constrained modulus and the flexural modulus and the constrained modulus based on 7-day-old specimens. These parameters are commonly tested for the evaluation of soil behavior. In addition to CBA, roller-compacted concrete (RCC) is also used in pavement construction. Regarding its mechanical properties, this material occupies a place between CBA and concrete. Prediction models have also been developed for such materials, which predict the compressive strength based on the rebound number and the UPV [15]. The relationship between the rebound number and the compressive strength is established by the power law, while the relationship between the UPV and compressive strength is established by the exponential law. Additionally, the authors developed a logarithmic relationship between the dynamic modulus of the elasticity and the compressive strength. Furthermore, Rao et al. [16] developed an equation for estimating the E_{dyn} of the RCC based on the fly ash content, UPV and curing period, which agrees with experimental tests. The RCC in these research works is combined with crumb rubber, nano-silica and fly ash, making UPV a universal tool for model development in different materials. The UPV mainly served as a compressive-strength-prediction tool in the research on stabilized granular materials and no research deals with the issue of the static modulus of elasticity. Furthermore, all the models of CBA developed are nonlinear. There is a consensus on the utility of using the UPV technique when evaluating the mechanical characteristics of coherent [9–11,17] and incoherent materials [18]. The reliability of all the developed models is based on the coefficient of determination, which is not a reliable parameter for evaluating nonlinear models in the sphere of statistical inference. Therefore, the need for a more detailed statistical data analysis in this area is emphasized.

The non-destructive nature of UPV measurement also applies to asphalt mixtures. Norambuena-Contreras et al. [19] state that the dynamic modulus measured by the UPV can replace the low-frequency standard dynamic test. They also emphasize this method as cheaper, faster and easier to implement. In determining the E_{dyn} of asphalt mixtures by UPV, Majhi et al. [20] concluded that more reliable modulus values are obtained by considering the bulk density rather than the geometric density of asphalt specimens. The testing of the moisture sensitivity of asphalt mixtures using the UPV in [21] resulted in a linear equation between the seismic modulus and the UPV with a good coefficient of determination. Using this model, the moisture susceptibility of asphalt specimens can be predicted.

In addition to the desire for non-destructive testing methods, the trend of the circular economy has also been expressed in recent times. There are increasing numbers of applications of different waste materials in composite materials used in construction. Some of these are used as aggregates, while those with pronounced pozzolanic properties are used as binders. For example, Jackowski et al. [22] investigated the possibility of using different additives to cement and different fibers in the production of concrete bricks, while Ramadani et al. [23] investigated the possibility of using glass powder in combination with waste rubber in concrete. Rubber has also showed potential in increasing the resistance of concrete structures to the impact of earthquakes [24]. Guided by the desire to preserve the environment, rubber was used as a waste material in this work, since, due to its pronounced elastic properties, it can affect the reduction in the high stiffness of cement-stabilized aggregates and, as a waste material, it is very easily available on the market, considering the large consumption of tires. In addition, in most countries in Europe, the collection and processing of tires is very well organized [25]. Furthermore, the possibilities of using waste rubber in road construction is highlighted [24,26]. However, prior to waste rubber's incorporation in pavement materials, it has to pass through a certain separation process, in which steel fibers are separated from the rubber. These steel fibers are applied as reinforcements in concrete [27].

The aim of this research is to develop reliable prediction models for f_c and E_{st} estimation based on the measured UPV of CBA modified with waste rubber. Such a model would ensure a simple, fast, non-destructive approach to characterizing CBA by adding waste rubber. Furthermore, based on the literature review, it is concluded that none of the prediction models developed to date consider both the UPV and the length of the curing regimes of specimens for the prediction of mechanical properties, which would greatly facilitate the application of such models. Furthermore, a complete lack of prediction models for estimating the static modulus of elasticity was observed. Considering the difficulty of precisely determining this parameter, such a model would contribute significantly to this field.

2. Materials and Methods

Within this research, 15 cement-stabilized mixtures were tested. The materials used were natural river sand and gravel, waste granulated rubber, Portland cement of grade 32.5R (CEM II B/M (P-S) 32.5R) as a binder and the optimal amount of water determined according to standard [28]. The density of used materials is presented in Table 1, while the physical and mechanical properties of used binder are presented in Table 2.

Table 1. Densities of used material.

Aggregate	Sand		Gravel		Rubber	Cement
Size	0–2 mm	0–4 mm	4–8 mm	8–16 mm	0–0.5 mm	
Density (g/cm^3)	2.86	2.96	2.63	2.70	1.12	2.92

Table 2. Physical and chemical properties of cement.

Physical Properties		Chemical Properties	
Start of binding (min)	200	SO$_3$ (%)	3.2
Volume stability acc. to Le Chatelier (mm)	0.4	Cl (%)	0.009
Pressure strength after 2 days (MPa)	16		
Pressure strength after 28 days (MPa)	42		

The granulometric composition of the aggregates was determined using the European standard EN 933-1 [29] and is presented in Figure 1. The composition of the mixture shown in Figure 1 is tailored to the inclusion of rubber as per the flexibility allowed in the fifth category of the EN 14227-1 [30]. Cement was used as a binder in proportions of 3%, 5% and 7% of the aggregate mass. Due to their similar granulometric curves, fine-granulated rubber (0–0.5 mm) derived from end-of-life (ELT) car and truck tires was used as a volume replacement for sand in amounts of 10%, 20%, 30% and 40%. The detailed composition of the tested CBA mixtures is presented in Table 3. The rubber content is defined according to previous results, indicating 60% replacement, causing extremely high strength loss [31,32].

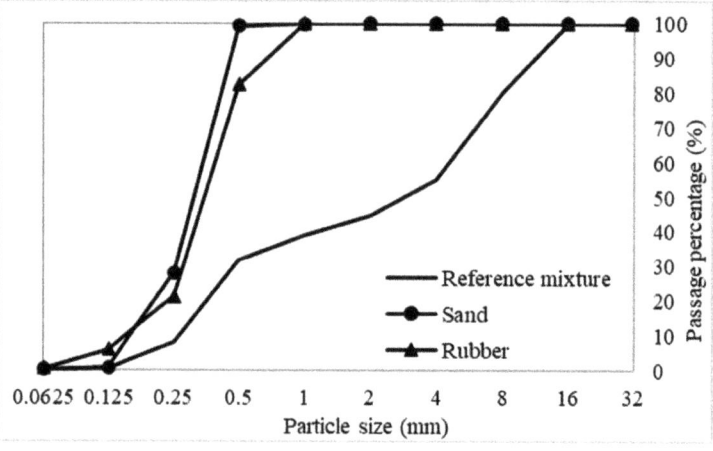

Figure 1. Granulometric composition.

Table 3. CBA mixture composition.

Mixture	Cement (%)	Sand (%)	Rubber (%)
C3R0		100	0
C3R10		90	10
C3R20	3	80	20
C3R30		70	30
C3R40		60	40
C5R0		100	0
C5R10		90	10
C5R20	5	80	20
C5R30		70	30
C5R40		60	40
C7R0		100	0
C7R10		90	10
C7R20	7	80	20
C7R30		70	30
C7R40		60	40

Rubber replacing the fine fraction complies with several other research papers [33–38]. Each rubber proportion was added to each cement amount, resulting in 12 rubberized mixtures. Three standard cylindrical specimens measuring Ø100 mm and with heights of 120 mm of each mixture were compacted by a vibrating hammer according to the procedure prescribed in EN 13286-51 [39].

The specimens were produced in order to test their compressive strength and determine their static (E_{st}) and dynamic (E_{dyn}) moduli of elasticity. The modulus of elasticity is the slope of a material's stress—strain curve with its elastic region. Before destructive testing, the non-destructive method for determining E_{dyn} was employed according to standard EN 12504-4 [40]. This method is carried out on specimens of known dimensions and density, with two transducers applied to opposite bases of cylindrical specimens, emitting ultrasonic waves and measuring the duration of their passage, which is used to calculate the E_{dyn}. Poisson's coefficient is needed for the calculation, for which the value 0.25 was adopted in this research as a typical value for CBA. Poisson's ratio usually ranges from 0.15 to 0.30 for cement-stabilized materials; the value of 0.25 was adopted in previous papers [41,42]. In addition, to neutralize the imperfect contact between the transducer and the rough specimen surface, a gel was applied. The procedure for determining the dimensions and mass of the specimen and the apparatus required for the UPV test are shown in Figure 2.

The measurement of E_{st} was carried out during the compressive strength test. The test was carried out according to EN 13286-43 [43] from the stress-and-strain relationship. Due to the inaccuracies in using LVDT for strain measurement, caused by the setting of the sample and the breakage of aggregates during the test, such results may be unreliable [6]. Therefore, in this research, a 3D DIC method was used to monitor the displacement of the characteristic points of the specimens. This is an optical non-contact method for monitoring the changes on the observed surface, in this case, vertical displacements. More details on the 3D DIC method used in this research and its applicability are presented in [44]. The procedure for E_{st} testing is shown in Figure 3. Testing of the compressive strength was carried out according to EN 13286-41 [45], exposing the specimen to a compressive load with the input force such that the fracture of the specimen occurred between the 30th and 120th second from the commencement of the load; specific experience is needed to conduct this test. The compressive strength was calculated from the peak force, i.e., the force at which the fracture occurred in the area on which the load was applied.

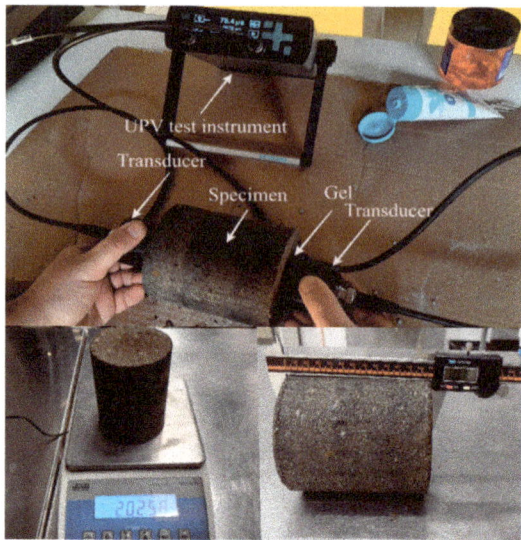

Figure 2. UPV testing.

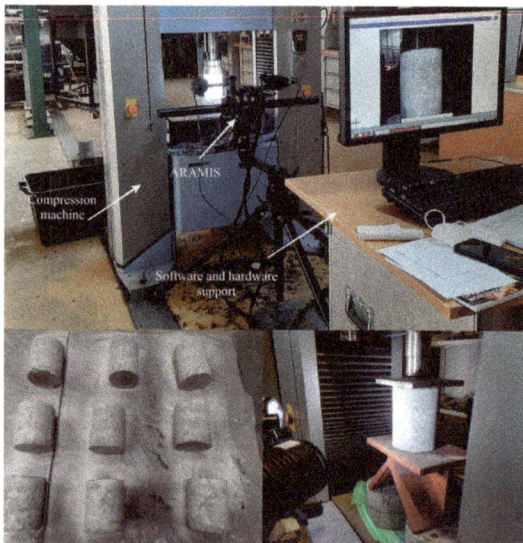

Figure 3. The f_c and E_{st} testing.

3. Results

The results obtained for the dynamic modulus of elasticity (E_{dyn}), compressive strength (f_c) and static modulus of elasticity (E_{st}) measurements are shown in Table 4. The presented results were calculated as the average value of the three tested specimens for each mixture and curing period. Values that deviated by over 20% from each other were discarded according to the standard EN 14227-1 [30]. The table contains the results of the mechanical characteristics for the curing periods, 7, 28 and 90 days, and express the standard deviation (St.dev.). The mixtures were divided into groups (columns) according to the amount of rubber and, additionally, the results were divided according to the proportion of cement in the mixture. For example, the third, fourth and fifth columns in the table show the results of mixtures with 0% rubber and 3%, 5% and 7% cement, respectively. From the plotted results (Table 4), it can be concluded that an increase in strength occurs with an increase in cement and in the duration of the curing period. The highest strength values were reached for the 90-day curing period and with 7% cement for each rubber content. On the other hand, increasing the amount of rubber in the mixture causes a decrease in the compressive strength. An increase in the E_{dyn} and E_{st} accompanies the increase in strength. The decrease in E_{din} and UPV with a decrease in f_c due to rubber incorporation complies with the findings in [16]. The authors of that paper state that a decrease in UPV is correlated with a decrease in compressive strength due to the incorporation of fly ash instead of cement, which has less early-age pozzolanic activity. In this case, the rubber is the reason for the compressive strength decrease. Observing the modulus values, one can conclude that there is no rapid modulus-growth phase, as stated in [8], because of the use of rapid hardening cement. The rapid growth in the strength and modulus occurred in the first seven days of the specimen curing. The same can be concluded from Figure 4, which shows the development of the UPV over time for the mixture with the highest cement content. These mixtures are shown because they are expected to have the most significant influence on the development of cement stiffness. It can be seen in Figure 4 that the difference between the UPV for 7 and 28 days increases with the amount of rubber because the hydration slows down due to the reaction of the Zn from the rubber with C3S [46]. However, for mixtures with up to 20% rubber, the UPV increases almost linearly until the 90th day of curing. It is impossible to determine the phase of the UPV's rapid growth as it is detected in rubberized mortars, i.e., the rapid growth phase occurs in the first 7 days. At the same time, the amount of 20% rubber was shown to reduce the initial development of the stiffness of the mixture. In general, the use of 20% rubber as a sand replacement reduces the rate of stiffness development and linearizes the stiffness development over time. This means that there is no sudden development of strength and stiffness and, consequently, no sudden development of internal stresses. Furthermore, by observing the static modulus of elasticity, it can be concluded that for reference mixtures, the E_{st} values stagnate for longer curing periods, and that this is more pronounced with higher cement contents. On the other hand, with the incorporation of rubber, the E_{st} develops with age and it is more pronounced with mixtures with higher cement contents. Greater changes in mixtures with higher proportions of cement and rubber directly indicate the interaction of rubber and cement.

Table 4. Results of f_c (MPa), E_{dyn} (GPa) and E_{st} (GPa) for three curing periods (7, 28 and 90 days) and corresponding standard deviations for f_c, E_{dyn} and E_{st}.

		R0			R10			R20			R30			R40		
		C3	C5	C7	C3	C5	C7	C3	C5	C7	C3	C5	C7	C3	C5	C7
7 days	f_c	1.73	4.11	6.82	1.31	3.54	6.04	0.94	2.20	3.69	0.60	1.56	2.84	0.51	1.32	1.66
	St. dev. (f_c)	0.07	0.37	0.16	0.07	0.22	0.30	0.07	0.07	0.13	0.02	0.07	0.03	0.01	0.04	0.02
	E_{dyn}	11.39	20.04	27.14	9.35	17.79	23.22	3.23	10.93	14.76	1.76	5.66	9.05	0.62	4.18	4.19
	St. dev. (E_{dyn})	0.07	0.94	0.80	0.27	0.57	0.66	0.28	0.89	0.74	0.05	0.26	0.60	0.01	0.17	0.12
	E_{st}	2.27	4.61	10.76	2.11	5.46	7.25	1.63	3.79	4.94	0.73	2.03	3.78	0.59	1.53	1.80
	St. dev. (E_{st})	0.03	0.37	0.50	0.16	0.38	0.17	0.03	0.22	0.26	0.06	0.13	0.00	0.03	0.08	0.10
28 days	f_c	2.69	6.45	8.89	2.07	3.99	7.81	1.15	3.01	4.36	0.85	1.94	3.32	0.59	1.51	1.09
	St. dev (f_c)	0.31	0.07	0.27	0.09	0.09	0.22	0.02	0.03	0.29	0.02	0.11	0.03	0.01	0.04	0.03
	E_{dyn}	15.64	27.05	31.33	11.90	20.43	28.10	6.72	13.88	18.67	4.58	8.59	12.31	1.80	6.43	4.24
	St. dev. (E_{dyn})	1.00	1.00	0.27	0.54	0.41	1.26	0.12	0.34	1.42	0.16	0.14	0.53	0.07	0.35	0.40
	E_{st}	3.46	10.03	11.96	3.27	6.18	8.61	2.51	4.31	5.51	1.76	3.24	3.59	0.91	2.55	2.73
	St. dev. (E_{st})	0.026	0.53	0.12	0.04	0.41	0.17	0.17	0.01	0.46	0.03	0.05	0.20	0.05	0.13	0.08
90 days	f_c	3.32	7.55	11.57	3.88	7.67	12.95	2.24	4.85	7.85	1.13	2.78	4.30	0.84	2.13	2.34
	St. dev. (f_c)	0.11	0.53	0.38	0.05	0.00	0.46	0.06	0.12	0.20	0.05	0.13	0.20	0.01	0.00	0.03
	E_{dyn}	18.77	25.70	31.59	24.23	27.60	31.29	15.50	18.89	26.14	7.19	12.94	15.95	2.26	8.01	6.70
	St. dev. (E_{dyn})	1.11	0.98	0.45	1.30	0.74	1.04	0.99	1.72	0.66	0.42	0.26	0.45	0.03	0.39	0.19
	E_{st}	4.66	9.54	12.69	5.26	11.42	13.42	5.80	8.48	12.92	3.06	4.72	5.96	0.92	3.07	3.27
	St. dev. (E_{st})	0.35	0.57	0.79	0.23	0.39	0.40	0.15	0.09	0.61	0.10	0.11	0.42	0.02	0.10	0.19

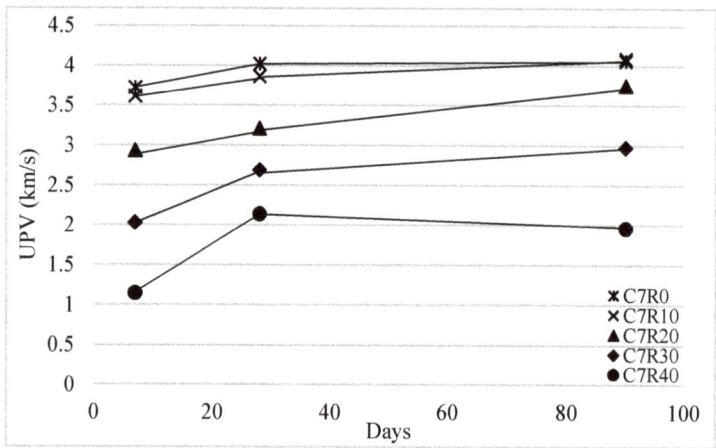

Figure 4. UPV development over time.

Furthermore, there is a strong linear correlation between these two moduli of the examined mixtures, with a coefficient of determination of $R^2 = 0.88$, as presented in Figure 5. There was inevitably a connection between these two material properties, which was expected, since these two parameters describe the same material property, its stiffness. The results are more homogeneous for lower elasticity modulus values, i.e., mixtures of lower strength and with a higher proportion of rubber. This is due to the more elastic behavior of these mixtures, the more uniform development of deformations during loading and, thus, the possibility of a more precise E_{st} determination. Considering that different methods are used to measure these two values and describe the same material characteristic, stiffness, their linear relationship is proof of the applicability of these two methods.

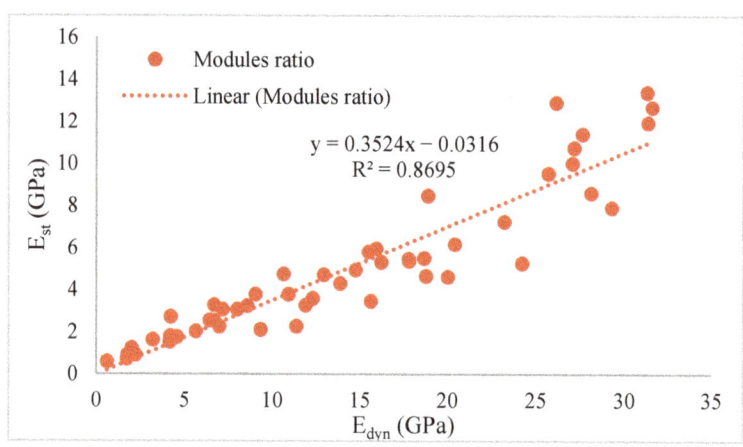

Figure 5. E_{dyn} and E_{st} relationship.

In this paper, we try to understand the relationship between the measured UPV and the mechanical characteristics to establish prediction models. Firstly, the impact of the amount of cement and rubber and the curing period on the UPV is analyzed. The UPV values for each mixture are presented in Figure 6a for 7 days, Figure 6b for 28 days and Figure 6c for 90 days. As shown in Figure 6, the ultrasonic pulse travels faster through mixtures with higher proportions of cement, resulting in a very rigid matrix, which is more pronounced for shorter curing periods. That is, mixtures with higher proportions of cement have more cement paste, which has a significantly higher stiffness than other mixture constituents. In Figure 6c, one can see the decrease in the UPV for the C7R40 mixture, which is attributed to the large amount of fine particles of rubber and cement, resulting in the filling of all the pores and the grouping of the rubber in clusters that form an obstacle to the passage of ultrasonic pulses [35]. The obtained results comply with [12,14], whose authors state that increased cement contents and curing times result in higher UPV and elasticity moduli. With higher amounts of rubber, a decrease in the UPV is also apparent. As expected, the ultrasonic pulse passes through the rubber particles more slowly due to their lower density. The rubber particles have a lower specific density and a porous structure filled with air [47,48], which slows down the ultrasonic pulse.

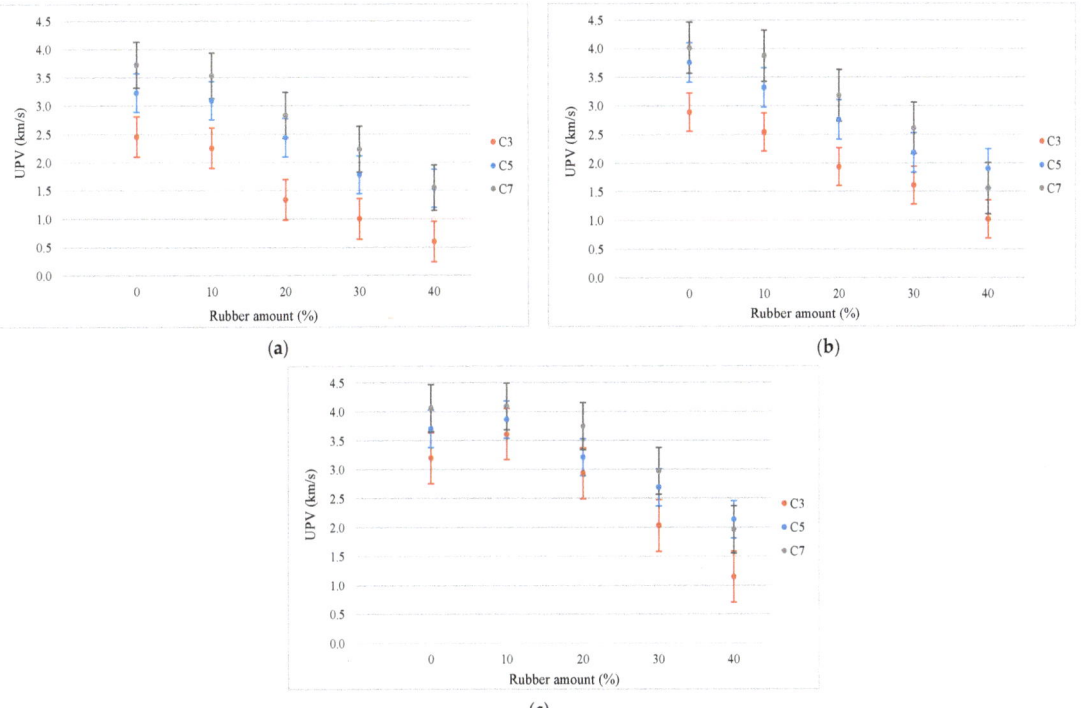

Figure 6. Impact of cement and rubber on UPV for curing periods of (**a**) 7 days, (**b**) 28 days and (**c**) 90 days.

4. Prediction Models

In order to make the results of the research conducted usable in practice, two prediction models were created by regression analysis. One was designed to predict the compressive strength (f_c) and the other was designed for the modulus of elasticity (E_{st}) prediction. In both models, the UPV was used as a predictor. This analysis was carried out based on raw pairs of data (three pairs for every mixture and curing period) of the UPV–f_c and UPV–E_{st} results. The R programming language was used to build the model. To build acceptable models, the predictors UPV and curing period were used in a linear relationship with the logarithm of the response variable.

The curing period was statistically insignificant in the model for f_c prediction based on the measured UPV values. Hence, the developed model takes only the ultrasonic pulse velocity as an input parameter. This means that the same model can be used for all three curing periods, simplifying the prediction of f_c. The plotted data and the regression line obtained are presented in Figure 7.

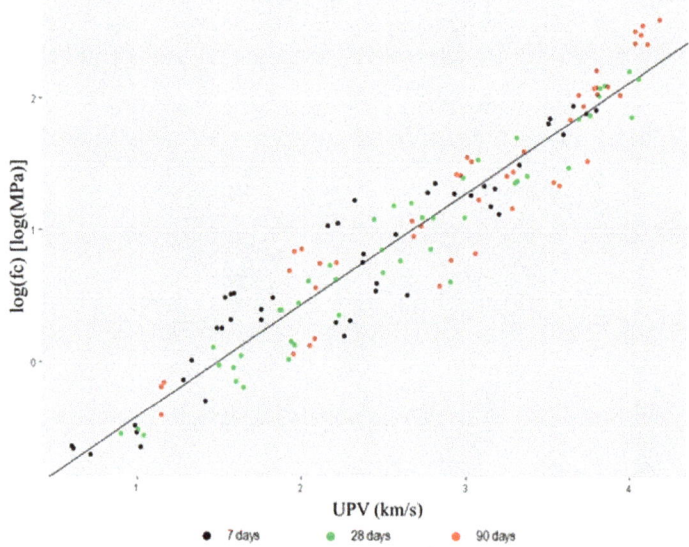

Figure 7. UPV-ln(f_c) relationship.

The model is homoscedastic (non-constant variance score test p-value = 0.84747), and errors are normally distributed (Shapiro–Wilk test p-value = 0.2124). The 95% confidence intervals for the intercept and UPV coefficient are (−1.3783653, −1.1341046) and (0.8035473, 0.8915409), respectively.

Given that this is a linear log(f_c) model, it must be transformed to obtain an expression for the prediction of f_c. The model for f_c is therefore:

$$f_c(\text{UPV}) = e^{(-1.25623 + 0.84754 \times \text{UPV} + \varepsilon)} = e^{\varepsilon} \times e^{(-1.25623 + 0.84754 \times \text{UPV})} \tag{1}$$

An adjusted coefficient of determination for this model equals 0.9154, which characterizes a very strong relationship. The ε is the zero mean model error, with an estimated standard deviation of 0.2414. As the hypothesis of normality was accepted, the mean prediction and prediction intervals were calculated based on the lognormal distribution, with the parameters $\mu = 0$ and $\sigma^2 = 0.2414^2 = 0.05828$. For instance, the mean prediction can be calculated by the formula:

$$\text{mean}(f_c(\text{UPV})) = 1.0296 \times e^{(-1.25623 + 0.84754 \times \text{UPV})} \tag{2}$$

The mean prediction line is with the 95%-prediction-interval boundaries presented in Figure 8.

The situation is more complicated in the case of E_{st} prediction based on the measured UPV values. The UPV and curing period were statistically significant, so a model with two predictors was developed. This means that the prediction of log(E_{st}) depends, apart from the UPV, on the duration of the curing period.

The model for log(E_{st}) is also homoscedastic (non-constant variance score test p-value = 0.38964), but the errors are not normally distributed (Shapiro–Wilk test p-value = 0.002291). The estimated model coefficients, intercept, UPV and days are −0.7216, 0.7585 and 0.0019, respectively, with p-values of 2×10^{-16}, 2×10^{-16} and 2.59×10^{-3}, respectively. Based on the asymptotic regression theory of 95% confidence intervals for the intercept, the UPV and days coefficients are (−0.8406, −0.6025), (0.7132, 0.8038) and (0.0007, 0.0031), respectively.

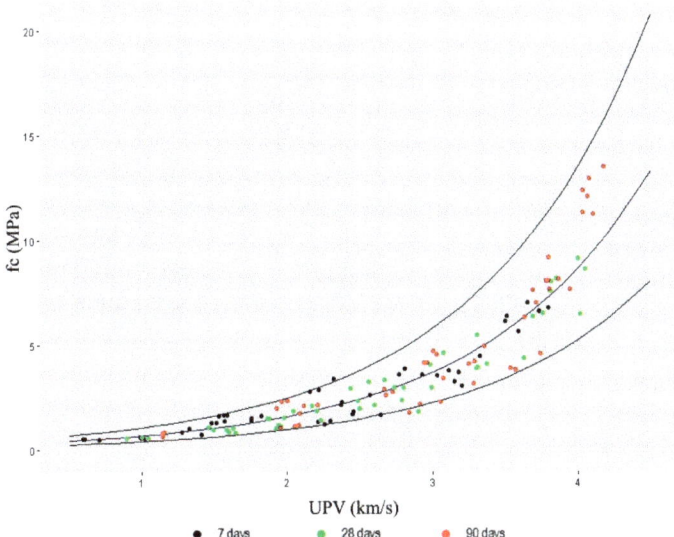

Figure 8. Mean predictions (blue line) and 2.5–97.5%-prediction-interval boundaries.

One can see that the longer curing period significantly affects the increase in E_{st}. The plotted data and the linear models obtained are presented in Figure 9. As the curing period is also a predictor, the presented lines differ for different values of the curing period.

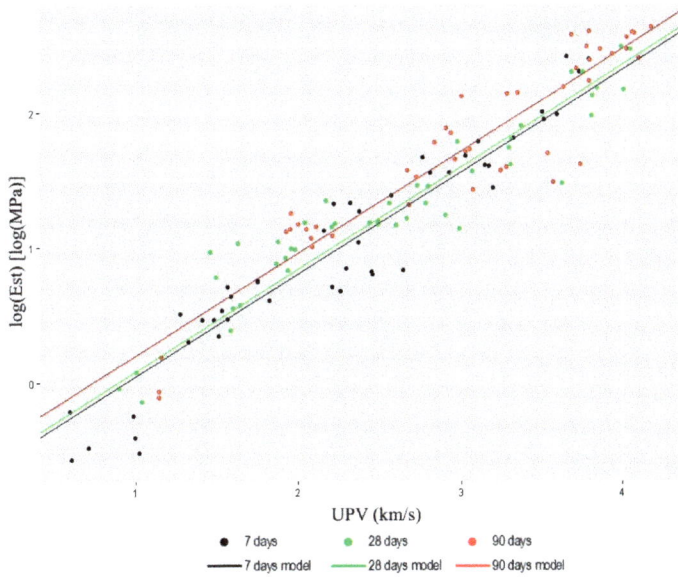

Figure 9. UPV-E_{st} relationship.

The established model is as follows:

$$E_{st}(UPV, days) = e^{-0.7216 + 0.7585 \times UPV + 0.0019 \times days + \varepsilon} = e^{\varepsilon} \times e^{-0.7216 + 0.7585 \times UPV + 0.0019 \times days} \qquad (3)$$

An adjusted coefficient of determination for this model equals 0.9077, which characterizes a very strong relationship. The ε is the zero mean model error, with an estimated standard deviation of 0.2345. Extensive simulations based on the empirical error distribution showed that the e^{ε} part in the model is negligible for practical purposes, so the resulting formula can be used to discuss the behavior of the mean E_{st}, depending on UPV and days. For a complete understanding of this model, we provide the following example: for the same UPV value, a change in the curing period from 0 days to 7 days would affect an increase in E_{st} of 1.013 (GPa), a change in curing period from 0 days to 28 days would affect an increase in E_{st} of 1.055 (GPa), while changing the curing period for the same UPV from 0 to 90 days would increase E_{st} of 1.186 (GPa). The mean prediction line and 95%-prediction-interval boundaries obtained by these simulations are shown in Figure 10.

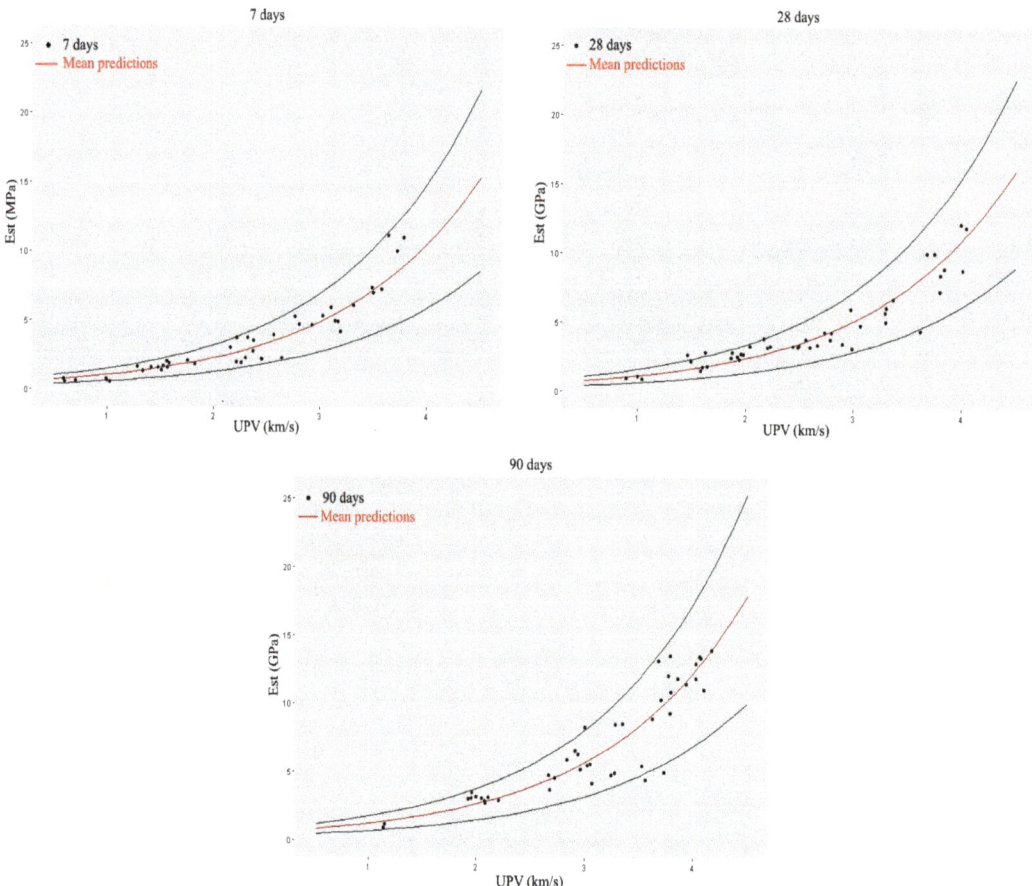

Figure 10. Mean predictions (red line) for E_{st} and 2.5–97.5%-prediction-interval boundaries.

The development of such models is significant in material testing and greatly facilitates the testing process, enabling the determination of more mechanical characteristics on the same specimen. The developed models enable reliable results, as shown by the distribution of the residuals. It was shown that the prediction of the compressive strength does not depend significantly on the length of the curing period. In contrast, the curing period is statistically significant for predicting the static modulus of elasticity. This is in accordance with [8]. It can be concluded that, during strength development, the stiffness passes through certain phases, which cannot be determined based on the results obtained. This represents the motivation for further research and the description of the development of stiffness in cement-bound mixtures with the addition of waste rubber.

5. Conclusions

This research includes the testing of the compressive strength (f_c) and dynamic (E_{dyn}) and static (E_{st}) moduli of elasticity of cement-bound aggregate modified with waste rubber to determine the inter-relationships of these characteristics and their time dependence. Reliable results were obtained through laboratory research, enabling the development of the prediction model. These are the main contributions of this research. Furthermore, a detailed statistical analysis of nonlinear relationships, which was not found in the available research for these materials, also contributes significantly to the non-destructive testing of pavement materials. Two models were developed: one for the estimation of the f_c and the other for estimating the E_{st} based on the measured UPV and the duration of specimen curing. From the results obtained, the following conclusions can be drawn:

- Increases in the amount of cement and in the curing period positively affect the f_c, E_{dyn} and E_{st}. The addition of rubber decreases these mechanical characteristics.
- The reliability of the modulus of elasticity results obtained by the two methods is supported with a strong linear correlation ($R^2 = 0.88$).
- A detailed statistical analysis of the obtained data resulted in two simple linear prediction models. One of these models serves for the estimation of the f_c based on UPV, while the other serves for the E_{st} estimation based on the UPV and curing period.
- An inter-relationship between rubber and cement was observed, especially in the mixtures with higher proportions of cement. The recommendation for further research is to analyze this influence through more mechanical properties and on a chemical and micro level.
- The increase in the UPV in the first 7 days and its linearization for a longer period of time indicates that the CBA stiffness passes through certain phases that cannot be precisely determined from the obtained results. To determine the stage of development of the stiffness, it is recommended to carry out tests in short time intervals between the first and seventh day of care.

The presented prediction models were developed on limited data and are valid only for the tested materials. As a recommendation for further research, the verification the developed models on a more significant number of specimens and other materials is indicated to prove their general applicability. It is recommended to limit the application of these models to gravel materials, considering the shape of the grains and the manner of their entrapment.

Author Contributions: Conceptualization, M.Z. and M.B.; methodology, M.Z. and M.B.; software, M.B.; validation, M.B., I.B. and T.D.; investigation, M.Z., I.B. and T.D.; writing—original draft preparation, M.Z.; writing—review and editing, M.B., I.B. and T.D.; funding acquisition, I.B. All authors have read and agreed to the published version of the manuscript.

Funding: This research was funded by Croatian Science Foundation (HRZZ), UIP-2019-04-8195, "Cement stabilized base courses with waste rubber for sustainable pavements—RubSuPave".

Institutional Review Board Statement: Not applicable.

Informed Consent Statement: Not applicable.

Data Availability Statement: Data available on request due to restrictions eg privacy or ethical. The data presented in this study are available on request from the corresponding author. The data are not publicly available due to ongoing research financed by HRZZ here].

Conflicts of Interest: The authors declare no conflicts of interest.

References

1. Zvonarić, M.; Dimter, S. Prevention and remediation measures for reflective cracks in flexible pavements. *J. Croat. Assoc. Civ. Eng.* **2022**, *74*, 189–197. [CrossRef]
2. *Development of New Bituminous Pavement Design Method*; European Commission: Brussels, Belgium, 1999.
3. Garber, S.; Rasmussen, R.O.; Harrington, D. *Guide to Cement-Based Integrated Pavement Solutions*; Portland Cement Association: Skoike, IL, USA, 2011.
4. Halsted, G.E.; Luhr, D.R.; Adaska, W.S. *Guide to Cement-Treated Base (CTB)*; Portland Cement Association: Skoike, IL, USA, 2006.
5. Marques, A.I.; Morais, J.; Morais, P.; Veiga, M.d.R.; Santos, C.; Candeias, P.; Gomes Ferreira, J. Modulus of elasticity of mortars: Static and dynamic analyses. *Constr. Build. Mater.* **2020**, *232*, 117216. [CrossRef]
6. Barišić, I.; Dokšanović, T.; Draganić, H. Characterization of hydraulically bound base materials through digital image correlation. *Constr. Build. Mater.* **2015**, *83*, 299–307. [CrossRef]
7. Jodhani, J.; Handa, A.; Gautam, A.; Ashwni Rana, R. Ultrasonic non-destructive evaluation of composites: A review. *Mater. Today Proc.* **2022**, *78*, 627–632. [CrossRef]
8. Zhao, G.; She, W.; Yang, G.; Pan, L.; Cai, D.; Jiang, J.; Hu, H. Mechanism of cement on the performance of cement stabilized aggregate for high speed railway roadbed. *Constr. Build. Mater.* **2017**, *144*, 347–356. [CrossRef]
9. Raavi, S.S.D.; Tripura, D.D. Ultrasonic pulse velocity and statistical analysis for predicting and evaluating the properties of rammed earth with natural and brick. *Constr. Build. Mater.* **2021**, *298*, 123840. [CrossRef]
10. Martin-del-Rio, J.J.; Canivelli, J. The use of non-destructive testing to evaluate the compressive strength of a lime-stabilised rammed-earth wall: Rebound index and ultrasonic pulse velocity. *Constr. Build. Mater.* **2021**, *242*, 118060. [CrossRef]
11. Zhou, T.; Zhang, H.; Li, B.; Zhang, L.; Tan, W. Evaluation of compressive strength of cement-stabilized rammed earth wall by ultrasonic-rebound combined method. *J. Build. Eng.* **2023**, *68*, 106121. [CrossRef]
12. Barišić, I.; Dimter, S.; Rukavina, T. Characterization of cement stabilized pavement layers with ultrasound testing. *Tech. Gaz.* **2016**, *23*, 447–453. [CrossRef]
13. Liu, H.; Ye, R.; Chen, L.; Ouyang, Z.; Chu, C.; Yu, H.; Lv, S.; Pan, Q. Characterization of strength, modulus, and fatigue damage properties of cement stabilized macadam based on the double modulus theory. *Constr. Build. Mater.* **2022**, *353*, 106121. [CrossRef]
14. Mandal, T.; Tinjum, J.M.; Edil, T.B. Non-destructive testing of cementitiously stabilized materials using ultrasonic pulse velocity test. *Transp. Geotech.* **2015**, *6*, 97–107. [CrossRef]
15. Mohammed, B.S.; Adamu, M.; Liew, M.S. Evaluating the effect of Crumb rubber and Nano silica on the properties of High volume fly Ash Roller compacted concrete pavement using Non-destructive Techniques. *Case Stud. Constr. Mater.* **2018**, *8*, 381–391. [CrossRef]
16. Rao, S.K.; Rao, T.C. Experimental studies in Ultrasonic Pulse Velocity of Roller compacted concrete pavement containing Fly Ash and M-sand. *Int. J. Pavement Res. Technol.* **2016**, *9*, 289–301. [CrossRef]
17. Subramanian, S.; Qasim, K.; Ku, T. Effect of sand on the stiffness characteristics of cement-stabilized clay. *Constr. Build. Mater.* **2020**, *264*, 120192. [CrossRef]
18. Norambuena-Contreras, J.; Castro-Fresno, D.; Vega Zamanillo, A.; Celaya, M.; Lombillo-Vozmediano, I. Dynamic modulus of asphalt mixture by ultrasonic direct test. *NDT E Int.* **2010**, *43*, 629–634. [CrossRef]
19. Gheibi, A.; Hedayat, A. Ultrasonic Investigation of granular materials subjected to compression and crushing. *Ultrasonics* **2018**, *87*, 112–125. [CrossRef] [PubMed]
20. Majhi, D.; Karmakar, S.; Roy, T.K. Reliability of Ultrasonic Pulse Velocity Method for Determining Dynamic Modulus of Asphalt Mixtures. *Mater. Today Proc.* **2017**, *4*, 9709–9712. [CrossRef]
21. Sarsam, S.I.; Kadium, N.S. Verifying Moisture Damage Impact in Asphalt Concrete with the Aid of Nondestructive Test NDT. *Adv. Sci. Eng.* **2020**, *12*, 13–20. [CrossRef]
22. Jackowski, M.; Małek, M. A multi-site study of a new cement composite brick with partial cement substitutes and waste materials. *Case Stud. Constr. Mater.* **2023**, *18*, e01992. [CrossRef]
23. Ramadani, S.; Abdelhamid, G.; Benmalek, M.L.; Aguiar, J.L.B. Physical and mechanical performance of concrete made with waste rubber aggregate, glass powder and silica sand powder. *J. Build. Eng.* **2019**, *21*, 302–311. [CrossRef]
24. Karalar, M.; Ozturk, H.; Özkılıç, Y.O. Experimental and numerical investigation on flexural response of reinforced rubberized concrete beams using waste tire rubber. *Steel Compos. Struct.* **2023**, *48*, 43–57. [CrossRef]
25. ETRMA. *End-of-Life Tyre. REPORT 2015*; European Tyre and Rubber Manufacturer's Association: Bruxelles, Belgium, 2015.
26. Shu, X.; Huang, B. Recycling of waste tire rubber in asphalt and portland cement concrete: An overview. *Constr. Build. Mater.* **2014**, *67 Pt B*, 217–224. [CrossRef]
27. Zeybek, Ö.; Özkılıç, Y.O.; Çelik, A.İ.; Deifalla, A.F.; Ahmad, M.; Sabri Sabri, M.M. Performance evaluation of fiber-reinforced concrete produced with steel fibers extracted from waste tire. *Front. Mater.* **2022**, *9*, 1057128. [CrossRef]
28. EN 13286-4:2021; Unbound and Hydraulically Bound Mixtures—Part 4: Test Methods for Laboratory Reference Density and Water Content—Vibrating Hammer. European Committee for Standardization: Brussels, Belgium, 2021.
29. EN 933-1:1997; Test for Geometrical Properties of Aggregates—Part 1: Determination of Particle Size Distribution—Sieving Method. European Committee for Standardization: Brussels, Belgium, 1997.

30. *EN 14227-1:2013*; Hydraulically Bound Mixtures—Specifications—Part 1: Cement Bound Granular Mixtures. European Committee for Standardization: Brussels, Belgium, 2013.
31. Zvonarić, M.; Barišić, I.; Galić, M.; Minažek, K. Influence of Laboratory Compaction Method on Compaction and Strength Characteristics of Unbound and Cement-Bound Mixtures. *Appl. Sci.* **2021**, *11*, 4750. [CrossRef]
32. Barišić, I.; Zvonarić, M.; Netinger Grubeša, I.; Šurdonja, S. Recycling waste rubber tyres in road construction. *Arch. Civ. Eng.* **2021**, *67*, 499–512. [CrossRef]
33. Farhan, A.H.; Dawson, A.; Thom, N. Behaviour of rubberised cement-bound aggregate mixtures containing different stabilisation levels under static and cyclic flexural loading. *Road Mater. Pavement Des.* **2020**, *21*, 2282–2301. [CrossRef]
34. Farhan, A.H.; Dawson, A.R.; Thom, N.H. Effect of cementation level on performance of rubberized cement-stabilized aggregate mixtures. *Mater. Des.* **2016**, *97*, 98–107. [CrossRef]
35. Farhan, A.H.; Dawson, A.R.; Thom, N.H. Compressive behaviour of rubberized cement-stabilized aggregate mixtures. *Constr. Build. Mater.* **2020**, *262*, 120038. [CrossRef]
36. Farhan, A.H.; Dawson, A.R.; Thom, N.H. Characterization of rubberized cement bound aggregate mixtures using indirect tensile testing and fractal analysis. *Constr. Build. Mater.* **2015**, *105*, 94–102. [CrossRef]
37. Farhan, A.H.; Dawson, A.R.; Thom, N.H. Recycled hybrid fiber-reinforced & cement-stabilized pavement mixtures: Tensile properties and cracking characterization. *Constr. Build. Mater.* **2018**, *179*, 488–499. [CrossRef]
38. Sun, X.; Wu, S.; Yang, J.; Yang, R. Mechanical properties and crack resistance of crumb rubber modified cement-stabilized macadam. *Constr. Build. Mater.* **2020**, *259*, 119708. [CrossRef]
39. *EN 13286-51:2004*; Unbound and Hydraulically Bound Mixtures—Part 51: Methods for the Manufacture of Test Specimens of Hydraulically Bound Mixtures Using Vibrating Hammer Compaction. European Committee for Standardization: Brussels, Belgium, 2004.
40. *CEN/TC104 TC. EN 12504-4:2021*; Testing Concrete—Part 4: Determination of Ultrasonic Pulse Velocity. European Committee for Standardization: Brussels, Belgium, 2021.
41. Barišić, I.; Dimter, S.; Rukavina, T. Cement stabilizations—Characterization of materials and design criteria. *J. Croat. Assoc. Civ. Eng.* **2011**, *63*, 8.
42. Đoković, K.; Tošović, S.; Janković, K.; Šušić, N. Physical-Mechanical Properties of Cement Stabilized RAP/Crushed Stone Aggregate Mixtures. *Tech. Gaz.* **2019**, *26*, 385–390. [CrossRef]
43. *EN 13286-43:2003*; Unbound and Hydraulically Bound Mixtures—Part 43: Test Method for the Determination of the Modulus of Elasticity of Hydraulically Bound Mixtures. European Committee for Standardization: Brussels, Belgium, 2003.
44. Barišić, I.; Dokšanović, T.; Zvonarić, M. Pavement Structure Characteristics and Behaviour Analysis with Digital Image Correlation. *Appl. Sci.* **2023**, *13*, 664. [CrossRef]
45. *EN 13286-41:2021*; Unbound and Hydraulically Bound Mixtures—Part 41: Test Method for the Determination of the Compressive Strenght of Hydraulically Bound Mixtures. European Committee for Standardization: Brussels, Belgium, 2021.
46. Rodrígues, R.S.L.; Domínguez, O.; Díaz, A.J.H.; García, C. Synergistic effects of rubber-tire-powder and fluorogypsum in cement-based composite. *Case Stud. Constr. Mater.* **2021**, *14*, e00471. [CrossRef]
47. Ul Islam, M.M.; Li, J.; Wu, Y.-F.; Roychand, R.; Saberian, M. Design and strength optimization method for the production of structural lightweight concrete: An experimental investigation for the complete replacement of conventional coarse aggregates by waste rubber particles. *Resour. Conserv. Recycl.* **2022**, *184*, 106390. [CrossRef]
48. Ul Islam, M.M.; Li, J.; Roychand, R.; Saberian, M. Investigation of durability properties for structural lightweight concrete with discarded vehicle tire rubbers: A study for the complete replacement of conventional coarse aggregates. *Constr. Build. Mater.* **2023**, *369*, 130634. [CrossRef]

Disclaimer/Publisher's Note: The statements, opinions and data contained in all publications are solely those of the individual author(s) and contributor(s) and not of MDPI and/or the editor(s). MDPI and/or the editor(s) disclaim responsibility for any injury to people or property resulting from any ideas, methods, instructions or products referred to in the content.

Mortars with Polypropylene Fibers Modified with Tannic Acid to Increase Their Adhesion to Cement Matrices

Joanna Julia Sokołowska, Paweł Łukowski * and Alicja Bączek

Department of Building Materials, Faculty of Civil Engineering, Warsaw University of Technology, 00-637 Warsaw, Poland; joanna.sokolowska@pw.edu.pl (J.J.S.); alicja.jaskulecka.stud@pw.edu.pl (A.B.)
* Correspondence: pawel.lukowski@pw.edu.pl

Featured Application: Fiber concrete, mortar and other cementitious composites reinforced with polypropylene macrofibers.

Abstract: The presented research's main objective was to evaluate the possibility of improving the adhesion between polypropylene fibers and mineral matrices in cementitious composites by modifying the fibers' surface with tannic acid (TA). This modifier was previously used for polyethylene fibers only. Cement mortar containing modified polypropylene fibers and mortar containing unmodified fibers were tested. The physical and mechanical properties (apparent density, compressive strength, flexural strength and modulus of elasticity) were determined, and the fibers' morphology after the specimens' destruction was observed. No adverse effect of the modification was found. The elastic modulus was 6% lower after 28 days, enabling the formation of a less stiff composite material. The integrity of the specimens after mechanical damage was improved, confirming the increased adhesion between the polypropylene fibers and the hardened cement paste. The results of the introductory tests are promising; however, further research is needed in the field.

Keywords: adhesion; cementitious composite; fiber-reinforced concrete; modification; polypropylene fibers; tannic acid

1. Introduction and Scope

Fiber concrete (or fiber-reinforced concrete) is a type of concrete that contains fibers to enhance its mechanical strength and resistance to cracks and, generally, its durability [1]. The fibers are made mainly of two materials, steel and synthetic polymers (e.g., polypropylene, polyethylene or polyamide); however, other types of fibers are also used, like cellulose, glass, carbon or basalt. The effectiveness of fibers in concrete depends on their strength, shape and dimensions, but also their content, dispersion and adhesion to the cementitious matrix. The last issue is crucial in the context of using polymer fibers, the adhesion of which to mineral matrices is much lower than those made of steel due to their smooth surfaces [2–4]. Therefore, to fully exploit the potential of polymer fibers, it is necessary to improve their adhesion and compatibility with cement matrices.

1.1. Polymer Microfibers and Macrofibers Used in Concrete and Other Cementitious Materials

Generally, there are two basic types of polymer fibers used in ordinary concretes and other concrete-like materials with cementitious matrices: micro- and macrofibers.

Microfibers are fibers with very small cross-sections (diameters of several to a dozen micrometers) and lengths usually not exceeding 12 mm (see Figure 1). The dosing of such fibers is 0.2–2.0 kg per 1 m^3 of the composite, depending on the intended use. Microfibers are intended mainly for non-structural applications and can be considered for microreinforcement, which reduces plastic shrinkage and prevents the appearance of microcracks in cement matrices in the early stage. It is advisable to use them in various composites (concretes; thin-layer screeds, including ones for underfloor heating and plastering; masonry mortars; repair mortars; adhesives; insulating coatings; etc.), especially if they contain high-class Portland cements with high early strength, i.e., with an increased tendency to shrinkage and crack development, or, in general, in advanced cementitious matrices that are characterized by a refined microstructure and low porosity (e.g., a water/binder ratio lower than about 0.3). Due to their small size, microfibers can easily be evenly distributed in cement paste and create a spatial mesh that can additionally reduce water absorption and increase the water tightness and frost resistance of modified composites. The authors confirmed in previous research investigations [5] that it was possible to obtain a seamless polymer–cement insulating coating with polyacrylonitrile microfibers that showed excellent elasticity and water resistance and at the same time remained water-vapor permeable. Also, when mortars with PP microfibers (presented in Figure 1b) were tested, no negative impact on the mixes' consistency was observed, while hardened composites were characterized by assumed compressive strength and flexural strength. Therefore, it was not necessary to analyze or to improve the adhesion between the used microfibers and the composite matrices.

(a) (b)

Figure 1. Commercial polypropylene microfibers of similar geometry (both with a length of 12 mm and a diameter of 18 µm) from different manufacturers: (**a**) fibers without an additional coating; (**b**) fibers additionally covered with a water-soluble coating.

The second type of polymer fibers used for concrete, i.e., macrofibers or structural fibers (sometimes referred to as "traditional fibers"), have much larger cross sections (diameters of 0.1 mm and more) and greater lengths. Bentur and Mindess [1] distinguished a third type of fiber, i.e., mesofibers, which have cross-sectional dimensions in the range of 0.1 to 0.3 mm; however, in most studies in the scientific and industrial literature and most technical data reports (including most of the mentioned references), fibers with such cross sections are classified as fine macrofibers. Unlike the situation with polymer micro-fibers, in the case of macrofibers made of polymers, the issue of adhesion is much more problematic. Polymer macrofibers are added to cementitious mortars and concretes (structural and non-structural) for several reasons:

- To reduce plastic shrinkage and the formation of microcracks in cement matrices;
- To increase the mechanical strength of composites (including flexural and tensile strength, as well as the level of fracture energy absorption and impact strength);
- To reduce the possibility of crushing and spalling at the edges of composite elements.

The positive effects of the presence of polymer macrofibers mentioned above are basically the same as the effects of steel fibers, but they cannot be treated as a direct replacement for steel fibers of the same geometry, mainly due to their much lower adhesion to cement binders.

The authors have reviewed the range of polymer macrofibers currently available on the Polish market and used in concrete production. The main commercially available polymer macrofibers are polypropylene, polyethylene and polyolefin copolymer fibers with lengths between 24 and 54 mm. Some fibers are crimped (Figure 2a) and others are twisted (Figure 2b) to provide better anchoring in cement matrices. However, despite these geometric adjustments, the adhesion of all polymer fibers to cement paste is much lower compared to steel fibers, and if the reinforced element is damaged, the polymer fibers can easily be pulled out from the matrix. The additional lengthening of fibers and increase in their cross-sections (and thus extension of the anchoring zones and the contact surfaces) do not significantly improve the integrity of the composites at failure [6]. A better solution seems to be the modification of polymer fibers in order to roughen their initially very smooth surface and to expand the contact zones with hardened cement paste. This can be achieved either by mechanical modification (the surface of the fibers is mechanically modified to create a rough texture by subjecting the fibers to abrasion or friction, e.g., by sanding) or chemical modification, including coating (i.e., applying chemical agents or rough materials onto the surface of the polymer fibers [6,7]) and chemical surface treatments (e.g., with acids [8,9]). Selected examples of chemical surface treatments for polymer fibers are given in the next section.

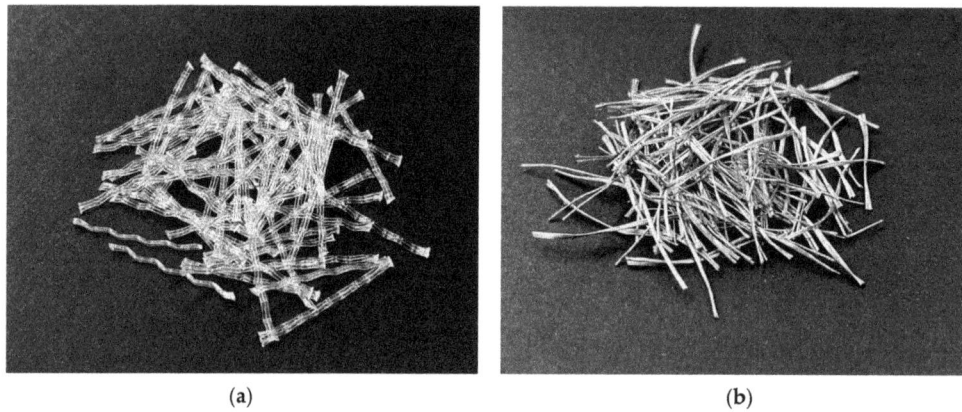

Figure 2. Examples of macrofibers with various geometries that are commercially available in Poland: (**a**) crimped polypropylene macrofibers (length of 25 mm); (**b**) polyolefin copolymer fibers twisted in bundles (length of 24 mm).

1.2. Chemical Surface Modification of Polymer Fibers—Selected Methods

The scientific and technical literature contains descriptions of numerous attempts to improve polymer fiber quality and durability by chemical modification. For instance, the authors themselves successfully performed the surface modification of fibers made from recycled PET with ethylene-vinyl acetate copolymer (EVA) [6] to prevent hydrolysis of ester linkages of poly(ethylene terephthalate) in the highly alkaline environment of fresh cement paste (for detailed information on PET degradation in alkaline environments, see also: Silva et al. [10], Pelisser [11] and Won et al. in [12]).

In the case of polypropylene fibers, Wiliński et al. [8] modified the fibers using chrome acid to roughen their surface, and they concluded that non-polar polymer surface oxidation could be an efficient way to improve the fiber-reinforced concrete mechanical properties. In the case of polyethylene fibers, Bashiri Rezaie et al. [7] applied the approach used earlier, among others, by Shanmugam et al. [4], Changani et al. [13] or Xi et al. [14], for other non-polar polymers to test the possibility of improving the adhesion between polyethylene fibers and cementitious matrix by modifying the surface of the fiber with polydopamine. Using dopamine through oxidative self-polymerization reactions applied by a simple waterborne deposition process, which forms a thin reactive polymeric layer comprising hydroxyl and amino groups on diverse inorganic and organic substances [13,15], enabled imparting a hydrophilic nature of the modified polymer, thus increasing its surface polarity and hydrophilicity [16]. They concluded that by dopamine surface modification, it was possible to increase the tensile strength of polyethylene fibers, pull-out strength and interfacial shear strength, thus enhancing the bonds between the initially hydrophobic polymer fibers and cementitious matrix.

Recently, Bashiri Rezaie et al. [9] researched the possibility of improving the adhesion between PE fibers and cementitious matrix using tannic acid. They obtained promising results, showing that this method allowed them to chemically roughen the fiber surface (Figure 3) and significantly increase the energy required to pull the modified fibers out of the matrix compared to unmodified fibers. They experimentally scaled the modification method, using different sequences of adding chemical reagents to the modifying solution and using different exposure times of the fibers to the modifying medium (compare Figure 3b–d), to finally indicate the most effective procedure for modifying polyethylene fibers assuming 3-h immersion of fibers in tannic acid supplemented with selected additional modifiers.

Figure 3. ESEM images of pristine PE fibers and various tannic acid-modified PE fibers: (**a**) pristine fibers; (**b**) fibers treated with tannic acid for 30 min; (**c**) fibers treated with tannic acid for 1 h; (**d**) fibers treated with tannic acid for 3 h (based on [9]).

The research by Bashiri Rezaie et al. [9] included only tests of individual fibers placed straight in a pure cement paste of high water/cement mass ratio of 2.0, i.e., a value several times higher than the ratio of actual scale composites used in construction. However, the authors decided to use the same chemical roughening mechanism in the context of polypropylene fibers available on the local market and check whether the method of rapid modification with tannic acid would also improve the adhesion of PP fibers to the cement matrix with a water/cement mass ratio of 0.50 (recommended by EN 206 standard [17] for concretes working in the aggressive environments described by selected exposure classes—including the risk of carbonation, chloride aggression, freezing/thawing or mechanical friction), and whether it would affect (positively or negatively) the mechanical properties of mortars with such fibers.

While fibers of vegetable origins, like raffia, coconut or similar ones, can be successfully treated with alkaline products (e.g., sodium or potassium hydroxide), such modification is not recommended for polypropylene fibers. Synthetic polymers have weak alkali resistance; instead, they are acid-resistant [1]. As mentioned above, tannic acid was employed to modify polyethylene fibers [9]. The authors intended to assess whether this treatment method could also improve the performance of polypropylene fiber-reinforced cement composite.

2. Materials and Methods

2.1. Qualitative and Qualitative Composition of Tested Composites

The experiment involved comparing the behavior of surface-modified and unmodified polypropylene fibers applied to a composite with a cement matrix, which was then subjected to destructive mechanical tests. To eliminate the influence of other components on the test results, the composite selected as the base one, i.e., reinforced with dispersed reinforcement in the form of fibers, was the so-called comparative mortar, i.e., standard mortar made of high-early strength Portland cement class 42.5, i.e., CEMI 42.5 R (CEMEX, Chełm, Poland). The applied cement fulfilled the requirements of European Standards EN 197-1 [18] and EN 196-1 [19] in terms of the composition (at least 95% Portland clinker) and properties. The aggregate used in the mortars was standard sand (meeting requirements of EN 196-1). It was natural quartz sand with rounded grains, SiO_2 content of at least 98%, and grain size up to 2 mm. The size distribution of sand grains recommended by the EN 196-1 standard is presented in Table 1. The mixing water was tap water (meeting the requirements of EN 1008 [20]).

Table 1. Size distribution of standard sand grains according to EN 196-1.

Square Mesh Size, mm	2.00	1.60	1.00	0.50	0.16	0.08
Total residue on the sieve, %	0	7 ± 5	33 ± 5	57 ± 5	87 ± 5	99 ± 1

The applied fibers (presented in Figure 2a) were pure polypropylene macrofibers of a length of 25 mm (DIIF, Dnepr, Ukraine) designated to be used in structural concrete and mortars. Table 2 contains the basic physical properties of the fibers listed by the manufacturer [21]. The fibers were also characterized by high acid, alkali, and salt resistance and low thermal and electrical conductivity.

Table 2. Properties of macrofibers used in tested composites (manufacturer's data [21]).

Type of Polymer	Specific Gravity, kg/m³	Melting Point, °C	Ignition Point, °C
Polypropylene (PP)	910	162	593

Standard EN 196-1 includes the composition of the standard mortar, which assumes a water/cement mass ratio of 0.50 and a cement/aggregate mass ratio of 1:3. For the experiment, both a standard mortar (without fibers) and mortars with an identical composition of essential components, but supplemented with chemically modified/unmodified fibers, were prepared. The amount of added fibers was based on the manufacturer's recommendations [21]. The manufacturer recommended dosing fibers in various amounts depending on the potential use of composites, i.e., 2–4 kg per 1 m³ of the concrete mix in the case of the production of industrial floors, screeds or sprayed concrete, and 2–6 kg per 1 m³ of the concrete mix in the case of concrete structural elements of residential and industrial buildings. For the experiment, it was assumed that the fiber content would be close to the upper limit of the range recommended for structural concrete, i.e., 4.5 kg per 1 m^3 of the mortar mix. The compositions of the tested mortars (calculated per mix following the standard procedure, assuming a specific gravity of 2650 kg/m^3 for standard sand and of 3100 kg/m^3 for the cement) are listed in Table 3.

Table 3. Composition of tested composites (by mass and volume).

No/Code	Cement		Water		Aggregate		Non-Modified Fibers		Modified Fibers	
	g	cm^3	g	cm^3	g	cm^3	g	cm^3	g	cm^3
1/SM							0	0	0	0
2/NM	450	145	225	225	1350	509	5	5.5	0	0
3/M							0	0	5	5.5

2.2. Procedure of Fiber Surface Modification with Tannic Acid

The surface modification procedure for polypropylene fibers used in the tested composites was adopted from the experiment described in [9] (among the methods described here, the variant that gave the best roughening effect was selected). The modification involved immersing clean fibers in a solution with the predominance of tannic acid, TA (CAS-Number: 1401-55-4), and the addition of sodium periodate, SP (NaIO$_4$, CAS-Number: 7790-28-5) and ethanolamine, EA (CAS Number: 141-43-5) for 3 h, and then washing and drying the fibers in ambient temperature for later use. All chemicals were delivered by Linegal Chemicals, Blizne, Poland. The modifying solution was prepared using 1 dm^3 of distilled water, 4 g of tannic acid, 20 cm^3 of ethanolamine, and 8 g of sodium periodate. The whole was mixed using a magnetic stirrer for about 15 min. The fiber modification procedure was as follows:

1. Inserting the fibers in the aqueous solution (TA + EA + SP) so that all of the fibers are immersed in the solution.
2. Keeping the fibers in the solution (covered container) for 3 h at ambient temperature.
3. Removing the fibers from the solution and rinsing several times in distilled water.
4. Drying the fibers at room temperature.

It is important to emphasize that the tannic acid was used only during the stage of the modification of the fibers (i.e., before the preparation of the concrete mix). The tannic acid had no contact with the concrete matrix and had no possibility of causing any chemical corrosion of the concrete.

2.3. Testing Methods

For all composites, the set of technical properties was determined as apparent density, flexural strength, flexural elastic modulus and compressive strength. The properties of mortars with fibers were determined after 7 and 28 days of curing. The standard mortar was tested after 28 days of curing (all specimens were demolded after 24 h after casting and then kept in the water in laboratory conditions). Flexural strength and flexural modulus were tested on a set of three standard specimens in the shape of beams of size 40 mm × 40 mm × 160 mm in the three-point bending test (using Instron 5567 electromechanical testing machine, Canton, OH, USA, Figure 4). The compressive strength was tested on the halves of the prisms remaining after the bending test (using Controls MC66 hydraulic press, Milan, Italy). According to EN 197-1, the applied method excludes the influence of the bending test on the compression test result—although the second test is performed on halves of the bent and broken prism specimen. The compressive force is applied to the specimen far enough away from the broken edge so that the intact part of the prism half is compressed. Apparent density was determined on the same specimens (mass of the specimens divided by their measured volume) just before the destructive tests, while the fractures of the specimens and fibers were visually observed using optical microscopes

(Biolux AL and Carl Zeiss Jena Neophot 32, Jena, Germany, with the Nikon D300 digital recording system, Tokyo, Japan) on the specimens after the tests.

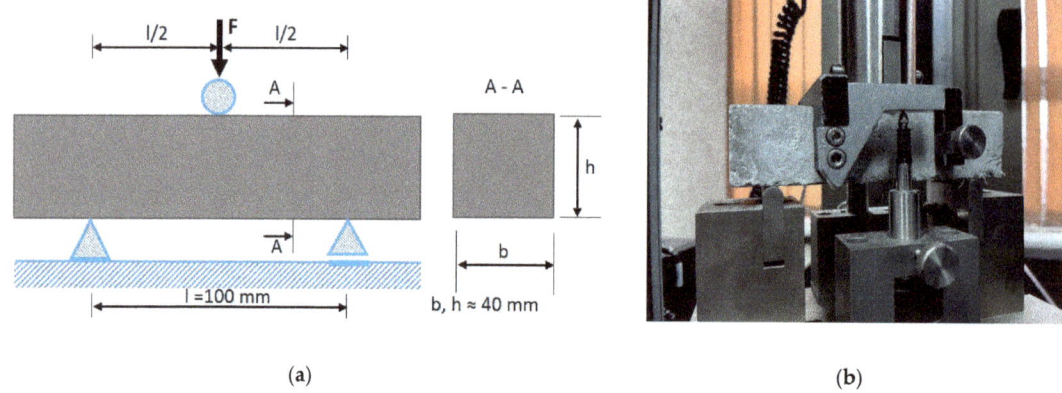

(a)   (b)

Figure 4. Flexural strength and modulus test performed on specimens in the shape of beams of size 40 mm × 40 mm × 160 mm in the three-point bending test acc. to EN 1015-11 [22]: (**a**) scheme of the test (specimen on two supports—one stationary, the other movable, so that the system is statically determinate, the force concentrated in the middle of the span); (**b**) specimen during the test in Instron 5567 electromechanical testing machine.

3. Results and Discussion

3.1. Physical and Mechanical Properties

The results for apparent density (average and basic statistic parameters: standard deviation, SD and coefficient of variation, CV) of mortars with non-modified and modified fibers after 7 and 28 days and the density of the standard mortar after 28 days are given in Table 4. As expected, mortars with fibers made of polypropylene (with a specific gravity of 910 kg/m^3—compare Table 2) were characterized by slightly lower (differences of up to 3%) apparent density compared to pure-mineral standard mortar. However, it is worth noting that the mortar containing non-modified fibers was characterized by a density with a 1.5 times higher coefficient of variation, while the mortar with fibers modified with tannic acid showed four times lower variability compared to standard mortar apparent density.

Table 4. Density of tested mortars: standard mortar (SM), mortar with non-modified polypropylene fibers (NM) and mortar with tannic acid-modified polypropylene fibers (M) determined after 7 and/or 28 days (SD—standard deviation, CV—coefficient of variation).

Composite Type	Density after 7 Days			Density after 28 Days			
	Mean, kg/m^3	SD, kg/m^3	CV, %	Mean, kg/m^3	SD, kg/m^3	CV, %	Relative Change to SM, %
Standard mortar (SM)	x	x	x	2261	37	1.6	-
Non-modified-fiber mortar (NM)	2186	25	1.1	2240	58	2.4	−0.9
Modified-fiber mortar (M)	2198	5	0.2	2198	8	0.4	−2.8

The results of the mechanical tests of mortars with fibers after 7 and 28 days are given in Figure 5. In the case of standard mortar, the results obtained after 28 days were as follows: flexural strength—7.46 MPa (on average) and compressive strength—45.16 MPa (on average), which stands in line with the expectation of standard EN 196-1 for such mortars with CEM I 42.5R binder (i.e., $f_{cm} \geq 42.5$ MPa). Both mortars with fibers had lower strength after 28 days, and none exceeded 42.5 MPa in compressive strength. The difference compared to the standard mortar was 5.2–5.6 MPa (11.5–12.3% reduction). In the case of 28-day flexural strength, the difference compared to the standard mortar was more significant—reduction by approx. 35% was observed. However, such weakening of the cementitious mortar after adding polypropylene fibers is unsurprising, as the connection between the mineral matrix and the polypropylene fiber is weakened. Moreover, a slight correlation was observed between the decrease in mortar density and mechanical strength.

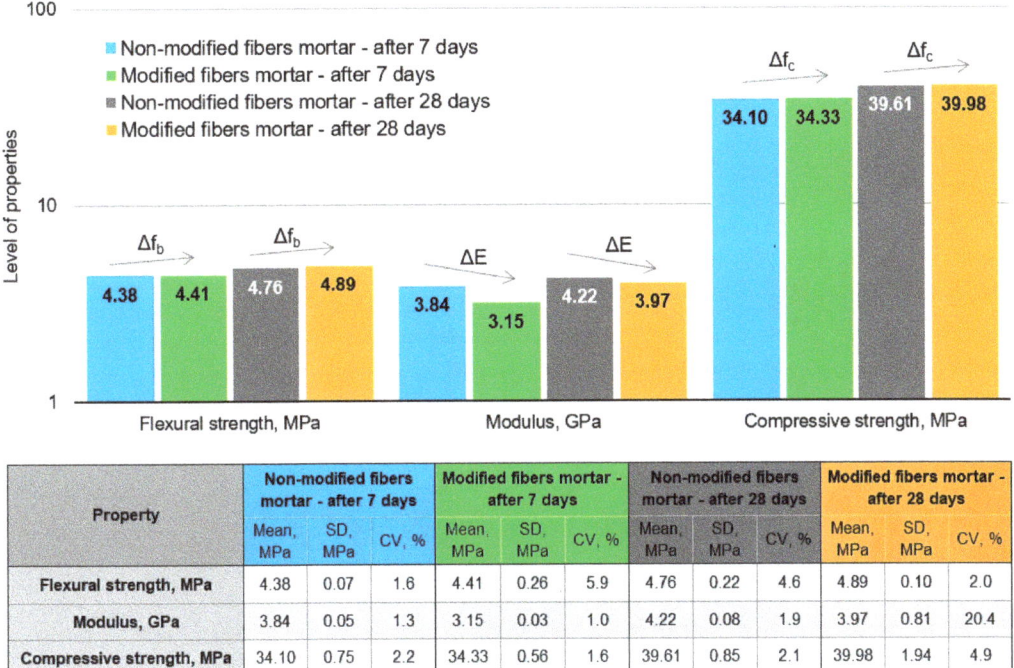

Figure 5. Mechanical properties of mortars with fibers after 7 days and 28 days of curing (SD—standard deviation, CV—coefficient of variation).

An interesting observation was that in each case—both in tests carried out after 7 days and after 28 days—higher flexural strength and compressive strength values were noted when using fibers modified with tannic acid. The differences were insignificant—they amounted to a maximum of 3%. However, authors expect that the effect of fiber modification on mechanical strength might be more remarkable in the case of composites of lower water/cement ratio and when adding plasticizing admixture that might provide a better coating of the fibers with a fluidized cement paste [23,24].

The crucial mechanical property when assessing the effect of introducing polymer fibers into the brittle mineral composite is the elastic modulus. When comparing the modulus of mortars with non-modified fibers and modified fibers, the latter was characterized by a significantly lower flexural elastic modulus. After 7 days of curing, the modulus of the mortar was lower by 700 MPa (18%), and after 28 days of curing, it was 240 MPa (6%). Thus, modification of fibers enabled obtaining less stiff composite material that can deform more easily under stress, making it more flexible.

3.2. Visual Inspection of Composite Fractures and Fibers

The polymer fiber modification was not expected to drastically improve the mechanical properties of the mortars, as such an effect would require changes in the composition of the target composite. The main aim of the performed investigation was to determine whether the modified fibers were better anchored in the cement matrix, as that would be a sign that the tannic acid modification was effective in roughening the surface of the polypropylene fibers and improving their adhesion to the mineral matrix. Here, the visual inspection of the damaged specimens was helpful—visual inspection of the surfaces of scratched specimens (after bending), as well as the fractures of the broken specimens and the conditions of fibers themselves.

In Figure 6, one can see the comparison of the specimen's outer surfaces with both fiber variants after the flexural strength test. The test procedure did not assume loading the bending elements until the specimen wholly disintegrated, but until the stress dropped, which meant damage to the brittle material structure. In the case of standard mortar testing, as a result of exceeding the maximum stresses, the brittle cracking occurred in the area of the cementitious matrix and the aggregate–matrix interface, and the visible crack appeared, leading to the separation of the halves of the bent prism. However, in the testing of mortars with polymer fibers, the polymer fibers ensured the stability of the specimen even after the failure of the brittle matrix. Despite the destruction and significant decrease in the load capacity, the specimens retained their integrity. However, the nature of the destruction differed in the case of mortars with unmodified and modified fibers. The specimens with unmodified fibers were cracked across the entire thickness of the bent element (Figure 5, upper specimen). However, the specimens with modified fibers were only scratched (up to a maximum of 2/3 of the element thickness), and the crack width did not exceed 0.3 mm (thus fulfilling requirements of Eurocode EN 1992-1-1 [25] in terms of carbonation, chlorides other than from seawater and most of the chlorides from seawater exposure classes). Specimens with modified fibers clearly showed significantly improved integrity after failure.

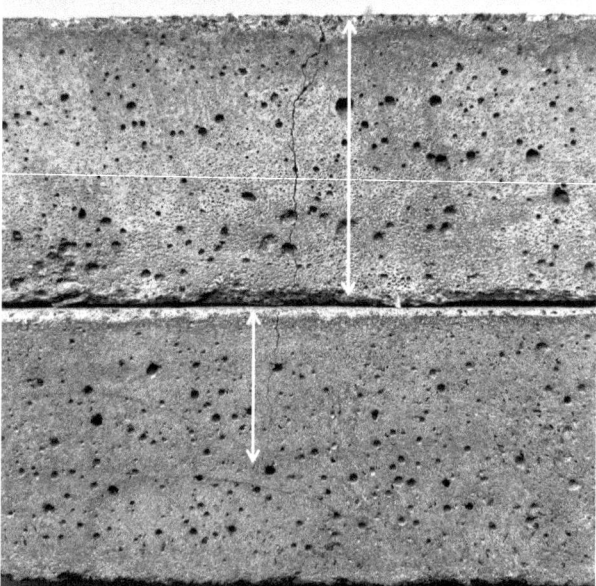

Figure 6. The outer surface and the mechanical damage (the length of the crack corresponds to the entire thickness of the specimen, i.e., 40 mm—see upper arrow) of the mortars with non-modified PP fibers (upper specimen) and PP fibers modified with tannic acid (lower specimen; crack range is 24.9 mm—see lower arrow) after flexural strength test (three-point bending test).

In order to measure compressive strength, which was the next phase of the experiment, the prisms had to be fully broken, which required additional loading of the specimens. After separating the prisms' halves, the fibers' fractures and condition were inspected. It was observed that the unmodified fibers were not broken but pulled out, often almost entirely, from the cementitious matrix (Figure 7a). Meanwhile, the tannic acid- modified fibers remained anchored in the matrix, and their breaks occurred close to the specimen fracture surface. Also, their ends were frayed, confirming breaking when their tensile strength was exceeded (Figure 7b). The above observation proves they better adhered to the matrix, which the authors attribute to the roughened surface of the fibers. The last conclusion was also supported by microscopic observations that showed that the surface of unmodified polypropylene fibers was smooth before their implementation into the mortar mix and after their pull-out from the broken specimen. The fibers subjected to the tannic acid modification showed a rough, more developed surface, i.e., predisposed to better adhesion to the cement paste, and a much more damaged surface after the destruction of the bent element (Figure 7c,d).

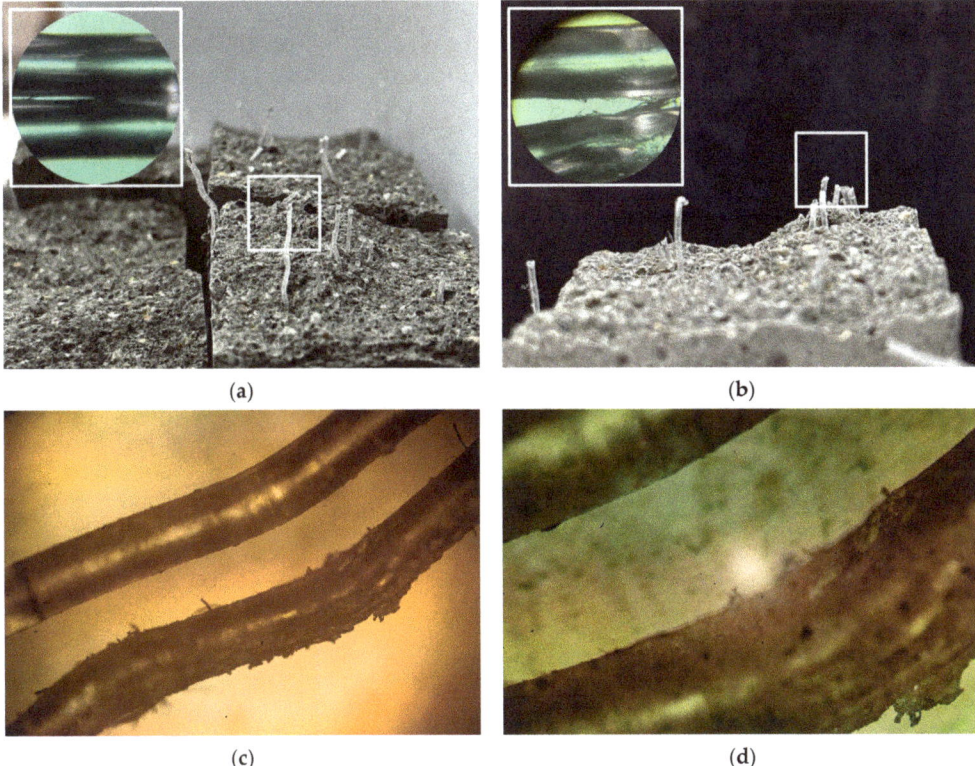

Figure 7. Fractures of PP fiber mortar specimens after flexural strength test and the image of the fibers after their pull-out from the cementitious matrix: (**a**) mortars with non-modified fibers; (**b**) mortars with TA-modified fibers (macro-observation and micro-observation under magnification: 50×).; (**c**) non-modified (upper) and modified (lower) fibers (magnification: 50.4×); (**d**) non-modified (upper) and modified (lower) fibers (magnification: 120×).

4. Conclusions

The presented research is a preliminary phase of work on improving the properties of polymer fibers for concrete available on the Polish market. Only one level of fiber dosing was used, and the modified composite was a standard mortar (the so-called comparative mortar) with Portland cement (although there is a tendency to reduce consumption of high-emission binders). Nevertheless, the results obtained in the presented research are promising. Firstly, after adding polypropylene fibers to mortars in a significant amount, recommended by the manufacturer as reinforcement for concrete for structural applications, there was no significant deterioration in compressive strength (only 11.5–12.3% reduction). Secondly, the modification with tannic acid did not lead in any case (neither after 7 days nor after 28 days) to a deterioration in compressive strength or flexural strength (even a slight improvement was noted). It significantly improved the flexural modulus of the tested fiber mortars. Thirdly, it was confirmed that the surface modification with tannic acid, which was earlier used in the case of different polymers, positively affected the polypropylene fibers, improving their adhesion and anchoring in the cementitious matrix.

The study aimed to evaluate the possibility of improving the performance of the polypropylene fibers in the cement composite with the method previously employed for the polyethylene fibers. The results show that the surface modification method with tannic acid can effectively improve the bonding between a cementitious matrix and the various polymer fibers.

In the subsequent investigation phase, the authors intend to extend the research to include composites with lower water/cement ratio values and fluidizing admixtures. They also intend to shift the scale of the experiment from mortars to concretes with coarse aggregate and to determine properties such as those assessed with the wedge splitting test (WST), which is considered a good measure of adhesion [26–28]. Moreover, they want to investigate the effectiveness of tannic acid modification on other polymer fibers for concrete available on the Polish market, such as copolymer polyolefin fibers.

Author Contributions: Conceptualization, J.J.S. and P.Ł.; methodology, J.J.S.; investigation, J.J.S. and A.B.; writing, J.J.S.; visualization, J.J.S.; supervision, P.Ł.; project administration, P.Ł. All authors have read and agreed to the published version of the manuscript.

Funding: This research received no external funding.

Institutional Review Board Statement: Not applicable.

Informed Consent Statement: Not applicable.

Data Availability Statement: The data presented in this study are available on request from the corresponding author. The data are not publicly available due to the policy of Warsaw University of Technology.

Conflicts of Interest: The authors declare no conflicts of interest.

References

1. Bentur, A.; Mindess, S. *Fibre Reinforced Cementitious Composites*, 2nd ed.; Taylor & Francis: London, UK; New York, NY, USA, 2007.
2. Brandt, A.M. Fibre reinforced cement-based (FRC) composites after over 40 years of development in building and civil engineering. *Compos. Struct.* **2008**, *86*, 3–9. [CrossRef]
3. Curosu, I.; Liebscher, M.; Mechtcherine, V.; Bellmann, C.; Michel, S. Tensile behavior of high-strength strain-hardening cement-based composites (HS-SHCC) made with high-performance polyethylene, aramid and PBO fibers. *Cem. Concr. Res.* **2017**, *98*, 71–81. [CrossRef]
4. Shanmugam, L.; Feng, X.; Yang, J. Enhanced interphase between thermoplastic matrix and UHMWPE fiber sized with CNT-modified polydopamine coating. *Compos. Sci. Technol.* **2019**, *174*, 212–220. [CrossRef]
5. Czarnecki, L.; Sokołowska, J. Optimization of polymer-cement coating composition using material model. *Key Eng. Mater.* **2011**, *466*, 191–199. [CrossRef]
6. Wiliński, D.; Łukowski, P.; Rokicki, G. Application of fibres from recycled PET bottles for concrete reinforcement. *J. Build. Chem.* **2016**, *1*, 1–9. [CrossRef]
7. Bashiri Rezaie, A.; Liebscher, M.; Ranjbarian, M.; Simon, F.; Zimmerer, C.; Drechsler, A.; Frenzel, R.; Synytska, A.; Mechtcherine, V. Enhancing the interfacial bonding between PE fibers and cementitious matrices through polydopamine surface modification. *Compos. Part B Eng.* **2021**, *217*, 108817. [CrossRef]
8. Wiliński, D.; Kraiński, Ł. Oxidation modification of surface of polypropylene fiber reinforcement. *Mater. Bud.* **2017**, *1*, 24–26. (In Polish) [CrossRef]

9. Bashiri Rezaie, A.; Liebscher, M.; Mohammadi, M.; Mechtcherine, V. Fast tannic acid surface modification for improving PE fiber-cement matrix bonding performances. In Proceedings of the ICPIC 2023–17th International Congress on Polymers in Concrete, Warsaw, Poland, 17–20 September 2023.
10. Silva, D.A.; Betioli, A.M.; Gleize, P.J.P.; Roman, H.R.; Gomez, L.A.; Ribeiro, J.L.D. Degradation of recycled PET fibers in Portland cement-based materials. *Cem. Concr. Res.* **2005**, *35*, 1741–1746. [CrossRef]
11. Pelisser, F.; Montedo, O.R.K.; Gleize, P.J.P.; Roman, H.R. Mechanical properties of recycled PET fibers in concrete. *Mater. Res.* **2012**, *15*, 679–686. [CrossRef]
12. Won, J.P.; Jang, C.I.; Lee, S.W.; Lee, S.J.; Kim, H.Y. Long-term performance of recycled PET fibre-reinforced cement composites. *Constr. Build. Mater.* **2010**, *24*, 660–665. [CrossRef]
13. Changani, Z.; Razmjou, A.; Taheri-Kafrani, A.; Warkiani, M.E.; Asadnia, M. Surface modification of polypropylene membrane for the removal of iodine using polydopamine chemistry. *Chemosphere* **2020**, *249*, 126079. [CrossRef] [PubMed]
14. Xi, Z.-Y.; Xu, Y.-Y.; Zhu, L.-P.; Wang, Y.; Zhu, B.-K. A facile method of surface modification for hydrophobic polymer membranes based on the adhesive behavior of poly (DOPA) and poly (dopamine). *J. Membr. Sci.* **2009**, *327*, 244–253. [CrossRef]
15. Sa, R.; Wei, Z.; Yan, Y.; Wang, L.; Wang, W.; Zhang, L.; Ning, N.; Tian, M. Catechol and epoxy functionalized ultrahigh molecular weight polyethylene (UHMWPE) fibers with improved surface activity and interfacial adhesion. *Compos. Sci. Technol.* **2015**, *113*, 54–62. [CrossRef]
16. Silva, C.; Simon, F.; Friedel, P.; Pötschke, P.; Zimmerer, C. Elucidating the chemistry behind the reduction of graphene oxide using a green approach with polydopamine. *Nanomaterials* **2019**, *9*, 902. [CrossRef] [PubMed]
17. EN 206+A1:2016-12; Concrete—Specification, Performance, Production and Conformity. European Committee for Standardization: Brussels, Belgium, 2016.
18. EN 197-1:2011; Cement—Part 1: Composition, Specifications and Conformity Criteria for Common Cements. European Committee for Standardization: Brussels, Belgium, 2011.
19. EN 196-1:2016; Methods of Testing Cement—Part 1: Determination of Strength. European Committee for Standardization: Brussels, Belgium, 2016.
20. EN 1008:2002; Mixing Water for Concrete—Specification for Sampling, Testing and Assessing the Suitability of Water, Including Water Recovered from Processes in the Concrete Industry, as Mixing Water for Concrete. European Committee for Standardization: Brussels, Belgium, 2002.
21. Technical Data Sheet for "PoliArm" TM "FIBER" 25 mm Fibers (in Accordance with EN 14889-2:2006). Available online: https://fiber.ua/pl/fiber-poliarm/poliarm-25mm (accessed on 30 October 2023).
22. EN 1015-11:2019; Methods of Test for Mortar for Masonry—Part 11: Determination of Flexural and Compressive Strength of Hardened Mortar. European Committee for Standardization: Brussels, Belgium, 2019.
23. Ohama, Y. Polymer-based admixtures. *Cem. Concr. Comp.* **1998**, *20*, 189–212. [CrossRef]
24. Materak, K.; Wieczorek, A.; Łukowski, P.; Koniorczyk, M. The influence of internal hydrophobization using organosilicon admixture on the performance and the durability of concrete. *Cem. Lime Concr.* **2023**, *28*, 225–237. [CrossRef]
25. EN 1992-1-1:2023; Eurocode 2—Design of Concrete Structures—Part 1-1: General Rules and Rules for Buildings, Bridges and Civil Engineering Structures. European Committee for Standardization: Brussels, Belgium, 2023.
26. Chmielewska, B. Adhesion Strength and Other Mechanical Properties of SBR Modified Concrete. *Int. J. Concr. Struct. Mater.* **2008**, *2*, 3–8. [CrossRef]
27. Pawelska-Mazur, M.; Kaszyńska, M. Mechanical Performance and Environmental Assessment of Sustainable Concrete Reinforced with Recycled End-of-Life Tyre Fibres. *Materials* **2021**, *14*, 256. [CrossRef] [PubMed]
28. Han, X.; Li, P.; Liu, J. Application of a Closed-Form Model in Analyzing the Fracture of Quasi-Brittle Materials. *Materials* **2024**, *17*, 282. [CrossRef] [PubMed]

Disclaimer/Publisher's Note: The statements, opinions and data contained in all publications are solely those of the individual author(s) and contributor(s) and not of MDPI and/or the editor(s). MDPI and/or the editor(s) disclaim responsibility for any injury to people or property resulting from any ideas, methods, instructions or products referred to in the content.

Article

Effects of Flaxseed Mucilage Admixture on Ordinary Portland Cement Fresh and Hardened States

Haris Brevet [1,2], Rose-Marie Dheilly [1], Nicolas Montrelay [1], Koffi Justin Houessou [1], Emmanuel Petit [2] and Adeline Goullieux [1,*]

[1] EPROAD–UR 4669, Université de Picardie Jules Verne, 80000 Amiens, France; harisbrevet@reseau.ispa.asso.fr (H.B.); rose-marie.dheilly@u-picardie.fr (R.-M.D.); nicolas.montrelay@u-picardie.fr (N.M.); houesfr@yahoo.fr (K.J.H.)

[2] BIOPI–UMRT BioEcoAgro, Université de Picardie Jules Verne, 80000 Amiens, France; emmanuel.petit@u-picardie.fr

* Correspondence: adeline.goullieux@u-picardie.fr

Citation: Brevet, H.; Dheilly, R.-M.; Montrelay, N.; Houessou, K.J.; Petit, E.; Goullieux, A. Effects of Flaxseed Mucilage Admixture on Ordinary Portland Cement Fresh and Hardened States. *Appl. Sci.* 2024, *14*, 3862.

https://doi.org/10.3390/app14093862

Academic Editors: Mouhamadou Amar and Nor Edine Abriak

Received: 20 March 2024
Revised: 19 April 2024
Accepted: 25 April 2024
Published: 30 April 2024

Copyright: © 2024 by the authors. Licensee MDPI, Basel, Switzerland. This article is an open access article distributed under the terms and conditions of the Creative Commons Attribution (CC BY) license (https:// creativecommons.org/licenses/by/ 4.0/).

Abstract: France is Europe's leading producer of flaxseed. This seed is rich in omega-3, energy, and protein for animals, but it also contains anti-nutritional factors such as mucilage. Thus, mucilage must be removed and could be used as a bio-admixture in cementitious materials development, reducing the environmental impact of cementitious materials. This study aims to valorize the usage of flaxseed mucilage (FM) in ordinary Portland cement. FM caused macroscopic and microscopic changes in the materials studied. The higher the concentration, the greater the changes were. The admixed samples showed an exponentially concentration-dependent delay in setting. FM degradation products induced by the cementitious conditions accentuated the delay. However, this delay in setting did not affect the hydrates' growth in the material. In fact, FM showed a "delay accelerator" behavior, meaning that once hydration began, it was accelerated as compared to a reference. Macroscopically, FM induced significant flocculation, increasing material porosity and carbonation. Consequently, bulk density and thermal conductivity were reduced. At the highest amount of FM admixture (0.75% w/w), FM allowed bridge formation between $Ca(OH)_2$ crystals, which can improve the mechanical properties of mortars. Because FM is highly hygroscopic, it has the capability to absorb water and subsequently release it gradually and under controlled conditions into the cement matrix. Therefore, regulation of water diffusion from the mucilage may induce the self-healing properties responsible for mechanical properties similar to that of the reference in the medium to long term.

Keywords: flaxseed mucilage; OPC; hydration; mechanical strength; FTIR; calorimetric analysis; SEM; alkaline degradation

1. Introduction

A sustainable agricultural and food approach aims at improving the nutritional quality of human food by balancing animal feed with forages and seeds naturally rich in omega-3s. Flaxseed is high in omega-3, energy, and protein, but it also contains anti-nutritional factors such as mucilage, which reduces nutrient digestibility and impacts broiler chicken growth [1]. France is Europe's leading producer of flaxseed, so its animal feed manufacturers want to improve the nutritional quality of the flaxseed by removing mucilage [2]. Therefore, it is necessary to find ways of adding value to mucilage, which is a by-product of the seed-dehulling process. One possibility would be to use it as a bio-admixture in cementitious materials.

Flaxseed mucilage (FM) is present in anhydrous form in flaxseed before hydration by contact with water. FM is a compound rich in polysaccharides (50–80%) but also in proteins (4–20%) and minerals (3–9%). FM is composed of two fractions of water-soluble heteropolysaccharides: the acidic rhamnogalacturonans type I (RG-I) fraction (17%) and the neutral arabinoxylans (AX) fraction (83%) [3]. The bone composition of the mucilage varies according to the extraction conditions [4].

Rhamnogalacturonans are pectic acidic polysaccharides that can impact cement hydration and cementitious properties. Shanmugavel et al. [5] observed a consistency decrease and longer setting times with the increase in pectin percentage. They stated that during the hydration of cement paste, the galacturonic acid can form strong intermolecular association among the galacturonan chains by forming calcium bridges, with the effect of enhanced viscosity. According to Thomas and Birchall [6], natural polymers have the inherent characteristic of surface absorbency of organic molecules and subsequent development of protective polymeric film on the cement particles and hydration products. This film would restrict further hydration of cement particles. Hazarika et al. [7] also observed the viscosity-enhancing property of the addition of acidic heteropolysaccharides but a setting time reduction. As pectin incorporates some amount of calcium ions in its structure during the hydration of cement, the Ca^{2+} concentration in pore fluid would decrease. To balance the Ca^{2+} concentration, the rates of the hydration of cement minerals would increase, and greater amounts of hydration products would be formed, resulting in a decrease in setting times. Pan et al. [8] used carrot extract, whose main polysaccharide is rhamnogalacturonan. They observed a delay in setting and an enhanced compressive strength of the mortars. The positive effect on the compressive strength would be due to greater development of Portlandite and C-S-H in the cementitious system.

Concerning the arabinoxylans, a neutral fraction of polysaccharides, Girones et al. [9] observed a retardation of cement hydration with 2% (w/w) of AX. The AX polymers might slow down not only the formation but also the growth of the C-S-H nuclei. The impact of rhamnogalacturonans and arabinoxylans on cement seem to be dependent on the polysaccharide concentration and other compounds present in the extracts.

One of the plants used in its extract form as a replacement for mixing water is *Opuntia ficus indica* (OFI)—its availability, low operating cost, and presence in arid areas where water is most needed make it an important element in cement admixture. OFI cladodes are waterlogged and contain a huge amount of polysaccharides and some proteins that probably can interact with the complex hydration mechanism of Portland cements [10,11]. But this extract used for water mixing in cement contains a percentage of mucilage and proteins but also some other natural compounds (cellulose, fats, hemicellulose, starch, ashes, etc.) [5,12]. Chandra et al. [11] indicated that the incorporation of OFI extract increases the plasticity of the cement paste with a consequent decrease in water absorption by the mortar and an improvement in the freeze–thaw resistance of the mortar. The author also showed a possible interaction between the polysaccharides and the Portlandite formed. The formation of complexes influences the crystallization process, interfering with the size of the Portlandite crystals and making them more amorphous. The ability of OFI polysaccharides to interfere with the growth of mineral species and the crystal microstructure was confirmed using several polysaccharides [13]. Finally, the addition of polysaccharides within a cementitious matrix reduces the carbonation phenomenon of the material by acting as a barrier property to gases and water, a property conferred by the viscous extract of OFI [10,11]. Not all polysaccharides have the same resistance to the highly alkaline cementitious environment. Polysaccharides added to the cementitious matrix degrade more or less easily into hydroxycarboxylic acids or smaller entities.

The objective of the present study was to investigate the impact of the admixture of FM at different concentrations in a cement matrix. An evaluation of the FM degradation in an alkaline environment is herein discussed. A fresh state study of Portland cement provided some understanding of the hydration properties of cement in the presence of FM polysaccharides. The mechanism and products of hydration were investigated in the fresh state and in the hardened state after different curing times. The impact of the admixture rate on the macroscopic and microscopic structure was also explored in this study to obtain a more comprehensive knowledge of the impact of the addition of mucilage on the growth of hydration products. To this end, mortars were manufactured, and their mechanical strengths and thermal conductivities were evaluated.

2. Materials and Methods

2.1. Materials

The cement used in this study was an Ordinary Portland Cement (OPC), CEM I 52.5 N CE PM-CP2 NF, commercialized by Calcia, Courbevoire, France. The clinker is the main component ($\geq$95% w/w) of this cement, and no fillers were applied in this study, so result dispersion was avoided. Its composition in weight % is 74.8, 3.7, 8.2, and 8.3 for C_3S, C_2S, C_3A, and C_4AF, respectively. The Bogue approximation gives 66.48, 20.97, 4.84, 2.73, 3.4, and 0.21% (w/w) for CaO, SiO_2, Al_2O_3, Fe_2O_3, SO_3, and Na_2O, respectively. It exhibits a Blaine fineness of 4360 $cm^2 \cdot g^{-1}$. The different mortars were elaborated with a 0/4 mm sand, according to the NF EN 12620 standard. With respect to the French standard NF EN 1008, the mixing water used in this study was tap water at a temperature of 20 °C $\pm$ 2 °C. Flaxseed mucilage (FM) used as admixture originated from a gold flaxseed cultivar (Eurodor) and is available in a lyophilized form after the extraction procedure described by Brevet et al. [14].

2.2. Flaxseed Mucilage Characterization

2.2.1. Determination of FM Proximate Composition

The protein content of FM was quantified according to the Kjeldahl method described by AOAC 954.01 [15] on the determination of total nitrogen content in the samples. A conversion factor of 6.25 was used to calculate the crude protein content from the nitrogen content.

The AOAC Official Method 942.05 was used to determine the ash content of FM. An approximate 2 g of raw materials was weighed in porcelain crucibles and then calcined at 600 °C for 2 h in a pre-heated muffle furnace. The cooling step of the calcined samples was different from the AOAC method. The calcined materials were cooled in an oven at 70 °C for 2 h to avoid any moisture regain before being weighed.

High-performance anion-exchange chromatography with pulsed amperometric detection (HPAEC-PAD) was used to determine the FM monosaccharides composition and content. Results are expressed in grams of carbohydrates/100 g of FM. The analysis was performed as described by Roulard et al. [16].

2.2.2. Alkaline Solubilization of FM

The solubilization of FM in alkaline solutions allows the evaluation of the impact of the pH and the presence of Ca^{2+} cations on the availability of the characteristic groups of polysaccharide chain length after solubilization. The mucilage was dissolved (2, 5, 10, 15, 20, 25, and 30 g/L) into two different solutions of pH = 12.6: NaOH 0.04 M and $Ca(OH)_2$ 0.02 M. Solubilization was carried out at a rotation speed of 140 rpm in a beaker with a paddle stirrer until complete mucilage solubilization occurred to simulate the slow speed of mixing in an alkaline environment of the standard EN 196-1 [17] relative to mortar development. Then, alkaline mucilage solutions were dried in an oven at 50 °C to avoid any degradation due to temperature. Fourier transform infrared spectroscopy (FTIR) and viscosity measurements were carried out to evaluate the alkaline environment impact on FM.

2.2.3. Apparent Viscosity of FM

The viscosity range of mucilaginous solutions was established using a rheometer (DVNext, Brookfield, Toronto, ON, Canada) with appropriate HA/HB spindles ranges. Mucilaginous solutions were prepared to obtain seven solutions at different concentrations (2, 5, 10, 15, 20, 25, and 30 g/L) by solubilizing FM into tap water, NaOH 0.04 M, and $Ca(OH)_2$ 0.02 M at 400 rpm until all the FM was solubilized. The viscosity values were obtained at different spindle rotational speeds from 20 to 200 rpm. The designated viscosity value corresponds to a constant rotational speed of 60 rpm, corresponding to the most adequate torque generated for all solutions. At 60 rpm, the torque value during the experiment was optimal (between 40 to 60% of the maximum torque tolerated by the instrument). At higher speeds, the torque was in the low range of the apparatus, while torque values that were too high were obtained at speeds lower than 60 rpm.

2.2.4. FT-IR Characterization

Fourier transform infrared spectroscopy was used to characterize the changes in FM following alkaline solubilization. Dried FM materials were analyzed. The FTIR spectra of FM were determined by FTIR spectrophotometer (IR-Prestige 21, Shimadzu, Noisiel, France). Approximately 1% (w/w) of FM was weighed and crushed on a KBr matrix pellet (200 mg). The FTIR spectra were obtained with 200 scans in a transmittance mode and a resolution of 2.0, where an Happ–Genzel apodization was applied in a range of 400–4000 cm^{-1}.

2.3. Mortars Preparation

The preparations are carried out in a standardized mortar mixer (EN 196-1) (Proviteq, Lisses, France). FM was incorporated by solubilizing within the mixing water to fully study the interaction between the polysaccharides and the cement. Mortars were elaborated with a W/C ratio of 0.5 according to EN 196-1. The mortar compositions are given in Table 1. The samples were molded without the use of an impact table. The objective was to study the macroscopic structural effect of mucilage on a cementitious mortar. The molds used have dimensions of 4 × 4 × 16 cm for the study of the mechanical strength and 10 × 10 × 2 cm^3 for the further thermal conductivity study of mucilage-admixed materials. Each sample for strength tests was then cured for 28, 60, and 90 days (Table 1) in a chamber with saturated humidity and at room temperature. The samples for the thermal conductivity measurements were cured for only 28 days. Before characterization, samples were dried in an oven at 50 °C until a constant mass was obtained.

Table 1. Polysaccharidic mortars compositions.

Sample	Water (g)	Solubilized Mucilage (g)	Admixture (% Cement)	Cement (g)	Sand (g)	Curing Time (Days)
OPC	225	0	0	450	1350	28, 60, 90
EURO2G	225	0.45	0.1	450	1350	28, 60, 90
EURO5G	225	1.125	0.25	450	1350	28, 60, 90
EURO15G	225	3.375	0.75	450	1350	28, 60, 90

2.4. Characterization of FM Admixtured Mortars

2.4.1. Setting Times Determination

The evaluation of the hydration heat release is an effective technique for monitoring the hydration process and determining the initial and final setting times of pastes containing different amounts of FM. The experiments were conducted on a Calvet calorimeter C80, Setaram. Pastes were prepared by mixing 100 g of cement with 50 g of the mixing tap water at different mucilage solution concentrations (0, 2, 5, 15, and 30 g/L). The cement and water mixture were homogenized with an electric whisk during a 30 s time minimum depending on how difficult it was to homogenize the mix. Then, an approximate 3000 mg of each paste was introduced into a stainless-steel capsule. The heat flow ran for a minimum time of 48 h in a 25 ± 0.5 °C regulated chamber.

2.4.2. Slump Test

The slump test was performed on a mini-Abrahams cone called MBE ("Mortier Béton Equivalent" translated as "Concrete Equivalent Mortar"), which has dimensions of 150 mm height and an upper and lower diameter of 50 and 100 mm, respectively. The test was carried out following the procedure described by Schwartzentruber et al. [18].

2.4.3. Hydration Degree

When the major cement compounds, i.e., tricalcium silicates (C_3S Equation (1)) and dicalcium silicates (C_2S Equation (2)), are in contact with water, there is formation in the first few hours of calcium silicate hydrate (also called C-S-H), ettringite, and Portlandite ($Ca(OH)_2$). Over time, it is possible to form calcium carbonate ($CaCO_3$ Equation (3)) in various forms—calcite, vaterite, and aragonite. These different compounds are dependent on the hydration of the cement grains. To quantify good cement hydration, Bhatty developed a method based on thermogravimetric analysis [19]. The thermogravimetry allows the degradation of the hydration products by dissociation reactions—dehydration for C-S-H (Equation (4)), dehydroxylation for Portlandite (Equation (5)), and finally, decarbonation for calcium carbonates (Equation (6)).

The hydration reaction of tricalcium silicates and dicalcium silicates is as follows:

$$Ca_3SiO_5 + 6H_2O \rightarrow CSH + 3Ca(OH)_2 \tag{1}$$

$$Ca_2SiO_4 + 4H_2O \rightarrow CSH + Ca(OH)_2 \tag{2}$$

That for Portlandite carbonation with ambient carbon dioxide is as follows:

$$Ca(OH)_2 + CO_2 \rightarrow CaCO_3 + H_2O \tag{3}$$

The first reaction of CSH dehydration is given below [20]:

$$(CaO)_a \cdot SiO_2 \cdot (H_2O)_b \stackrel{\Delta}{\leftrightarrow} (CaO)_a \cdot SiO_2 \cdot (H_2O)_{b-c} + c \cdot H_2O \tag{4}$$

The second reaction of dehydration or dehydroxylation of Portlandite is as follows:

$$Ca(OH)_2 \stackrel{\Delta}{\leftrightarrow} CaO + H_2O \tag{5}$$

The decarbonation reaction of the calcium carbonate phases occurring at higher temperature is given below:

$$CaCO_3 \stackrel{\Delta}{\leftrightarrow} CaO + CO_2 \tag{6}$$

The method developed by Bhatty [19] allows determining the degree of hydration (DH) of mortars and cement pastes, taking into account the different dissociation reactions. The Equation (7) is used to determine the DH, according to the author.

$$DH(\%) = \frac{W_b}{0.24} \times 100 \tag{7}$$

where W_b corresponds to the chemically bound water, and 0.24 corresponds to the part of chemically bound water that is combined with each part of cement. W_b (Equation (8)) quantifies the weight loss during the dehydration (Ldh), the dehydroxylation (Ldx), and the decarbonation (Ldc). The correction coefficient 0.41 is based on the assumption that carbonate is formed only by the reaction of CO_2 with $Ca(OH)_2$ [21].

$$W_b = (Ldh + Ldx) + 0.41(Ldc) \tag{8}$$

A differential scanning calorimetry coupled with thermogravimetric analysis (DSC-TGA) (Themys LV, Setaram, Caluire-et-Cuire, France) was carried out on cement mortars and pastes to determine their degree of hydration and thus to evaluate the FM action on the evolution of the hydration products. The mortars were powdered using a mixer mill (MM400, Retsch, Eragny, France). About 130 mg of this powder was introduced into porcelain crucibles. The analysis started at 25 °C and went up to 1100 °C at a rate of 5 °C/min under a helium atmosphere.

2.4.4. Mechanical Characterization of Mortars

The strengths of the mortars were determined in compression at different curing times (28, 60, and 90 days) on a machine from Proviteq, France, with a load cell threshold of 100 kN. The strengths were obtained at a load rate of 2400 N·s^{-1} on twelve replicates according to the EN 196-1 standard.

2.4.5. Microstructure Visualization

The morphology of the samples was studied using scanning electron microscopy (namely SEM). This analysis allowed visualizing the macro and microporosities of the different mortars. The micrographs were obtained using a PHILIPS FEG XL 30 microscope. To facilitate observation, the samples were first dried and then coated with a thin layer of spray-on gold to enhance their electric conductivities. SEM observations were coupled with an energy dispersive X-ray spectroscopy (EDS) analysis allowing the surface samples' elements identification.

2.4.6. Thermal Conductivities Determination of Mortars

The samples ($10 \times 10 \times 2$ cm^3) were cured for 28 days and then analyzed by transient plane source (TPS) 2500, HotDisk, to determine their thermal conductivities. Before measurements, the samples were dried in an oven at 50 °C for at least a week until a constant mass was obtained. Measurements were repeated three times.

3. Results and Discussion

3.1. FM Characterization

3.1.1. Proximate Composition of FMs

The proportion of protein is 10.3%, and the quantity of carbohydrates is 45.1% mass of mucilage. The extracted product is relatively clean since the ash content (3.8%) is relatively low compared with values found in the literature [3,4,16,22,23]. All these values are highly affected by the process parameters [4,24] and the flaxseed cultivar [3].

The polysaccharide composition is categorized into two distinct fractions. The neutral fraction consists of xylose and arabinose, forming arabinoxylans (AX). This fraction accounts for 72% of the total polysaccharides. The second fraction, constituting 28% of the polysaccharide content, is an acidic fraction primarily composed of rhamnose and galacturonic acid, resulting in type I rhamnogalacturonans (RG-I). Warrand [25] initially elucidated this mucilage separation into these two fractions. He noted that the neutral and acidic fractions make up approximately 75 and 25% of the polysaccharide content in mucilage, respectively.

Finally, it is interesting to look at one of the mucilages that was referenced as a bio-admixture and whose admixture effects on the characteristics of cementitious matrices are widely praised and positive. The Table 2 shows an interesting parallel between the monosaccharide composition of the FM and OFI cladode mucilage. It is clearly visible that the monosaccharide profiles are completely different, with a large presence of galacturonic and glucuronic acid for OFI. The large presence of a pectic fraction greatly differentiates it from FM, where 73% of the content is a neutral fraction. It is therefore possible that FM does not have the same effects on cement materials as OFI mucilage.

Table 2. Monosaccharides content (g sugar/100 g) of FM and OFI cladodes.

Monosaccharide	HPAEC-PAD FM (Raw) [1]	FM [3]	OFI [4]
Galacturonic acid	3.686 ± 0.708	8.17	23.2
Arabinose	8.580 ± 0.373	19.02	18.8
Galactose	5.857 ± 0.165	12.98	31.8
Glucose	1.910 ± 0.106	4.23	25.1
Fucose	1.673 ± 0.038	3.71	
Rhamnose	7.751 ± 0.169	17.18	

Table 2. *Cont.*

Monosaccharide	HPAEC-PAD FM (Raw) [1]	FM [3]	OFI [4]
Xylose	15.640 ± 0.984	34.67	1.1
Glucuronic acid	0.009 ± 0.003	0.020	23.2
TOTAL	45.105 ± 1.793 [2]	100	100

[1] Raw data obtained from the HPAEC-PAD. [2] The sum of compounds does not reach 100% due to carbohydrates losses during the HPAEC-PAD hydrolysis [26]. [3] First column monosaccharides content on 100% basis to compare with OFI. [4] Values from Lefsih et al. [27].

3.1.2. Impact of Alkaline Conditions on FM

Apparent viscosity: When the mucilage is dissolved in the mixing water, the pectic and neutral fractions interact and structure themselves to form a more or less viscous network, depending on the concentration. This viscosity was therefore determined at different mucilage concentrations and is used as a reference for determining the quantity added to the cementitious matrix. It is also possible to simulate the effect of the cementitious matrix on the composition and/or degradation of polysaccharides using viscometry. The viscosimetric behavior of polysaccharides dissolved in different alkaline solutions (NaOH and $Ca(OH)_2$) with a deliberately high pH, such as the cementitious medium, can be seen in Figure 1.

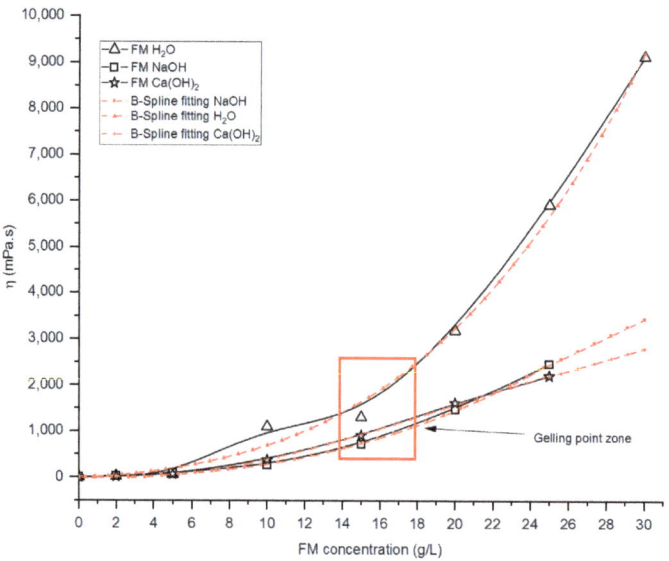

Figure 1. Apparent viscosity of FM solutions at 20 ± 2 °C (60 rpm).

Firstly, Figure 1 shows that the viscosity of the FM increases significantly as a function of the FM concentration. This increase is accelerated at around 15 g/L. This concentration was identified as a gelling point or a sol–gel transition [25]. Warrand [25] conducted a study on the viscous properties of FM and found that FM exhibits shear-thinning behavior. This means that the viscosity increases with increasing polysaccharide concentration, primarily due to interactions between the two distinct fractions present in FM. The pectic fraction that is anionic has minimal or no impact on determining the solution's physicochemical properties. Instead, it is mainly the neutral fraction, primarily composed of AX, that is responsible for the exceptionally high viscosity of the mucilage solution. The gel-forming characteristics of the AX solution arise from intermolecular hydrogen bonds and a relatively larger molecular weight compared to the acidic fraction. Several studies have reported a critical transition point in the apparent viscosity of the mucilage solution [4,25]. Like many plant extracts [28,29], this sol–gel transition point corresponds to the concentration or condition at which the solution ceases to be considered a true solution and instead behaves as a gel. This transition occurs due to the increasing entanglement of polymeric chain networks, which restricts the mobility of the polysaccharide chains and leads to a significant increase in viscosity [25].

FM does not exhibit the same behavior when added in alkaline conditions, as shown by the Figure 1. The viscosity curves part ways with the aqueous medium from a 5 g/L concentration. The apparent viscosity increases with the concentration but much less for the alkaline solutions than in water. There are two possible explanations for these observations.

First, the intermolecular network formed by hydrogen bond is reduced due to the action of NaOH and $Ca(OH)_2$ solutions. Chen et al. [30] worked on the gelation properties of flaxseed gum and confirmed the results shown in Figure 1. Flaxseed gum is characterized by its anionic polysaccharide nature, resulting from the presence of ionized carboxyl groups that generate a negative charge. The electrostatic forces between like-charged molecules cause the molecular chains to fully extend and intertwine, facilitating the formation of intermolecular cross-links that induce gelation at pH between 6 and 9. The authors explained that increasing the pH above 9 diminishes the gel strength and thus the apparent viscosity. Furthermore, they also mentioned that the Na^+ action on the zeta potential leads to a decrease in the intramolecular charge repulsions. The same mechanism of electrostatic repulsion exists with a large amount of divalent cations like Ca^{2+}, which inhibits the formation of a three-dimensional network [30]. The presence of divalent cations in a lower concentration can, however, induce a cross-linking or flocculation of the polysaccharides, as we observe a sedimentation because of agglomerates' formation. These phenomena are visible in Figure 1, as the NaOH and $Ca(OH)_2$ solution curves flatten compared to that for water. A higher flattening on the $Ca(OH)_2$ curve can be evidence of a divalent cation intermolecular cross-linking, shown by a flocculation visible during the material preparation.

The second hypothesis that can lead to the apparent lowering of viscosity concerns the possibility of polysaccharide degradation. In our conditions, molecular weight determination by SEC-MALS analysis does not demonstrate backbone hydrolysis (Figures S1 and S2). This hypothesis is not consistent with our results.

Thus, in NaOH and $Ca(OH)_2$ solutions, the decline in gel strength is linked to fewer junction zones due to high pH levels (pH > 9) and the presence of monovalent and divalent cations, as reported by several authors [31–33].

Compositional modifications—FTIR: The FM solubilized under different conditions were analyzed in FTIR to determine possible compositional modifications. The FTIR spectra are visible in Figure 2. First, there is an appearance of a doublet at 897 and 866 cm^{-1} when alkaline conditions are applied to the FM. These bands are characteristic of the co-existence of β- and α-glycosidic bonds [34], respectively. This new co-existence is possible evidence of an alkaline hydrolysis, as alkaline catalysts are often used to liberate carbohydrate chains from glycoconjugates [35]. The liberation of glycoconjugates such as proteins is visible on the NaOH solution spectrum with the disappearance of the N-H bond at 1541 cm^{-1}. Also, the C-H "hairy zone" and -CH_2 branched at 1413 cm^{-1} look different. The increase in the band and the apparition of a triplet at 1444, 1413, and 1382 cm^{-1} seem to confirm the hydrolysis of the side-branched conjugates. In the $Ca(OH)_2$ FM solution, this N-H bond seems to still be present, as the divalent cation can bind different polysaccharidic/protein compounds and then limit the impact of alkaline solution as characteristic groups are protected.

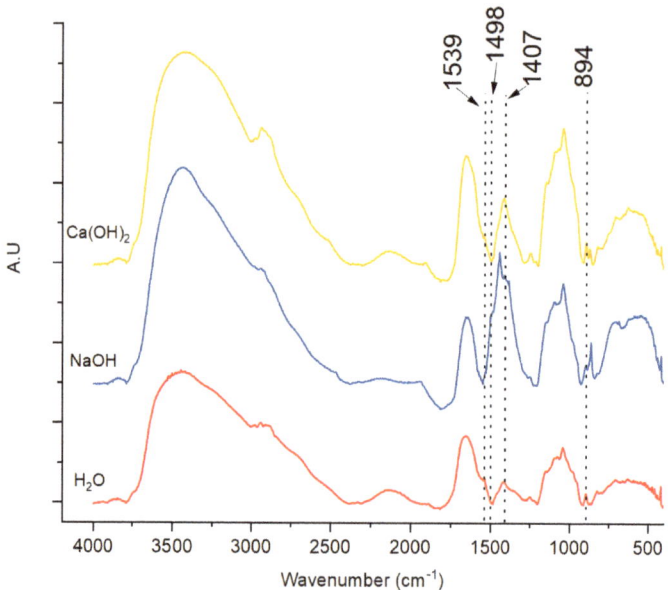

Figure 2. Infrared spectra of solubilized mucilages at different alkaline conditions.

The complexification and increase in organic species present in the cementitious medium after the FM degradation make it more complicated to conduct studies and to identify the mechanisms intrinsic to the initial polysaccharides. For example, Kang et al. [36] related that xylose, which is the main saccharide of the FM, degrades mainly to formic acid and furfural. Whistler and BeMiller [37] described the alkaline degradation process of polysaccharides as a peeling process in which the reducing end-group is liberated from a chain by elimination of the rest of the chain as a glycoxy anion.

3.2. Characterization of Cement Composites

3.2.1. Fresh State Characterization

Slump test: The elaboration of mortars requires characterization in the fresh state to determine the workability and viscosity of the hydraulic material. This measurement is all the more relevant when an admixture is incorporated. Figure 3a,b show the consistency of a EURO5G admixed mortar and the values obtained from the slump test, respectively. Figure 3a clearly shows the appearance given by the mucilage to the material. The addition of FM to a cementitious composite tends to form flocks whose size and shape depend on the mucilage concentration. The size decreases as the concentration increases, and the sphericity of these aggregates becomes more regular as the mucilage concentration increases. This particular granular structure suggests a coating of the cement grains/sand by the polysaccharide and not an increase in workability, as may be the case with OFI mucilage [11,38]. This important granular aspect makes the cement particularly dry (in conditions of W/C = 0.5) and difficult to work. During samples molding and demolding, this granular aspect, bordering on sandy, makes the material quite friable. As mucilage is a very hygroscopic material, the water available for hydration during mixing may be reduced initially. Dissolving the mucilage also allows the polysaccharide chains to unwind and ionize. This ionization makes the anionic groups available to bind with the metallic cations present in the cement, as shown previously with the FM solubilization in alkaline solutions and FTIR conclusions. This interaction at the surface of the cement grains has been shown to be an absorption or adsorption of the polysaccharide groups onto the cement grains, resulting in a flocculation process [39]. This phenomenon can be seen in Figure 4. Moreover, the kinetics of cement hydration makes this phenomenon even more plausible. Indeed, due to the substantial water absorption by the mucilage and the cement's affinity for water, the hydration kinetics of the cement enable the formation of this bond. Figure 4 shows that the greater the amount of mucilage, the greater the flocculation.

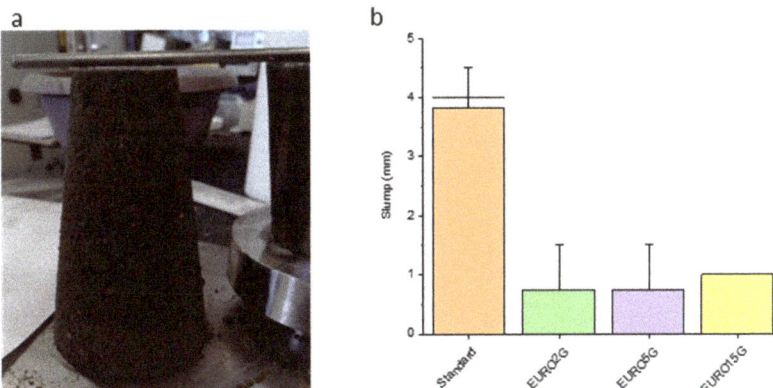

Figure 3. Slump test (**a**) of the mortars and the results (**b**).

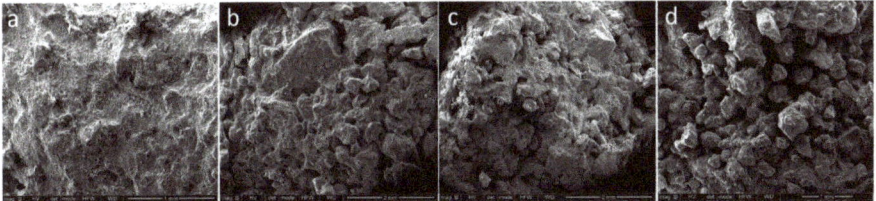

Figure 4. Granular shape of 28 days cured mortars—(**a**) reference, (**b**) EURO2G, (**c**) EURO5G, and (**d**) EURO15G.

Setting times determination: The polysaccharide has the ability to bind with cement cations and absorb an amount of mixing water necessary to the cement for hydration and hardening. The Figure 5 illustrates the isothermal calorimetry of the polysaccharide admixed cement pastes. Zhang et al. [40] worked on the retarding effect of saccharides on cement pastes and concluded that the initial and final setting times increased exponentially as a function of the sugar concentration. To confirm these results, an additional cement paste, with 30 g/L mucilage mixing water (1.5% w/w cement), was used. The results are shown in Figure 5a–c.

The Figure 5a clearly shows the impact of the polysaccharides on the cement hydration. The setting times, both initial and final, are increased as a function of rising polysaccharides concentration. The polysaccharides also influence the heat generated during the hydration and extend the induction period. The Figure 5c confirms the conclusions of Zhang et al. [40] that the setting is exponentially influenced by the polysaccharides concentration. Figure 5b shows a crossing point of the admixed cement pastes curves with the reference one around 32 h, indicating a higher hydration rate above that crossing point for the admixed cement pastes. Zhang et al. [40] mentioned this point. Moreover, after 48 h of hydration, excepted for EURO15G and EURO30G, the cumulated heat is at least similar or higher than that of the reference.

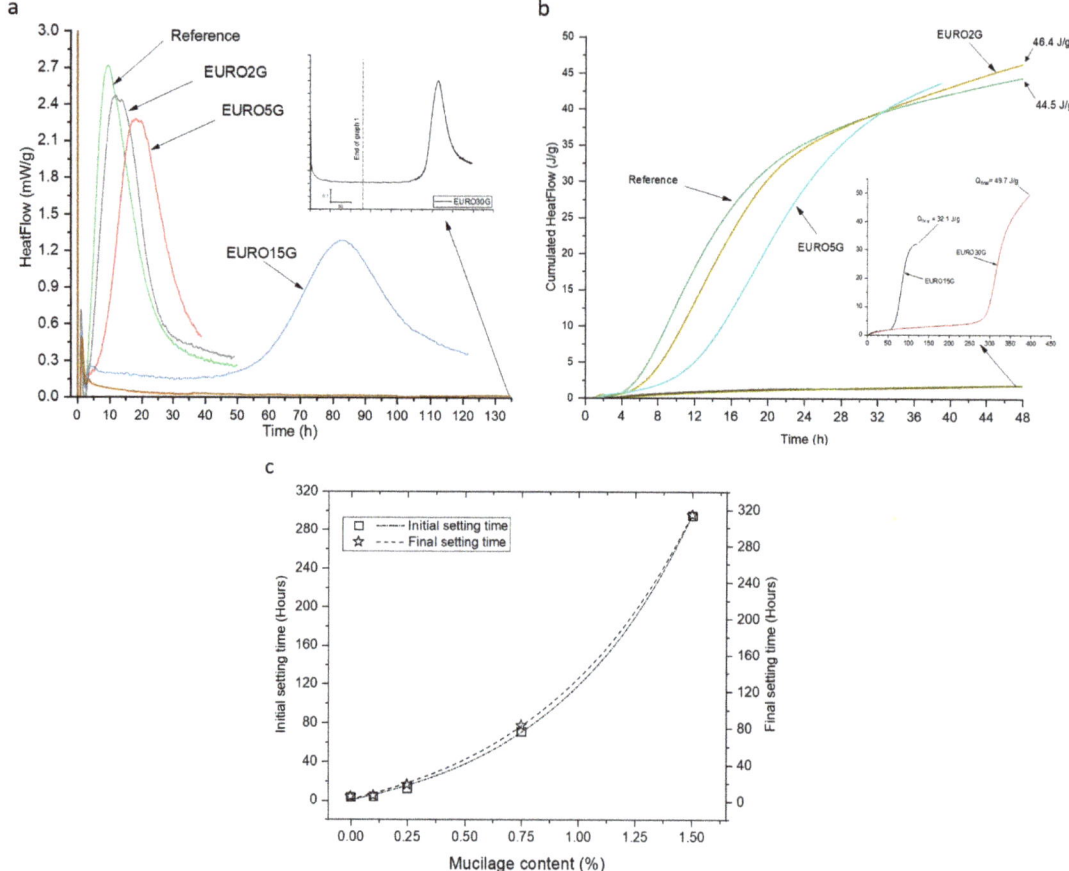

Figure 5. Setting times determination of mucilaginous cement pastes at W/C = 0.5 ((**a**) heat flow versus time according mucilage content, (**b**) cumulative heat flow versus time according mucilage content, and (**c**) initial and final setting times versus mucilage content).

Another evidence of the polysaccharide impact is on the mineral activity. The first part of the hydration process is characterized by a linear curve (Figure 5b). Before the deceleration of the hydration process, the slopes of the integrated curves are 2.49, 2.37, 2.19, 0.84, and 0.68 $h \cdot g \cdot J^{-1}$ for the standard, EURO2G, EURO5G, EURO15G, and EURO30G, respectively (upper and lower bounds are: X = [7.3; 16.7]; [9; 18.9]; [12.9; 24.4]; [67.2; 98]; [296.3; 329.4] for the standard, EURO2G, EURO5G, EURO15G, and EURO30G, respectively). All the evidence of the crossing point, the higher ending heat generated, and the lower slope during the induction period highlight the acceleration of the hydration process at the end of the 48 h, which was confirmed by the work of Zhang et al. [40]. For these authors, this behavior (called "delayed acceleration") is consistent with the fact that the induction period is controlled by slow formation or poisoning of the CSH nuclei [40] induced by the FM.

For EURO15G and EURO30G, the mineral activity is very low as the induction period is increased. FM is composed by AX, which is a neutral fraction, and by RG, which is the anionic one. In the neutral fraction, especially at the anomeric carbon (C1) of the end reducing units, the HO-C-C1=O groups of the arabinose and xylose [41] but also the α-hydroxylated acid as galacturonic acid units in the RG fraction [42] have the ability of binding with the cement dissolved ions, thus limiting the CSH nucleation at the surface of the cement grain. The increasing setting times and induction period as the polysaccharide concentration rises is the result of a co-action of chemisorption of the metallic ions from the cement and the increasing viscosity of the paste at higher polysaccharide concentrations. This high viscosity, provoked by an increasing amount of unwound polysaccharides, generates more cement grains ions absorption and an important reduction in mobility of those ions caused by a steric hindrance. It has the effect of lowering the rate of the CSH nucleation and $Ca(OH)_2$ precipitation (Figure 6). This point is confirmed by combining viscosity measurements of FM solutions and setting time measurements of the admixed cement pastes.

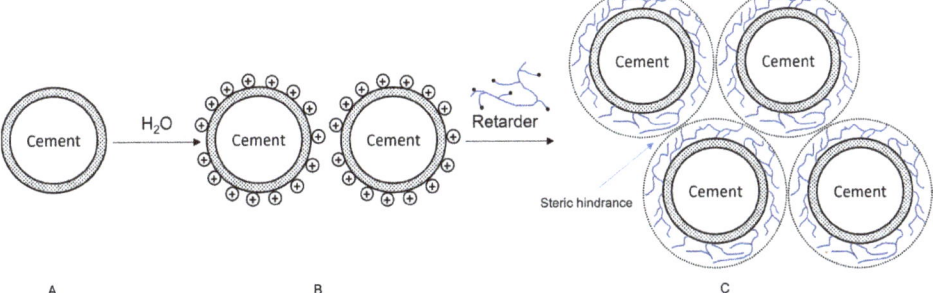

Figure 6. Schematic representation of the retardant barrier formation mechanism and the increase in steric hindrance. (**A**) Anhydrous cement grain, (**B**) formation of cationic sites on the surface of the cement grain, and (**C**) bonding of retarder and formation of a semi-permeable barrier around the cement grains.

3.2.2. Evaluation of the Hydrates Formed in Cement Mortars

Figure 7a shows the compressive strength and bulk density of mortars aged from 28 to 90 days, with their respective degrees of hydration shown in Figure 7b. It can be seen from this figure that the compressive strength evolves constantly over time and in the same way as the bulk density, which is not the case for the degree of hydration. The EURO2G and EURO5G mortars yield a strength that is identical or close to the reference, but the maximum strength seems to be reached after 60 days and then no longer evolves. Lastly, EURO15G shows weaknesses in compressive strength despite a significant change in strength over time compared to the other samples. For the same curing time, the higher the addition of FM, the lower the bulk density. The degree of hydration is at least identical to that of the reference but higher. This indicates good hydration of the hydraulic binder over time. These elements highlight the macroscopic granular, sandy shape of the mortar (Figure 4) caused by flocculation, which is responsible for the weakness of the material.

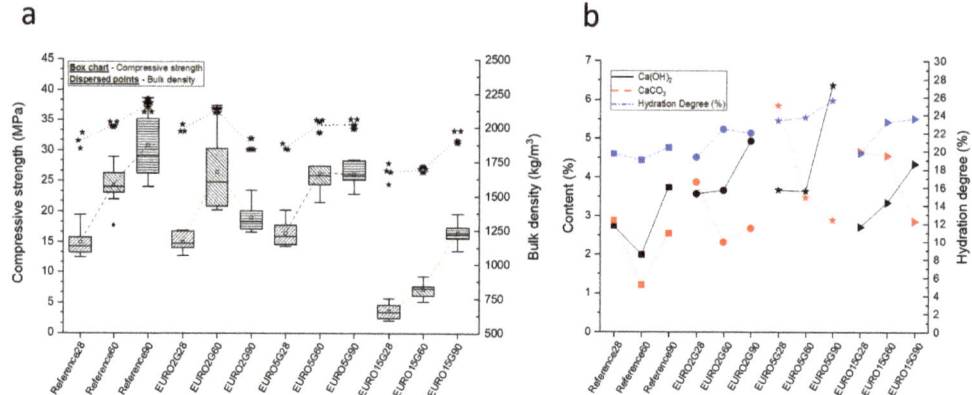

Figure 7. Compressive strength and bulk density (dispersed stars) (**a**) and hydration degree and hydrates contents (**b**) of 28-, 60-, and 90-day cured mortars.

For EURO2G and EURO5G, Figure 7b shows the quantity of hydrates developed at different curing times. The evolution of the quantity of Portlandite and $CaCO_3$ over time are the opposite, indicating a dissolution of the carbonates by prolonged hydration over time. This trend is not observed with the reference, in which the evolution of Portlandite and carbonates are linked. These results suggest that mucilage promotes carbonation at young ages and, further, the presence of Portlandite by prolonged hydration. This phenomenon has been described for certain polysaccharides in the case of self-healing concretes [43,44]. The increase in the carbonation of cement at young ages is favored by various factors, either jointly or independently:

- The degradation of mucilage into alcoholic saccharides, which promote the formation of calcium carbonate;
- The presence of -COO^- carboxyl groups within the mucilage and the accumulation of Ca^{2+} on the surface of the cement grains increase the conditions for the bond between CO_3^{2-} and Ca^{2+} ions. These conditions favor the precipitation and crystallization of $CaCO_3$ [45];
- The increase in mucilage concentration leads to an increase in grain flocculation and a decrease in the compactness of the cementitious matrix. The porosity created by the macro-structure of the composite can lead to a deeper access of ambient CO_2 into the material and thus to carbonation at an early age. This argument is validated by the work of Wang et al. [46];
- The viscosity confers a bubble trap characteristic to the mucilage. The higher the viscosity, the stronger this characteristic. During the mucilage solubilization and mixing phase, the air bubbles contained in the mucilage entrap a significant quantity of O_2 and CO_2 from the ambient air in particular. These gases find their way into the cementitious matrix at early ages, encouraging the formation of $CaCO_3$ [47,48].

When FM is added in the matrix, the high amount of calcium carbonates formed in the first ages is followed by a decrease. FM is able to capture water in the ambient air as a high hygroscopic material. The captured water will diffuse slowly and in a controlled manner over time into the cement matrix, which may lead to a long-term hydration. This long-term hydration induced by the FM corresponds to the rise in the amount of Portlandite in Figure 7b.

3.2.3. Microstructure Visualization—SEM

Figure 8 shows SEM images of the mortars at 28 days and 90 days. The matrices and the matrix/sand interfaces enable the evaluation of the evolution of the material over time as well as the adhesion of the matrix to the sand grain. According to these images, the adhesion of the matrix to the sand grain evolves negatively from the reference to EURO15G, with interfacial transition zones of 0 μm, 0.2–0.3 μm, 0.3 μm, and 0.5–1.4 μm on average for the reference, EURO2G, EURO5G, and EURO15G, respectively. These values are the second argument that the compressive strength of mortars decreases as the mucilage concentration increases after the flocculation by polysaccharides, as shown in Figure 4.

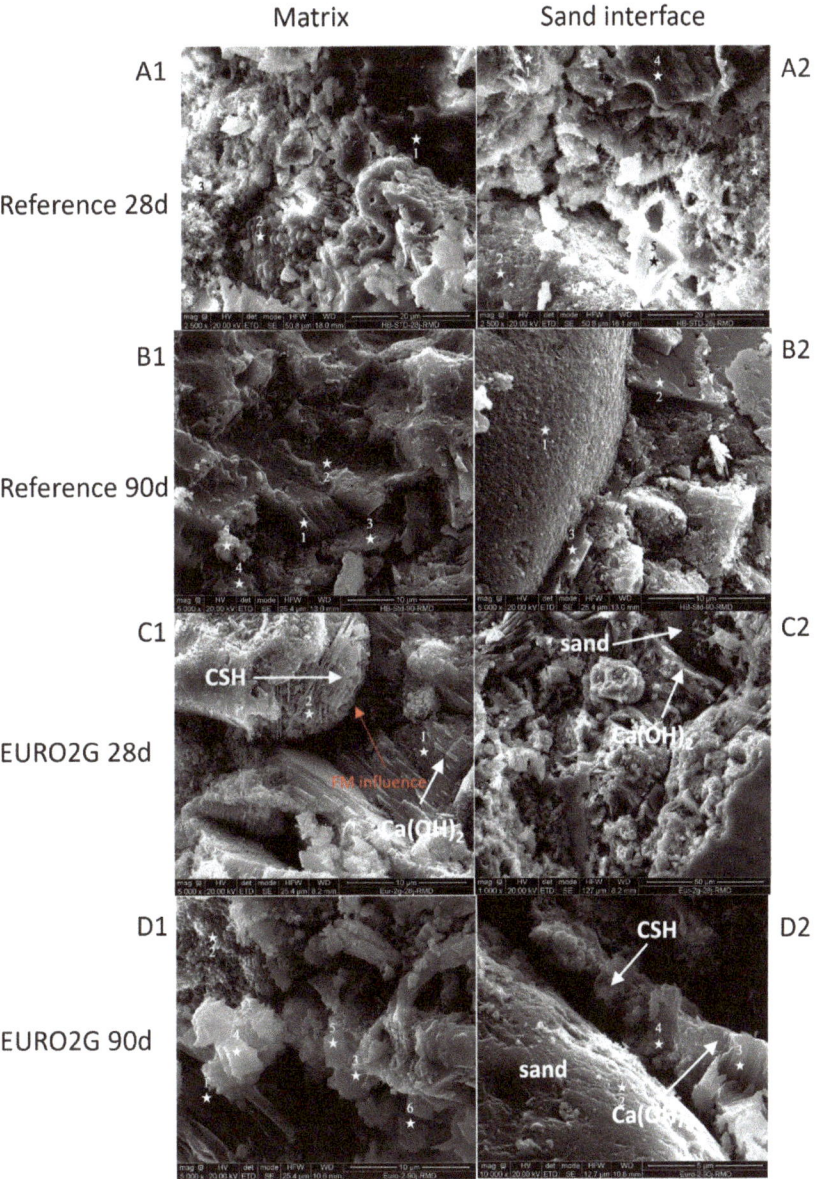

Figure 8. *Cont.*

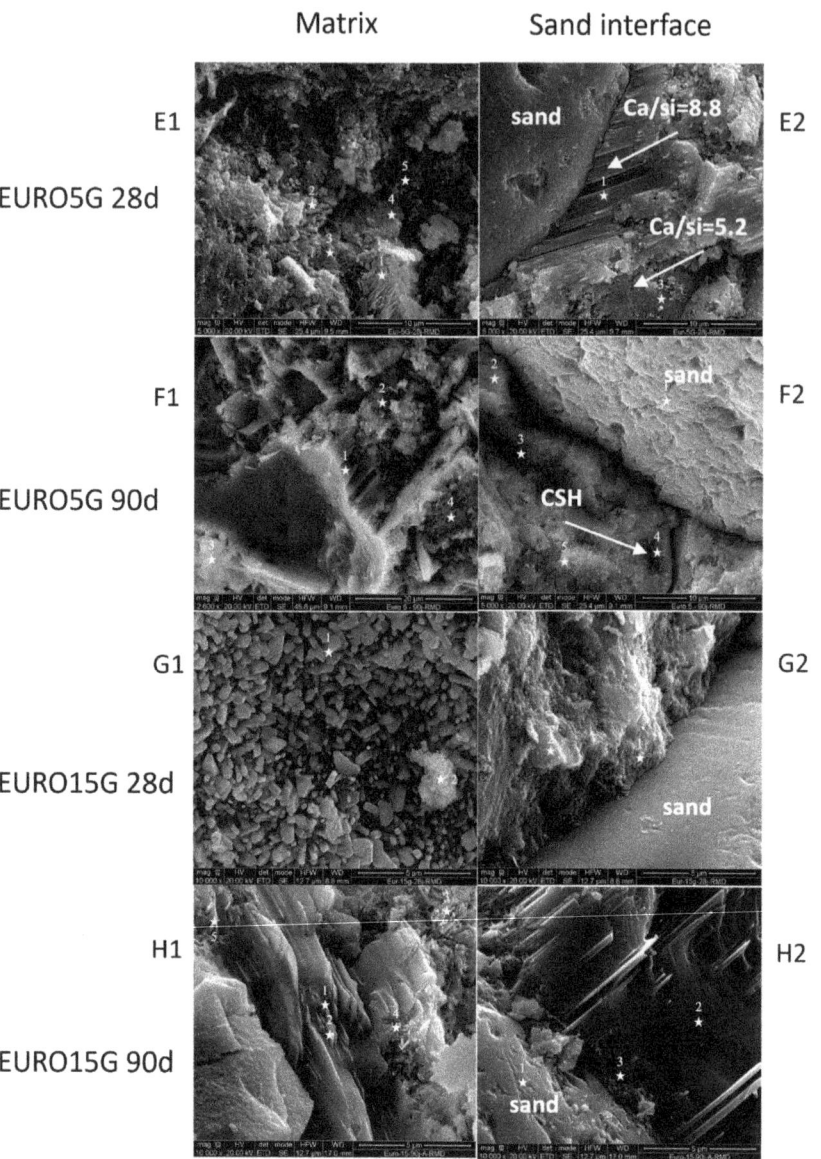

Figure 8. SEM observations of 28- and 90-day mortars matrices and sand/matrix interface. Stars on the micrographs correspond to EDS targets and arrows point to characteristic mineral species ((**C2**) Magnification (M) = 1000; (**A1,A2**) M = 2500; (**F1**) M = 2600; (**B1,B2,C1,D1,E1,E2,F2**) M = 5000; (**D2,G1,G2,H1,H2**) M = 10,000).

The reference sample (A1 to B2) shows a densification of the matrix with the presence of microporosities at 28 days and their disappearance at 90 days. In images A1 and B1, there is a good distribution of Portlandite and CSH. In the vicinity of the sand grain, weakly polymerized CSH and Portlandite evolve at 90 days into an adhesive interface of CSH gel with Ca/Si between 1.2 and 2.2. The presence of mucilage in small quantities in the EURO2G images perfectly illustrates the impact of the polysaccharide on the cement matrix. Image C1 shows a spherical geometric shape of CSH (Ca/Si of 2.6) surrounded by Portlandite that is particularly well ordered towards this sphere. This Portlandite organization is particularly visible and amplified on H1. This micrograph corresponds to a higher FM concentration, confirming the polysaccharides' influence on Portlandite structure. Knapen et al. [49] already observed this aligned arrangement of Portlandite because of polysaccharides. Image D1, despite the relatively long curing time, shows CSH and Portlandite forming, predominantly. The structure of the matrix remains particularly disordered and less dense than the reference. As far as the matrix around the sand grain is concerned, it consists mainly of Portlandite in the 28-day cure (C2), whereas in the 90-day cure, there is Portlandite and CSH at the same time (D2).

The increase in mucilage concentration (Figure 8) induces more and more morphological changes, confirming the arguments presented earlier. Image E1 shows that mucilage appears to influence CSH morphology since all the points analyzed in this image have Ca/Si values ranging from 2.2 to 2.9. At 90 days (F1), the evolution is notable, with Portlandite and CSH present. The matrix remains less dense than the reference matrix (B1) at the same age. The matrix/sand interface consists of Portlandite, as in mortar with less FM (C2). Indeed, image E2 shows Portlandite in two forms: one with a Ca/Si of 8.8 (close to the sand grain) and a Portlandite probably in the process of forming CSH, with a drop in Ca/Si = 5.2 (at distance of 6 µm from the sand grain). Finally, EURO15G shows significant carbonation (G1) not seen in the other samples. The matrix is filled with $CaCO_3$ at 28 days (G1), whereas in the 90-day cure, the matrix consists mainly of $Ca(OH)_2$, as the FM proportion is high. The observation of $Ca(OH)_2$ in the matrix of the 90-day cure, when at 28 days $CaCO_3$ was in the majority, has never been reported in the literature regarding low temperatures, but it is corroborated by Figure 7b showing that $CaCO_3$ disappears in favor of $Ca(OH)_2$.

The shape of the Portlandite is particularly ordered (H1) and has a veiled appearance (H2). The veiled appearance is the result of the visualization of mucilage, as reported by Knapen et al. [49]. These authors clearly showed the interaction of polysaccharides with Portlandite in particular. Some organic additives have the ability to structure and make Portlandite durable. Usually, Portlandite is not able to resist the stresses of early cement hydration, but the very visible bonds between Portlandite and polysaccharides create a layer-like development of the Portlandite platelets. According to Knapen et al. [49], the presence of polymer bridges between the $Ca(OH)_2$ crystals acts as an additional bond between the crystal layers and strengthens the crystal structure.

3.2.4. Thermal Conductivities of Cement Mortars

The thermal conductivities (Figure 9) decrease as the mucilage amount increases. It is also obvious that thermal conductivities are largely related to the bulk density of mortars. This decrease in the bulk density is related to an increase in microstructural flocculation and internal porosity, as observed by SEM. Finally, the increase in the bulk density of EURO15G is related to the elaboration of the mortar. It is necessary to apply a different compaction on the fresh material in order to obtain a non-friable hardened mortar. This compaction leads to an increase in the bulk density and thermal conductivity of the material. This manual compaction during the elaboration of the EURO15G mortars is observed by a significant increase in the standard deviation of the bulk density of EURO15G, as the process is hardly reproducible.

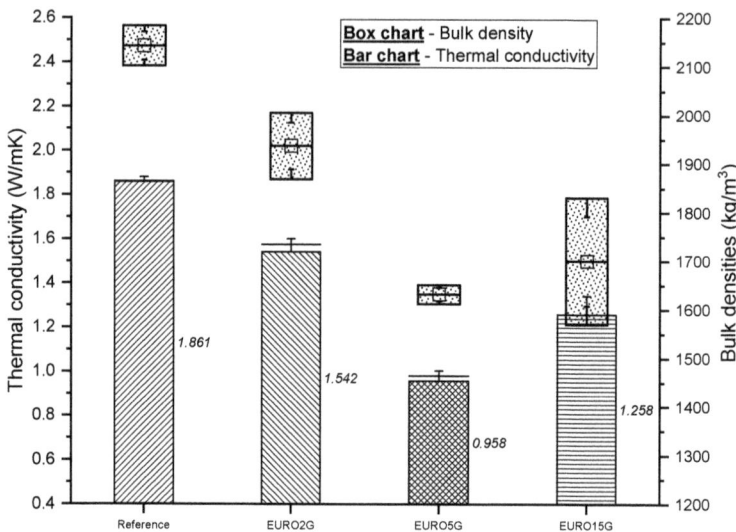

Figure 9. Thermal conductivities and bulk densities of 28-day cured mortars.

4. Conclusions

The paper discusses the influence of the concentration of a co-product from a local waste: flaxseed mucilage. The study shows that the mucilage of flaxseed seems to show signs of degradation in a simulated cementitious environment, with a decrease in the apparent viscosity. This may be due to the elimination of hydrogen bonds governing the viscosity of the solution and a degradation of the conjugated products of the polysaccharide main chain.

The presented mucilage-admixed cement pastes study reveals that increasing the mucilage concentration delays drastically, even exponentially, the beginning and ending of cement setting. This delay is probably due to two main phenomena: (i) an increase in steric hindrance caused by a mucilaginous solution that is all the more viscous as it is concentrated and (ii) an increase, proportional to the amount of mucilage, in the absorption and chelation of cement Ca^{2+} ions by FM. The simultaneous effects lead to the decrease in the transport of the metal ions of the cement responsible for the nucleation of the CSH and the precipitation of the Portlandite. This delay lasts throughout the induction period. Later, a higher hydration acceleration is observed compared to the reference. The hydration degrees of the mortars are equivalent, whatever the formulation, proving the non-inhibition but delaying effect of the FM.

FM leads to cement flocculation, and the higher the FM concentration, the higher the flocculation. This flocculation induces an increase in porosity and a greater carbonation as well as a decrease in the compressive strength and thermal conductivity of the material. As demonstrated by other work from our group, this disorder can be eliminated by the use of a W/C ratio > 0.5 at high admixture concentration (0.75% cement). Thus, not only the admixture rate but also the influence of the FM addition method (anhydrous and in-solution forms, a current work in progress) and the W/C ratio must be improved to better understand the potential of flaxseed mucilage in cementitious composites to target future applications.

As a highly hygroscopic material, FM can capture water and then release it gradually and under controlled conditions into the cement matrix over time. It would be interesting to verify that the regulation of water diffusion from the mucilage can induce self-healing properties. Effectively, by acting on cement grains not yet hydrated, this progressive release of water could produce new hydrates, thus reinforcing the mortar properties in the medium to long term. Another opportunity for valorizing mucilage would be to use a cement with a lower Blaine fineness, which has a lower water requirement and slows down the evolution of the heat of hydration. A design of experiments combining admixture rate, W/C ratio, and cement type will enable more defined recommendations for the formulation of this type of mortar.

Supplementary Materials: The following supporting information can be downloaded at: https://www.mdpi.com/article/10.3390/app14093862/s1, Figure S1: SEC-MALS analysis of flaxseed mucilage (1 g·L^{-1}) in NaOH solution (0.04 M) at different analysis times. (a) SEC-MALS profile (0 h: blue, 48 h: green and 72 h: red). Light curves are the molecular weight distribution, dark curves are the Refractive Index signal (RI). (b) SEC-MALS results (Mn: number average molecular weight, Mw: weight-average molecular weight, and polydispersity values (Mw/Mn)); Figure S2: SEC-MALS analysis of flaxseed mucilage (1 g·L^{-1}) in Ca(OH)$_2$ solution (0.02 M) at different analysis times. (a) SEC-MALS profile (0 h: blue, 48 h: green and 72 h: red). Light curves are the molecular weight distribution, dark curves are the Refractive Index signal (RI). (b) SEC-MALS results (Mn: number average molecular weight, Mw: weight-average molecular weight, and polydispersity values (Mw/Mn).

Author Contributions: Conceptualization, H.B., E.P. and A.G.; methodology, H.B.; validation, E.P. and A.G.; formal analysis, H.B.; investigation, H.B., R.-M.D., N.M. and K.J.H.; resources, R.-M.D., N.M. and K.J.H.; data curation, H.B. and A.G.; writing—original draft preparation, H.B.; writing—review and editing, E.P. and A.G.; supervision, E.P. and A.G. All authors have read and agreed to the published version of the manuscript.

Funding: The French Ministry of Higher Education and Research supported this work through Haris Brevet's doctoral grant.

Institutional Review Board Statement: Not applicable.

Informed Consent Statement: Not applicable.

Data Availability Statement: Data will be made available on request.

Acknowledgments: The authors would like to acknowledge the support from Calcia for providing the Portland cement used in the study at a time when it was extremely difficult to obtain (post lockdown in France) and especially Bailly from Calcia Company. The authors also extend their appreciation and gratitude to Romain Roulard for the sugar composition analysis with HPAEC-PAD.

Conflicts of Interest: The authors declare that they have no known competing financial interests or personal relationships that could have appeared to influence the work reported in this paper.

References

1. Pirmohammadi, A.; Khalaji, S.; Yari, M. Effects of Linseed Expansion on Its Dietary Molecular Structures, and on Broiler Chicks Digestive Enzymes Activity, Serum Metabolites, and Ileal Morphology. *J. Appl. Poult. Res.* **2019**, *28*, 997–1012. [CrossRef]
2. Chesneau, G.; Guillevic, M.; Germain, A.; Juin, H.; Lessire, M.; Enjalbert, F.; Burel, C.; Ferlay, A. Method for Treating Flax Seeds with a View to Improving the Value of Same. Patent WO 2019/101751 A1, 31 May 2019.
3. Kaur, M.; Kaur, R.; Punia, S. Characterization of Mucilages Extracted from Different Flaxseed (*Linum usitatsiumum* L.) Cultivars: A Heteropolysaccharide with Desirable Functional and Rheological Properties. *Int. J. Biol. Macromol.* **2018**, *117*, 919–927. [CrossRef]
4. Fedeniuk, R.W.; Biliaderis, C.G. Composition and Physicochemical Properties of Linseed (*Linum usitatissimum* L.) Mucilage. *J. Agric. Food Chem.* **1994**, *42*, 240–247. [CrossRef]
5. Shanmugavel, D.; Selvaraj, T.; Ramadoss, R.; Raneri, S. Interaction of a Viscous Biopolymer from Cactus Extract with Cement Paste to Produce Sustainable Concrete. *Constr. Build. Mater.* **2020**, *257*, 119585. [CrossRef]
6. Thomas, N.L.; Birchall, J.D. The Retarding Action of Sugars on Cement Hydration. *Cem. Concr. Res.* **1983**, *13*, 830–842. [CrossRef]
7. Hazarika, A.; Hazarika, I.; Gogoi, M.; Bora, S.S.; Borah, R.R.; Goutam, P.J.; Saikia, N. Use of a Plant Based Polymeric Material as a Low Cost Chemical Admixture in Cement Mortar and Concrete Preparations. *J. Build. Eng.* **2018**, *15*, 194–202. [CrossRef]
8. Pan, J.; Feng, K.; Wang, P.; Chen, H.; Yang, W. Retardation and Compressive Strength Enhancement Effect of Upcycling Waste Carrot as Bio-Admixture for Cement Mortars. *J. Build. Eng.* **2022**, *62*, 105402. [CrossRef]
9. Girones, J.; Vo, L.T.T.; Mouille, G.; Narciso, J.O.; Arnoult, S.; Brancourt-Hulmel, M.; Navard, P.; Lapierre, C. Impact of Miscanthus Lignin and Arabinoxylan on Portland Cement. *Ind. Crops Prod.* **2022**, *188*, 115585. [CrossRef]
10. Hernández, E.F.; Cano-Barrita, P.D.J.; Torres-Acosta, A.A. Influence of Cactus Mucilage and Marine Brown Algae Extract on the Compressive Strength and Durability of Concrete. *Mater. Constr.* **2016**, *66*, e074. [CrossRef]

11. Chandra, S.; Eklund, L.; Villarreal, R.R. Use of Cactus in Mortars and Concrete. *Cem. Concr. Res.* **1998**, *28*, 41–51. [CrossRef]
12. Azizi, C.E.; Hammi, H.; Chaouch, M.A.; Majdoub, H.; Mnif, A. Use of Tunisian *Opuntia ficus-indica* Cladodes as a Low Cost Renewable Admixture in Cement Mortar Preparations. *Chem. Afr.* **2019**, *2*, 135–142. [CrossRef]
13. Knapen, E.; Van Gemert, D. Cement Hydration and Microstructure Formation in the Presence of Water-Soluble Polymers. *Cem. Concr. Res.* **2009**, *39*, 6–13. [CrossRef]
14. Brevet, H.; Petit-Laignel, E.; Goullieux, A. Effects of Flaxseed Mucilage and Water to Cement Ratio on Mechanical and Hydration Characteristics of an OPC Mortar. *Acad. Mater. Sci.* **2023**, *1*, 1–10. [CrossRef]
15. Helrich, K. *Official Methods of Analysis of the AOAC*; AOAC: Rockville, MD, USA, 1990.
16. Roulard, R.; Petit, E.; Mesnard, F.; Rhazi, L. Molecular Investigations of Flaxseed Mucilage Polysaccharides. *Int. J. Biol. Macromol.* **2016**, *86*, 840–847. [CrossRef] [PubMed]
17. AFNOR NF EN 196-1; Methods of Testing Cement. Part 1: Determination of Strength. Normes et Normalisation Européenne: Brussels, Belgium, 2016.
18. Schwartzentruber, A.; Catherine, C. La Méthode Du Mortier de Béton Équivalent (MBE)–Un Nouvel Outil d'aide à La Formulation Des Bétons Adjuvantés. *Mater. Struct.* **2000**, *33*, 475. [CrossRef]
19. Bhatty, J.I. Hydration versus Strength in a Portland Cement Developed from Domestic Mineral Wastes—A Comparative Study. *Thermochim. Acta* **1986**, *106*, 93–103. [CrossRef]
20. Zhang, Q.; Ye, G. Dehydration Kinetics of Portland Cement Paste at High Temperature. *J. Therm. Anal. Calorim.* **2012**, *110*, 153–158. [CrossRef]
21. El-Jazairi, B.; Illston, J. The Hydration of Cement Paste Using the Semi-Isothermal Method of Derivative Thermogravimetry. *Cem. Concr. Res.* **1980**, *10*, 361–366. [CrossRef]
22. Hellebois, T.; Fortuin, J.; Xu, X.; Shaplov, A.S.; Gaiani, C.; Soukoulis, C. Structure Conformation, Physicochemical and Rheological Properties of Flaxseed Gums Extracted under Alkaline and Acidic Conditions. *Int. J. Biol. Macromol.* **2021**, *192*, 1217–1230. [CrossRef]
23. Hadad, S.; Goli, S.A.H. Fabrication and Characterization of Electrospun Nanofibers Using Flaxseed (*Linum usitatissimum*) Mucilage. *Int. J. Biol. Macromol.* **2018**, *114*, 408–414. [CrossRef]
24. de Paiva, P.H.E.N.; Correa, L.G.; Paulo, A.F.S.; Balan, G.C.; Ida, E.I.; Shirai, M.A. Film Production with Flaxseed Mucilage and Polyvinyl Alcohol Mixtures and Evaluation of Their Properties. *J. Food Sci. Technol.* **2021**, *58*, 3030–3038. [CrossRef] [PubMed]
25. Warrand, J. Etude Structurale et Propriétés en Solution des Polysaccharides Constitutifs du Mucilage de lin (*Linum usitatissimum* L.). Ph.D. Thesis, University of Picardie Jules Verne, Amiens, France, 2004.
26. Emaga, T.H.; Rabetafika, N.; Blecker, C.S.; Paquot, M. Kinetics of the Hydrolysis of Polysaccharide Galacturonic Acid and Neutral Sugars Chains from Flaxseed Mucilage. *Biotechnol. Agron. Soc. Environ.* **2021**, *16*, 139–147.
27. Lefsih, K.; Delattre, C.; Pierre, G.; Michaud, P.; Aminabhavi, T.M.; Dahmoune, F.; Madani, K. Extraction, Characterization and Gelling Behavior Enhancement of Pectins from the Cladodes of *Opuntia ficus indica*. *Int. J. Biol. Macromol.* **2016**, *82*, 645–652. [CrossRef] [PubMed]
28. Yu, L.; Stokes, J.R.; Yakubov, G.E. Viscoelastic Behaviour of Rapid and Slow Self-Healing Hydrogels Formed by Densely Branched Arabinoxylans from Plantago Ovata Seed Mucilage. *Carbohydr. Polym.* **2021**, *269*, 118318. [CrossRef] [PubMed]
29. Zhou, P.; Eid, M.; Xiong, W.; Ren, C.; Ai, T.; Deng, Z.; Li, J.; Li, B. Comparative Study between Cold and Hot Water Extracted Polysaccharides from Plantago Ovata Seed Husk by Using Rheological Methods. *Food Hydrocoll.* **2020**, *101*, 105465. [CrossRef]
30. Chen, H.-H.; Xu, S.-Y.; Wang, Z. Gelation Properties of Flaxseed Gum. *J. Food Eng.* **2006**, *77*, 295–303. [CrossRef]
31. Pourchez, J.; Govin, A.; Grosseau, P.; Guyonnet, R.; Guilhot, B.; Ruot, B. Alkaline Stability of Cellulose Ethers and Impact of Their Degradation Products on Cement Hydration. *Cem. Concr. Res.* **2006**, *36*, 1252–1256. [CrossRef]
32. Jayasingh, S.; Selvaraj, T. Influence of Organic Additive on Carbonation of Air Lime Mortar–Changes in Mechanical and Mineralogical Characteristics. *Eur. J. Environ. Civ. Eng.* **2022**, *26*, 1776–1791. [CrossRef]
33. Ponder, G.; Richards, G. Arabinogalactan from Western Larch, Part III: Alkaline Degradation Revisited, with Novel Conclusions on Molecular Structure. *Carbohydr. Polym.* **1997**, *34*, 251–261. [CrossRef]
34. Jing, Y.; Zhu, J.; Liu, T.; Bi, S.; Hu, X.; Chen, Z.; Song, L.; Lv, W.; Yu, R. Structural Characterization and Biological Activities of a Novel Polysaccharide from Cultured Cordyceps Militaris and Its Sulfated Derivative. *J. Agric. Food Chem.* **2015**, *63*, 3464–3471. [CrossRef]
35. Biermann, C.J. Hydrolysis and Other Cleavages of Glycosidic Linkages in Polysaccharides. *Adv. Carbohydr. Chem. Biochem.* **1988**, *46*, 251–271. [CrossRef]
36. Kang, X.; Wang, Y.-Y.; Wang, S.; Song, X. Xylan and Xylose Decomposition during Hot Water Pre-Extraction: A pH-Regulated Hydrolysis. *Carbohydr. Polym.* **2021**, *255*, 117391. [CrossRef] [PubMed]
37. Whistler, R.L.; BeMiller, J. Alkaline Degradation of Polysaccharides. In *Advances in Carbohydrate Chemistry*; Elsevier: Amsterdam, The Netherlands, 1958; Volume 13, pp. 289–329.
38. León-Martínez, F.; Cano-Barrita, P.D.J.; Lagunez-Rivera, L.; Medina-Torres, L. Study of Nopal Mucilage and Marine Brown Algae Extract as Viscosity-Enhancing Admixtures for Cement Based Materials. *Constr. Build. Mater.* **2014**, *53*, 190–202. [CrossRef]
39. Dickinson, E.; Eriksson, L. Particle Flocculation by Adsorbing Polymers. *Adv. Colloid Interface Sci.* **1991**, *34*, 1–29. [CrossRef]
40. Zhang, L.; Catalan, L.J.; Balec, R.J.; Larsen, A.C.; Esmaeili, H.H.; Kinrade, S.D. Effects of Saccharide Set Retarders on the Hydration of Ordinary Portland Cement and Pure Tricalcium Silicate. *J. Am. Ceram. Soc.* **2010**, *93*, 279–287. [CrossRef]

41. Bruere, G.M. Set-Retarding Effects of Sugars in Portland Cement Pastes. *Nature* **1966**, *212*, 502–503. [CrossRef]
42. Young, J. A Review of the Mechanisms of Set-Retardation in Portland Cement Pastes Containing Organic Admixtures. *Cem. Concr. Res.* **1972**, *2*, 415–433. [CrossRef]
43. Martinez-Molina, W.; Torres-Acosta, A.; Martínez-Peña, G.I.; Guzmán, E.A.; Mendoza-Pérez, I. Cement-Based, Materials-Enhanced Durability from *Opuntia ficus indica* Mucilage Additions. *ACI Mater. J.* **2015**, *112*, 165. [CrossRef]
44. Torres-Acosta, A.A.; González-Calderón, P.Y. *Opuntia ficus-indica* (OFI) Mucilage as Corrosion Inhibitor of Steel in CO_2-Contaminated Mortar. *Materials* **2021**, *14*, 1316. [CrossRef]
45. León-Martínez, F.; Cano-Barrita, P.D.J.; Castellanos, F.; Luna-Vicente, K.; Ramírez-Arellanes, S.; Gómez-Yáñez, C. Carbonation of High-Calcium Lime Mortars Containing Cactus Mucilage as Additive: A Spectroscopic Approach. *J. Mater. Sci.* **2021**, *56*, 3778–3789. [CrossRef]
46. Wang, J.; Xu, H.; Xu, D.; Du, P.; Zhou, Z.; Yuan, L.; Cheng, X. Accelerated Carbonation of Hardened Cement Pastes: Influence of Porosity. *Constr. Build. Mater.* **2019**, *225*, 159–169. [CrossRef]
47. Janotka, I.; Madejová, J.; Števula, L.; Frt'alová, D.M. Behaviour of $Ca(OH)_2$ in the Presence of the Set Styrene-Acrylate Dispersion. *Cem. Concr. Res.* **1996**, *26*, 1727–1735. [CrossRef]
48. Silva, D.A.; Roman, H.R.; Gleize, P.J.P. Evidences of Chemical Interaction between EVA and Hydrating Portland Cement. *Cem. Concr. Res.* **2002**, *32*, 1383–1390. [CrossRef]
49. Knapen, E.; Van Gemert, D. Polymer Film Formation in Cement Mortars Modified with Water-Soluble Polymers. *Cem. Concr. Compos.* **2015**, *58*, 23–28. [CrossRef]

Disclaimer/Publisher's Note: The statements, opinions and data contained in all publications are solely those of the individual author(s) and contributor(s) and not of MDPI and/or the editor(s). MDPI and/or the editor(s) disclaim responsibility for any injury to people or property resulting from any ideas, methods, instructions or products referred to in the content.

Article

Properties and Durability of Cementitious Composites Incorporating Solid-Solid Phase Change Materials

Yosra Rmili [1], Khadim Ndiaye [1,*], Lionel Plancher [1,2], Zine El Abidine Tahar [1], Annelise Cousture [1] and Yannick Melinge [3]

1. Laboratoire L2MGC, CY Cergy Paris University, 95031 Neuville-sur-Oise, France; yosra.rmili@etu.cyu.fr (Y.R.); lionel.plancher@u-cergy.fr (L.P.); zine-el-abidine.tahar@cyu.fr (Z.E.A.T.)
2. Laboratoire GEC, CY Cergy Paris University, 1 Rue Descartes, 95000 Neuville-sur-Oise, France
3. LRMH, CRC—MNHN, CNRS, Ministère de la Culture—UAR 3224, 77420 Champs sur Marne, France; yannick.melinge@culture.gouv.fr
* Correspondence: khadim.ndiaye@cyu.fr

Abstract: This paper investigates the properties and durability of cementitious composites incorporating solid-solid phase change materials (SS-PCM), an innovative heat storage material. Mortars with varying SS-PCM contents (0%, 5%, 10%, 15%) were formulated and characterized for rheological, structural, mechanical, and thermal properties. Durability assessment focused on volume stability (shrinkage), chemical stability (carbonation), and mechanical stability (over thermal cycles). Mortars with SS-PCM exhibited significant porosity and decreased mechanical strength with higher SS-PCM content. However, thermal insulation capacity increased proportionally. Notably, the material's shrinkage resistance rose with SS-PCM content, mitigating cracking issues. Despite faster carbonation kinetics in SS-PCM mortars, attributed to high porosity, carbonation appeared to enhance long-term mechanical performance by increasing compressive strength. Additionally, SS-PCM composites demonstrated superior stability over thermal cycles compared to reference mortars.

Keywords: cementitious materials; phase change materials; properties; durability; shrinkage; carbonation

1. Introduction

Energy consumption is a real challenge nowadays, particularly in the building sector, representing a significant part of it [1]. That is why many studies have been interested in and are still interested in finding ways and solutions to save energy and reduce greenhouse gas emissions. In fact, solar energy is one of these solutions, but the problem of intermittency is imposed.

In this regard, storing and using renewable energy become an urgent need in order to reduce energy consumption and then to improve indoor thermal comfort [1–3]. The heat storage is the key technology to streamline energy consumption and improve energy efficiency by reducing the gap between energy available and demand in buildings. There are different types of heat storage as sensitive, latent, thermochemical or chemical [1]. Therefore, phase change materials (PCM) have been advanced as one of the potential solutions to meet some of these expectations. Indeed, cementitious materials incorporating PCM, improve the thermal comfort of building, by increasing its inertia, thus avoiding sudden changes in the indoor temperature [4–10]. In the current market, there is a range of commercialized encapsulated phase change materials (solid-liquid) developed by leading companies such as BASF, Climator, Cristopia, Dupont de Nemours, Rubitherm, and Winco Technologies. All these technologies use encapsulated paraffin (micronal) as solid liquid phase change materials (SL-PCM). However, three significant drawbacks arise when utilizing SL-PCM. Firstly, the flammability of paraffin must be noticed. The associated risk is particularly unconvinient with fire safety requirements within buildings [11]. Secondly, because of a transition to a liquid state, such PCM require encapsulation to prevent against the leakage risk within the building material. The technologies for containment involve encapsulations, which can be intricate, costly, and energy-intensive. Thirdly, the presence of PCM in the cement matrix can cause sustainability problems such as the drop in their mechanical performance [2,4–16] and the appearance of cracking [1,3,14,17]. When PCM is added to the mix, it influences the mechanical and even thermal properties such as the thermal conductivity of concrete [16]. The decrease of compressive strength seems to be more pronounced than that of tensile strength [18]. The volume instability of these SL-PCM generated durability problems (cracking, etc.) constraining their use in the building sector despite their energy efficiency. To address these challenges, solid solid phase change material (SS-PCM) with crystalline to amorphous conversion while remaining at a solid state, have been developed [19–23]. Among these researchers, Harlé et al. [23] has been developing several SS-PCM including the innovative PUX1520 used in this work. This SS-MCP exhibits interesting and relevant physical and thermal properties (89 J.g^{-1} at 22 °C). for applications in the building sector. The solid liquid phase change of SL-PCM consists in endothermic melting during the heat charging phase and exothermic crystallization during the heat discharging phase. Solid solid phase change of SS-PCM undergoes a microstructural phase change from crystalline to amorphous while remaining at a solid state. This is referred to as the charging process. Conversely, when the temperature decreases, the SS-PCM releases heat as it crystallizes again, undergoing the discharging process (reversible conversion). The SS-PCM, more stable than SL-PCM, should be more suitable for building applications (masonry coating, bricks, roof coating, covering panels, etc.).

The innovative poly (ether urethane) based SS-PCM PUX1520 has interesting physical and thermal properties that are relevant for building applications [23,24]. Its integration in plaster has been the subject of preliminary studies [23], which have highlighted the interest and advantages of this technology SS-PCM compared with SL-MCP. They showed the high thermal inertia of Plaster/SS-PCM composites. The rheological properties in mortars are studied [24].

However, there is not study about the volume stability (shrinkage) and chemical stability (carbonation) of mortars incorporating this SS-MCP. Research on the durability of cementitious materials incorporating SS-PCM is very limited, which hinders their use in the building sector despite their energy efficiency. To use it as a construction material capable of storing heat, it is necessary to first study its properties and stability. The objective of this paper is to characterize and investigate the durability of cementitious materials incorporated the innovative SS-PCM to improve the thermal inertia of buildings. Furthermore, the effect of SS-PCM addition on the properties of cementitious materials (rheological, structural, thermal and mechanical properties) was studied on the one hand. On the other hand, since durability is a significant criterion for the use of building materials, the evaluation of their volume stability (drying shrinkage), chemical stability (accelerated carbonation) and finally mechanical stability (accelerated aging test over thermal cycles) was measured.

2. Materials and Methods

2.1. Materials

The SS-PCM composite mortars were prepared from white Portland CEM I 52.5 N cement containing 95%. The cement characterized by a Blaine specific surface 4200 cm^2.g^{-1} is chemically composed by 74% C_3S, 12% C_2S, 11% C_3A and 2.7% SO_3. The detail of the chemical composition is provided in the Table 1.

Table 1. Chemical composition of white cement.

Oxide	SiO_2	Al_2O_3	CaO	MgO	SO_3	K_2O	Na_2O	Fe_2O_3
%	21.53	3.59	65.47	0.70	3.49	0.26	0.6	0.22

The SS-PCM PUX1520 is an innovative poly (ether urethane) based SS-PCM developed by Harlé et al. [23,24]. Such home made SS-PCM is characterized by a high heat storage capacity of 89 J.g^{-1} with a low phase change temperature of 22.3 °C. Thanks to its high hardness (30 shore D), the SS-PCM seems to be suitable for cementitious materials applications. Nevertheless, the tested formula is water soluble. The thermal conductivity at room temperature is 0.231 W.m^{-1}.K^{-1} suggesting a good thermal insulation capacity [23]. The synthesized SS-PCM was crushed and sieved in white powder form with a grain size range between 300 µm and 600 µm (Figure 1). The SS-PCM has a molar mass of 1500 g.mol^{-1}, a bulk density of 1200 kg.m^{-3} and a true density of 450 kg.m^{-3}.

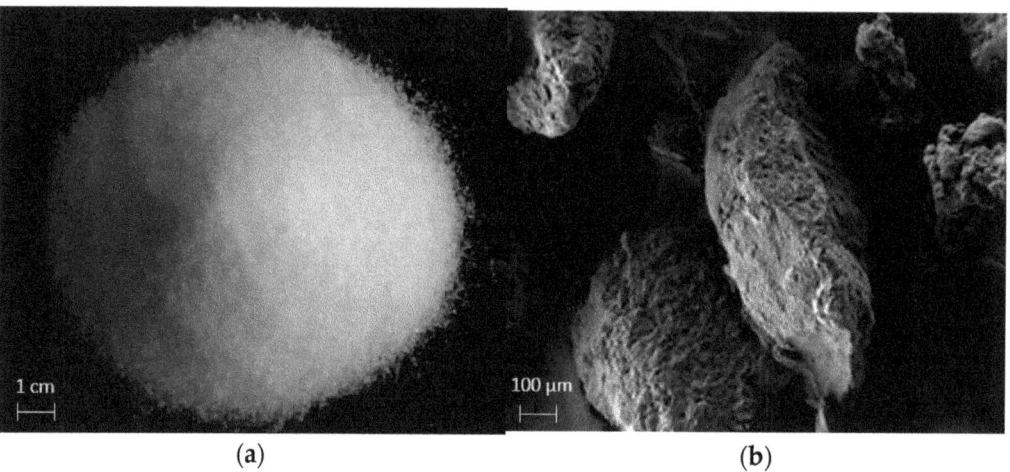

Figure 1. SS-PCM grains morphology: (**a**) macrostructure and (**b**) microstructure.

The SS-PCM composite mortars studied in the present work, are formulated as it reported in Table 2. The mortars were prepared according to the protocol described in Section 2.2. The cement-water ratio (W/C) has been fixed to 0.5 for all mix-designs. The mortar M0 is the reference sample without SS-PCM. The SS-PCM mortar composites M5, M10 and M15 contained addition of 5%, 10% and 15% of SS-PCM (SS-PCM in %wt mass of cement replaced in sand), respectively. The comparison between SS-PCM composite mortars and the reference mortar allows showing the SS-PCM addition effects.

Table 2. Composition of mortars.

Mortars	SS-PCM (%wt Cement)	Water/Cement	Cement (kg.m^{-3})	Water (kg.m^{-3})	Sand (kg.m^{-3})	SS-PCM (kg.m^{-3})
M0	0	0.5	512	256	1535	0
M5	5	0.5	503	251	1483	25
M10	10	0.5	494	247	1433	49
M15	15	0.5	486	243	1385	73

2.2. Methods

2.2.1. Sample Preparation

The mixture of the mortars was ensured by the standard mixer NF EN 196-1 [25]. The cement and SS-PCM were mixed at a slow speed during 60 s to obtain a homogenous mixture (the cement only was mixed for M0). Then, sand was added and mixed at a slow speed during 30 s. Water is then added according to the fixed cement-water ratio of 0.5. After mixing during 30 s (slow speed), a stop time of 60 s is taken to scrape the bottom of the mixer bowl and left to be mixed for the next 60 s at high speed. The prepared mortars were cast in prism molds (4 × 4 × 16 cm^3), and placed on a vibrating table. The vibration during 2 min allowed to ensure uniform thickness and to remove the air entrapped in the samples. This same mixing process was used for all formulations. The samples were stored in sealed bags at 20 °C and 100% relative humidity for 28 days. Finally, the mechanical and microstructure properties were determined after 28 days of moist curing.

2.2.2. Analytical Techniques

The mortar workability was studied using a mini-slump test [26]. It was carried out using a metal mold in the form of a truncated steel cone with a height of 60 mm, a base diameter of 100 mm, and a top diameter of 70 mm. The mold was placed on a smooth, level surface, then filled with mortar in 3 layers. Each layer was tamped with stokes throughout its depth. The steel mold was removed from the mortar immediately by raising it slowly and carefully in a vertical direction. This allows the mortar to spread over the level surface. Then, the slump and the spread diameter were measured.

The flexural and compressive strength were measured with a 3R Quantech device at 28 days using 3 samples (4 × 4 × 16 cm^3) for each mix-design mortar. The tests were performed in accordance with standard NF EN 196-1 [25], with loading speeds of 50 and 2400 N.s^{-1} for flexural and compressive strength tests, respectively. The resulting flexural and compressive strength values were the means of three and six individual values, respectively.

The microstructure of mortars was observed using a Gemini 300 (ZEISS) Scanning Electron Microscope (SEM) under high vacuum, with a working distance of 9 mm and a low voltage (2 kV) to avoid sample coating.

Mercury intrusion porosimetry (MIP) has been widely used to investigate the pore structure of cement-based materials. The pore distribution of composite samples (prism of 1 × 1 × 1 cm^3) was tested with the AutoPore IV 9500 device. Before testing, the samples were dried to remove air and water at 45 °C in a vacuum oven until a constant weight. The mercury pressure increasing allowed progressive access to the low porosity. The measured pore diameters were in the range of 0.003 µm to 358 µm.

The thermal conductivity and the thermal diffusivity have been identified with the use of a Hot-Disk device at 20 °C. The thermal properties was measured by the Transient Plane Source (TPS) methodology using prism samples (4 × 4 × 16 cm^3). The Hot-Disk probe was placed between two prepared surfaces of the prism samples. Finally, the specific heat was calculated from the measured thermal conductivity, thermal diffusivity and density. Before testing, the 28-day aged samples were oven-dried at 45 °C to remove the free water. Then, each sample was placed in a sealed bag and the cooling period was started at 20 °C. The free water has a significant influence on the measurements as water has its own thermal properties.

2.2.3. Durability

The drying shrinkage of the prepared mortar samples (same preparation as in Section 2.2.1 but with different storage conditions) is estimated by weighting the samples and measuring their length. The different mortars were cast in metallic molds of 4 × 4 × 16 cm^3 previously equipped with two screws (i.e., incorporated in the mortar sample). After 24 h of hydration under autogenous conditions (into sealed bags), the specimens were removed from the mold. The mass loss of each sample was followed during water evaporation. A comparator (precision ± 0.002 mm) is used to manually measure the sample length of the sample over time. The initial measurement corresponded to 24 h after the mortar casting.

The accelerated carbonation testing was performed in a chamber at 3% CO_2, 20 °C and 65% RH according to PERFDUB protocol [27]. A prior drying process of samples has been done before the accelerated carbonation. Indeed, the 28 days aged mortar samples (4 × 4 × 16 cm^3) were dried in 45 °C oven during 14 days. Then a cooling period allows to prepare the specimens at 20 °C and 65% RH during 7 days [27]. After this drying process, the samples were placed into the accelerated carbonation chamber. The samples were removed from the enclosure and weighed at 0, 14, 28, 48 and 360 days of carbonation. After that, the depth of carbonation was measured using a colored pH indicator (phenolphthalein). The phenolphthalein indicator was sprayed on the cutted surface (4×4 cm^2) to determine the carbonation depth.

The thermal stability of PCM mortars was followed by measuring their mechanical behavior after cycles thermally loading. The experimental protocol in terms of cycle time was previously set according to the thermal proprieties range of mortars in order to ensure the full spread of heat into the prism sample during a single considered cycle. The hygrometry was kept constant during the test. A monitored climate chamber is used to carry out the tests through 100 cycles. Each cycle of 5 h consisted of a heating step to 40 °C and a cooling step to 15 °C (Figure 2):

- Heating phase (heat charging): temperature increasing from 15 °C to 40 °C through a heating rate of 1 °C/min (for 25 min). After such heating period, the temperature is kept constant at 40 °C during 125 min.
- Cooling phase (heat discharging): temperature decreasing from 40 °C to 15 °C through a cooling rate of 1 °C/min (for 25 min). Finally after this cooling period, the temperature is kept constant at 15 °C during 125 min.

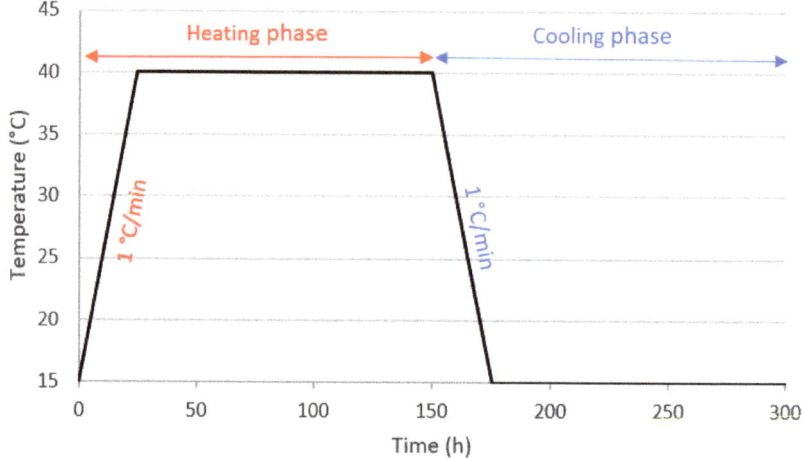

Figure 2. Set points of climate chamber temperature during a thermal cycle.

The mass variation of prism samples over time was followed by weighing specimens during cycles. The sample stability in terms of mechanical properties was evaluated by the measurement of their compressive strength over cycles.

3. Results

3.1. Mortar Proprieties

The mortars M0, M5, M10 and M15 were characterized by the measurement of rheological, structural, mechanical and thermal properties using experimental protocols described in Section 2.2.

3.1.1. Rheological Properties

In this study, the mini-slump test was used to evaluate the workability of cementitious materials incorporating SS-PCM and to get an idea of the influence of the different SS-PCM content on the rheological behavior of mortars (Figure 3). The reference mortar without SS-PCM shows lower workability with an average value of 3.3 cm for slump and about 13.3 cm for spread diameter. The slump and spread diameter increased with the added SS-PCM from 0% to 15%. That is to say, the SS-PCM added to the fresh mortar increased its workability by fluidifying the mixture. The results obtained in the present work are in good agreement of those of Plancher et al. [24]. The noticed fluidification is induced both by the solubility of SS-PCM and by the non-cohesive granular effect which contributes to reduce the energy needed to shear the fresh suspensions. This visco-plastic behaviour is well described by the Herschel–Bulkley model [24]. Finally, such results indicate an ability of the SS-PCM to be used as a fluidifying agent in cementitious materials.

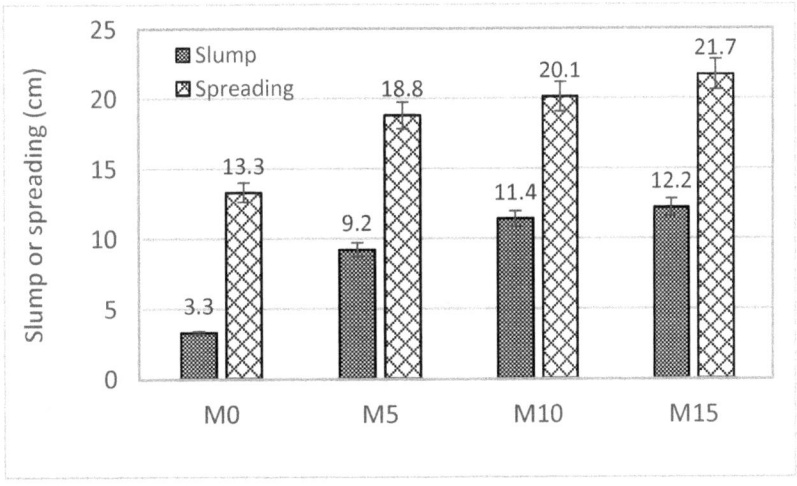

Figure 3. Slump and spread diameter of SS-PCM composite mortars.

3.1.2. Microstructure Properties

The effect of the SS-PCM incorporation on the hardened mortar microstructure was analyzed by scanning electron microscopic (SEM) observations (Figure 4). The morphology of 28 days aged mortars (M0, M5, M10 and M15) was observed for 2 scales: large field (a), and low field (b).

The observation of the large field images (a) shows clearly the macroporosity increasing with the SS-PCM incorporation. As a consequence, the SS-PCM addition should generate the modification of the pore size distribution. This should be confirmed by the mercury intrusion porosimetry test. The high macroporosity visible in M15 matrix (a) should have an effect on mechanical and thermal performance.

There is difference in morphology between M5, M10 and M15 noted in the low field images (b). Furthermore, Plancher et al. [24] observed the partial dissolution of SS-PCM in the cement paste during the mixing phase. This should generate chemical interactions between cement grains and soluble part of SS-PCM, which influences the cement hydration, hence the morphology. The centrifugation extraction of the interstitial fluid of the fresh mortar without (M0) and with SS-PCM (M5, M10 and M15), then ICP-OES analysis (Inductively Coupled Plasma Optical Emission spectroscopy) of extracted solutions would allow to quantity the dissolved part of SS-PCM in each fresh mortar.

The microstructure of M10 (low field), with needles sharped hydrates looking like ettringite, is very different to others mortars. There also seems to be C-S-H and ettringites phases visible in the M15 matrix. It is easier to distinguish hydrates in an aerated matrix, such as M15 matrix (less dense, less volume congestion). The hydrates in the other matrices (M0, M5, M10) are less visible because of their high density (volume congestion). Additional EDS on metallized samples are necessary to confirm the hydrate identification.

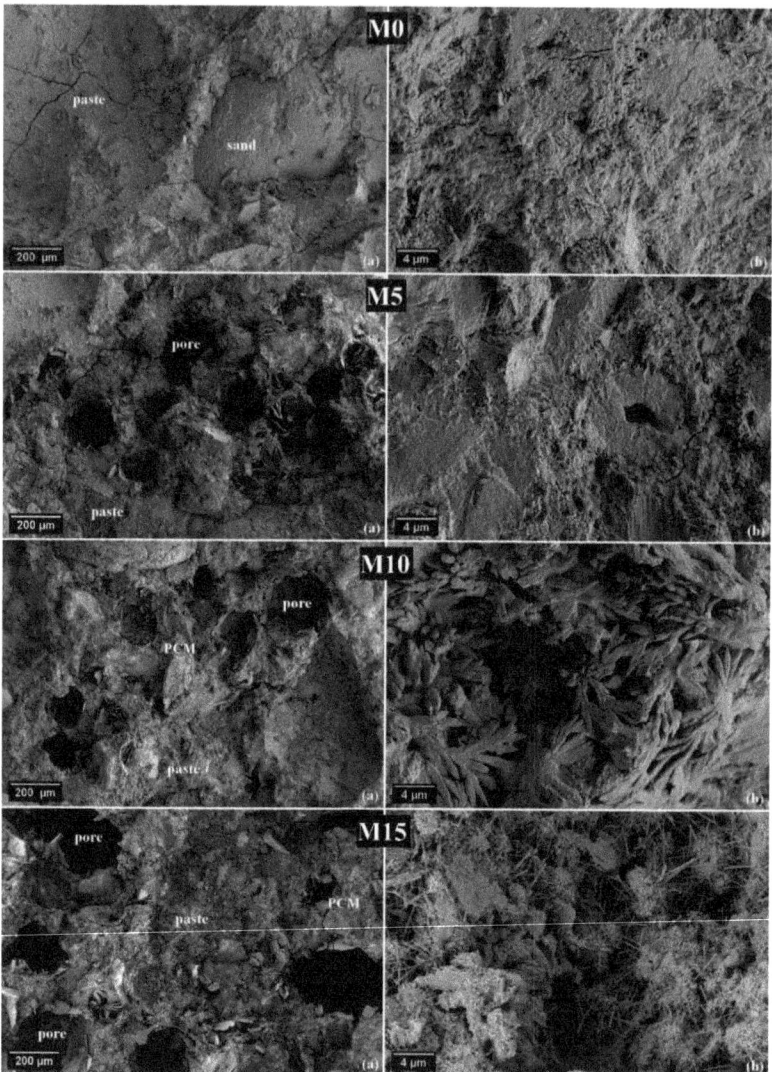

Figure 4. SEM Images of SS-PCM composite mortars after 28 days: (**a**) large field, and (**b**) low field.

3.1.3. Structural Properties

The porous structure network of mortars was investigated using Mercury Intrusion Porosimetry (MIP). It is by far the most widely used method to evaluate the size distributions of pores in cementitious materials, this is partially due to its wide range of pore size identification [28]. The Figure 5 shows the pore size structure of SS -PCM composite mortars with microporosity (≤ 0.01 µm), mesoporosity (0.01–1 µm) and macroporosity (≥ 1 µm). The mesoporosity of the reference mortar (1–100 µm) is represented by a single peak located at 10 µm. By considering the SS-PCM composite mortars, the access volume for the main mode (10 µm) is first reduced with the increasing incorporation of SS-PCM and a new pore size is created at around 60 µm. This new mode becomes more and more important via accessibility increasing induced by increasing the incorporation SS-PCM.

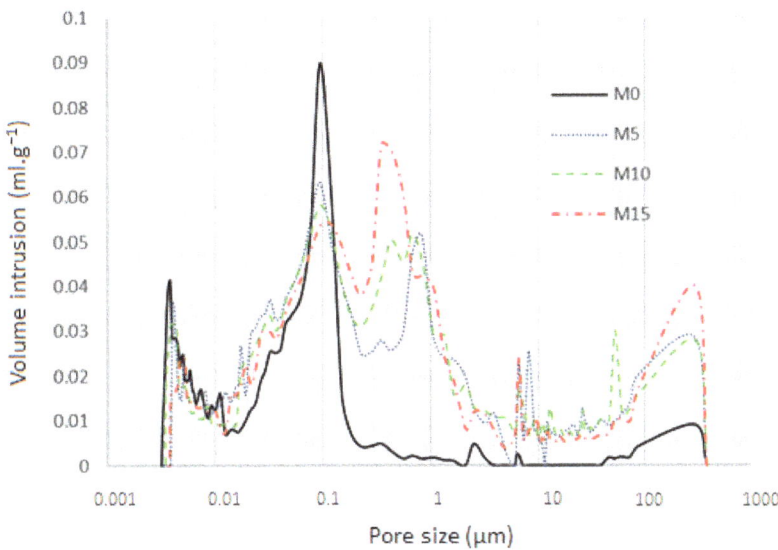

Figure 5. Pore size structure of SS-PCM composite mortars after 28 days.

The Figure 6 provides the pore size distribution of mortar porosity according to the mercury intrusion method. The total porosity of the mortars increased with the SS-PCM incorporation, because of the increase in macroporosity and mesoporosity, even if the microporosity decreased. The increase of macropores could be linked to the fact that the addition of SS-PCM leads to the decrease of the reactants content (cement and water), hence the drop of hydrates content into the mortar matrix leading to less dense matrix, and to more air content. As a consequence, for an equal amount of mortar, the cement paste content decreased with the SS-PCM addition. In addition, the sand grains were less bonded to each other (more pore voids between sand grains) because of the presence of SS-PCM grains without binding properties in the cement paste matrix. The decrease in microporosity should be related to the decrease of hydrate content with the SS-PCM addition (less cement and water content), knowing that the microporosity is related to the hydrate porosity. These results confirm the SEM observations in Section 3.1.2.

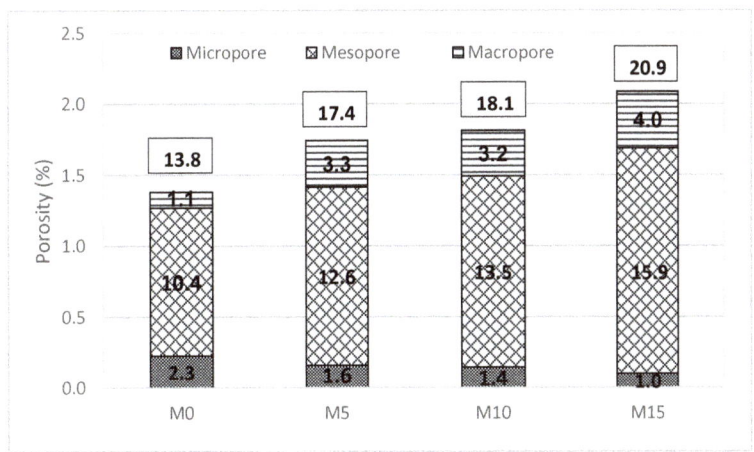

Figure 6. Pore size distribution of SS-PCM composite mortars after 28 days.

As expected, and in the opposite of porosity, the relative mass of SS-PCM increasing in the composites induced a functional decrease of the mortar density (Figure 7). This would induce modification of the construction material properties, such as mechanical or thermal performance.

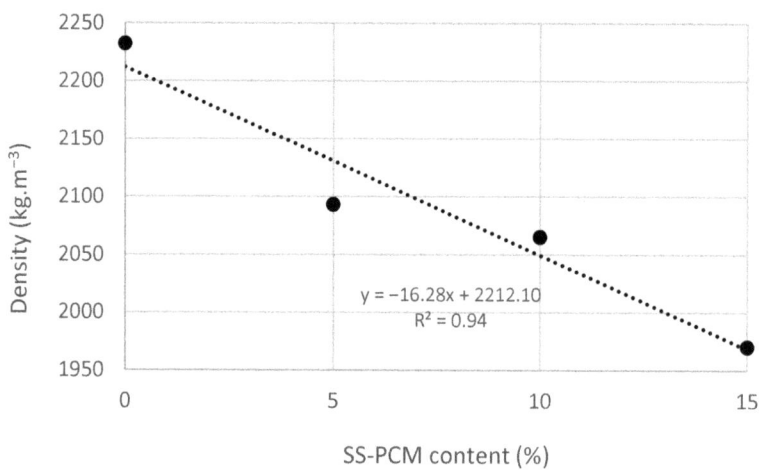

Figure 7. Density of mortars as a function of SS-PCM content of mortars.

3.1.4. Mechanical Properties

The flexural and compressive strengths were determined after 28 days of hydration to analyze the effects of SS-PCM incorporation on the mechanical properties of mortars. Figure 8 shows the results.

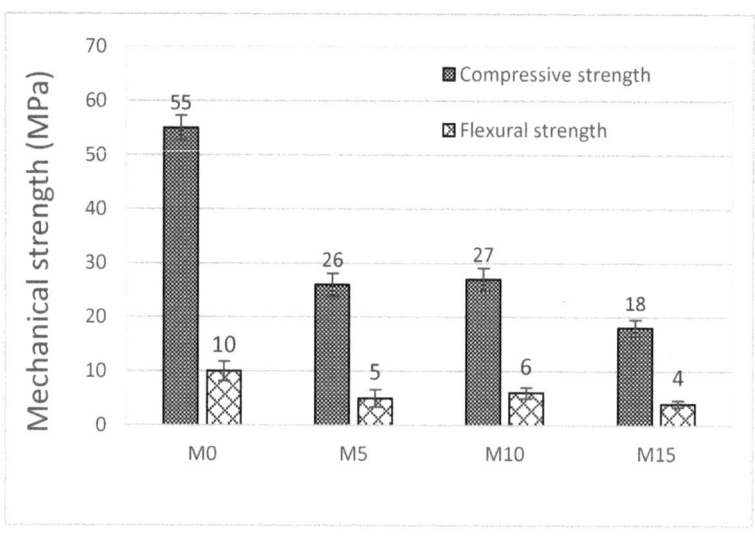

Figure 8. Mechanical strength of 28 days mortars as a function of SS-PCM content.

The flexural and compressive strengths of mortars incorporating SS-PCM were lower than those obtained for the reference mortar, and this reduction increased with the content of SS-PCM. There was a compressive strength drop of about 50% for the mortars M5 and M10, and 70% for M15%. These results are in agreement with the increase in mortar porosity previously observed (Figures 5 and 6). It should be noted that the compressive strength of the M5 mortar was similar to that of the M10 mortar. As well as the total porosity, the M5 compressive strength (26 MPa) is very close to that of M10 (27 MPa). The Figure 8 shows a reduction in flexural strength of 50% between the reference mortar (M0) and SS-PCM composite mortars (M5, M10 and M15), and a similar compressive strength (around 5 MPa) between SS-PCM composite mortars (M5, M10 and M15).

The mechanical strength drop is a disadvantage for construction materials. However, the mechanical performance of SS-PCM composite mortars remains sufficient for most building applications (26 MPa for M5, 27 MPa for M10 and 18 MPa for M15). For building applications, in which higher mechanical is not needed (bricks, masonry wall, coating wall, roof panels) [2], the SS-PCM could be added to cementitious materials to improve the thermal performance of the building.

Thanks to review articles [2,29–31], the mechanical resistance of lightweight cementitious material and PCM cementitious material according to its density was provided in the Figure 9. The results of the present work were added to the figure as red squares. All results in the literature show a trend for the compressive strength to increase with the material density. However, for a given density, the mechanical strength differs according to the binder composition, foaming method and the hydration cure. This explains the wide dispersal of compressive strength according to the studies of lightweight materials. Our results (red squares) remain in the range of compressive strengths for PCM materials. The mechanical strength of PCM composites was lower than that of lightweight material for a given density. This is due to the presence of PCM molecules without binding properties in the PCM composite matrix. The advantage of PCM over lightweight aggregates is its latent enthalpy (heat storage capacity) providing high thermal inertia of building.

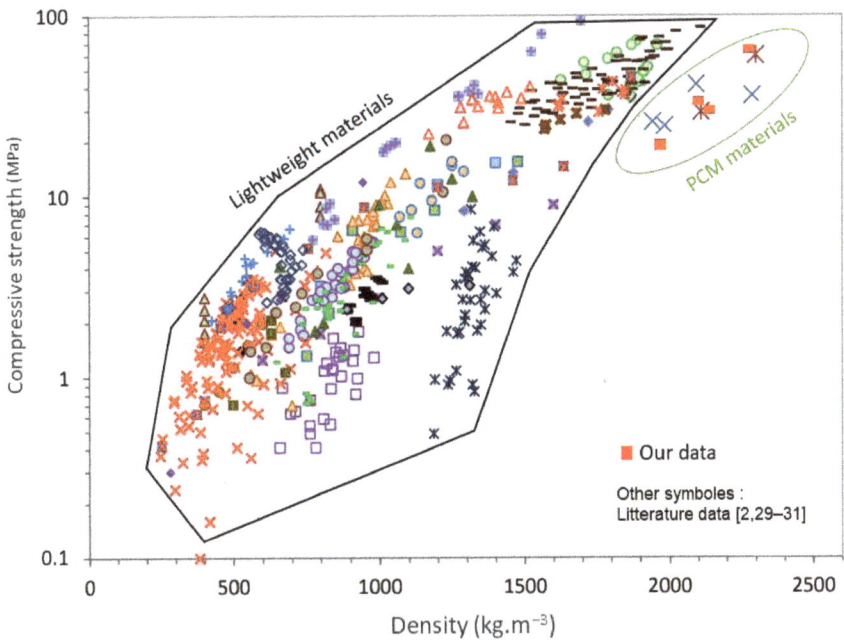

Figure 9. Compressive strength according to density [2,29–31].

3.1.5. Thermal Properties

The addition of 5% SS-PCM in the mortar led to a significant drop (26%) in terms of thermal conductivity. However, there was a slight variation in thermal conductivity among the SS-PCM composite mortars (Figure 10a). The Figure 10b shows the rise of thermal conductivity with the porosity of mortars. Indeed, the thermal conductivity of cementitious materials depends on their air content. In addition, it should be noted that the low thermal conductivity of the added SS-PCM (0.231 W.m^{-1}.K^{-1}) promotes the high insulation capacity of mortar composites. The SS-PCM conductivity was about 7 times lower than that of the reference mortar. Unlike the mechanical strength decreased, the thermal conductivity drop is an advantage for building applications. The SS-PCM composite mortars have better thermal insulating capability compared to reference mortar, and seem suitable for low-energy buildings.

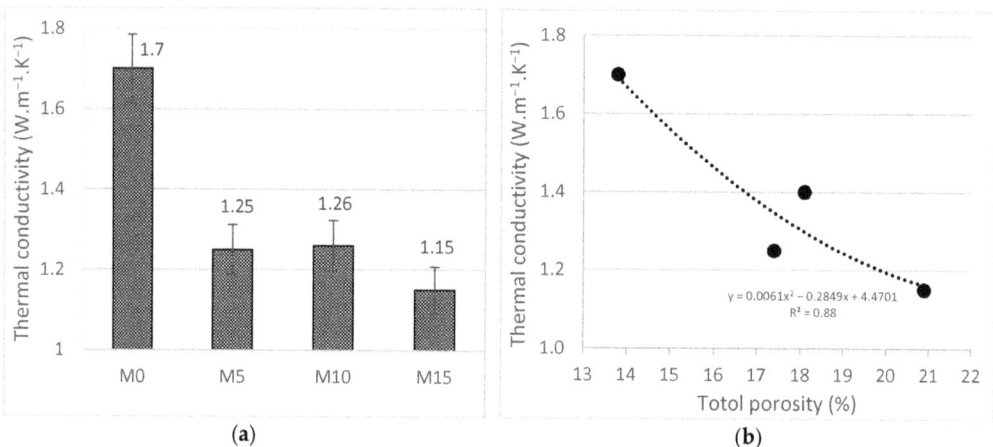

Figure 10. Thermal conductivity of mortars related to (**a**) SS-PCM content and to (**b**) porosity of mortars.

Unlike the compressive strength, the thermal conductivity of PCM composites was slightly lower than that of lightweight materials for a given density (Figure 11). This is due to the low thermal conductivity of the added PCM in the matrix. Our results (red squares) remain in the range of thermal conductivity for PCM materials. SS-PCM and lightweight materials are similar in terms of insulating capacity (similar thermal conductivity), but not in terms of thermal inertia. The advantage of SS-PCM over lightweight materials is its latent enthalpy, i.e., its high solar energy storage capacity providing high thermal inertia of building (unlike, the thermal inertia of lightweight material is low). In addition, the SS-PCM provides better durability and rheological properties. However, it should be noted that the lightweight materials are cheaper than PCM composites.

Unlike thermal conductivity, specific heat was higher when SS-PCM was added to mortar (Figure 12). In fact, the specific heat only depends on solid and phases of the mortar (i.e., cement paste, sand and SS-PCM), while the thermal conductivity takes into account all phases of the material (i.e., including the gas phase (porosity)). However, the specific heat of SS-PCM mortars was not a linear function of the SS-PCM amount.

The mechanical and thermal performances of SS-PCM composite mortars remain sufficient for most building applications. Moreover, its high thermal insulating capability and heat storage [23] are also relevant benefits for low-energy consumption buildings. This allows to improve the thermal inertia of the building, thus avoiding sudden changes in the indoor temperature, all while maintaining the supporting structure capacity. However, its higher porosity could lead to durability problems.

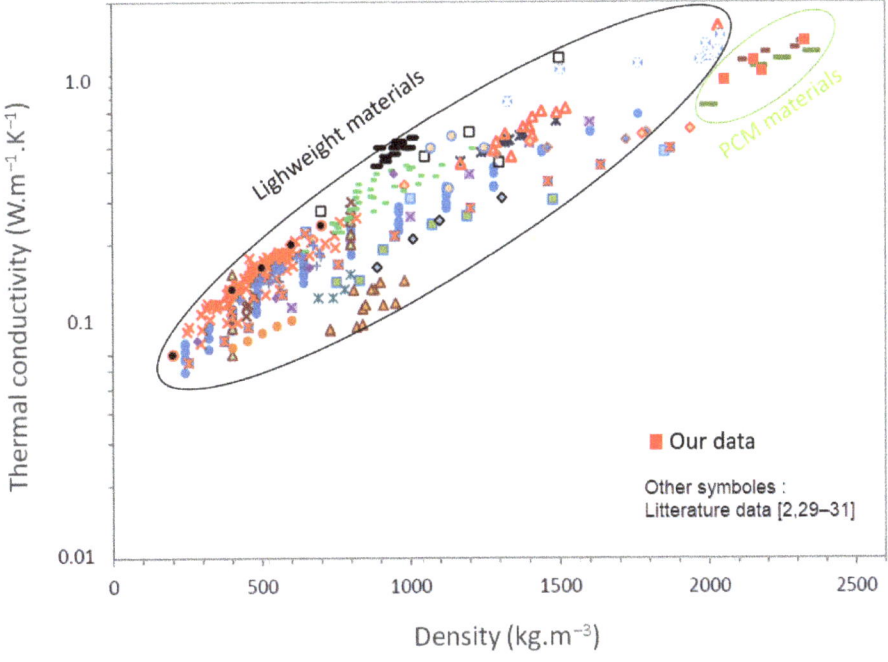

Figure 11. Thermal conductivity according to density [2,29–31].

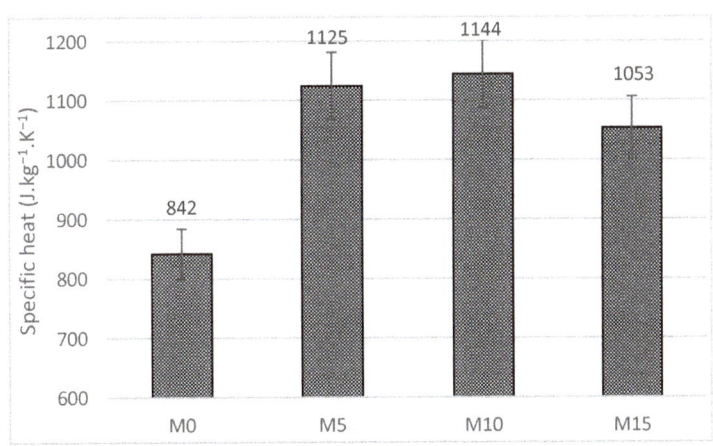

Figure 12. Specific heat of SS-PCM composite mortars.

3.2. Durability of Cementitious Composites Incorporating SS-PCM

As durability is an essential criterion for building materials, the SS-PCM composite mortars stability was investigated. This aims to evaluate volume stability (shrinkage), chemical stability (carbonation) and mechanical stability over thermal cycles of SS-PCM mortars.

3.2.1. Volume Stability: Drying Shrinkage

The drying shrinkage of cement-based materials can lead to structural cracking. In the literature, drying shrinkage is considered to be a consequence of changes in disjoining pressure, capillary pressure, and surface free energy or combinations thereof, accompanying a decrease of saturation and internal relative humidity [32–35]. The capillary pressure in the pore network is the mechanism that is usually employed to describe the hydric strains in porous materials [35]. It is related to the formation of water-air menisci in the partially empty pores, which induce isotropic compressive stress within the rigid solid skeleton that leads to bulk shrinkage. The essential cause of drying shrinkage is, of course, the evaporation of the capillary water from the material surface exposed to the ambient air. Evaporation occurs as soon as the relative humidity of the ambient air is lower than that prevailing in the capillary network. The tension forces developed inside the concrete lead to its contraction. The SS-PCM effect on drying shrinkage of SS-PCM mortars is investigated.

The specimens of 24 h ($4 \times 4 \times 16$ cm^3) are removed from the molds, and placed in a climate chamber (20 °C and 50% RH). The mass evolution of each sample is followed during the water evaporation. Figure 13 shows the mass loss over time for each mix-design mortar. The water desorption kinetics was rapid in the short term, then the kinetics slowed down for all mortar samples (M0, M5, M10 and M15) after 20 days. The mass loss of water into the mortars increased with the SS-PCM content. This was related to the pore diameter and porosity which increased with SS-PCM content.

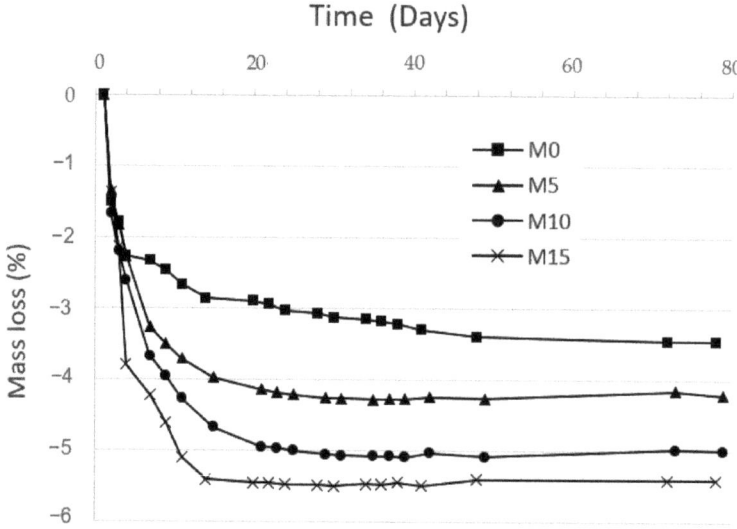

Figure 13. Mass loss of mortars over time.

Moreover, the capillary pressure, is related to the internal relative humidity according to the Kelvin–Laplace law [32]. So, the drying shrinkage amplitude should increase with the water loss [36]. However, it is noted that the SS-PCM addition generated more water loss (Figure 13), but less drying shrinkage (Figure 14). The incorporation of 5%, 10% and 15% of SS-PCM into the mortar allowed reducing the drying shrinkage amplitude by 7%, 42% and 53%, respectively. This should be related to the higher compressibility of the SS-PCM compared to cementitious materials able to amortize the drying shrinkage [19]. Therefore, the SS-PCM (grains size 300–600 μm) into the mortar matrix would amortize local stress due to the capillary pressure (generated by water loss) into the porous network, leading to less measurable bulk deformation of the skeleton. Hence, the addition of SS-PCM has the advantage of reducing macroscopic deformation linked to drying shrinkage.

Furthermore, it should be noted that shrinkage is probably the most common cause of structural cracking [37,38]. If not controlled, the deformation due to shrinkage can lead to durability problems and even shear stress [39,40]. The SS-PCM could be used as a mitigation agent to reduce the drying shrinkage of cementitious materials.

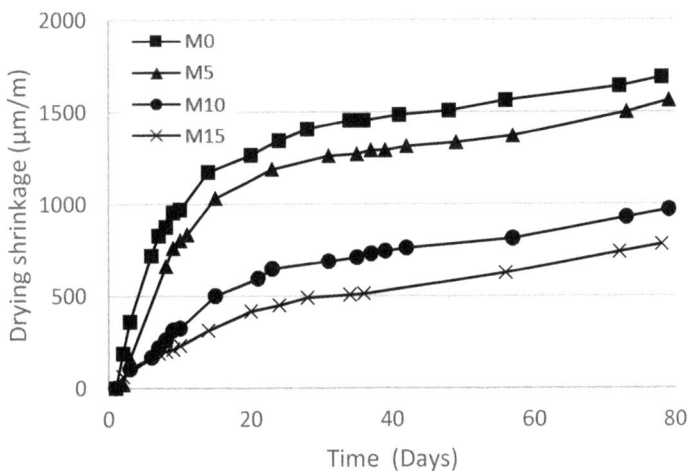

Figure 14. Drying shrinkage as a function of time.

3.2.2. Chemical Stability: Carbonation

Carbonation of concrete is a natural phenomenon that occurs in the long term in the ambient air. However, it is a slow process in atmospheric conditions taking a few years to reach high carbonation depth in the ambient air. The accelerated carbonation in a chamber at 3% CO_2 (20 °C and 65%RH) allows to speed up the process while maintaining stable environmental conditions according to the PERFDUB test method [27]. Figure 15 shows the evolution of the mass loss of mortar samples (4 × 4 × 16 cm^3) during the drying period at 45 °C before the carbonation test. The water loss is greater for mortars incorporating SS-PCM compared to the reference mortar. As seen before (Figure 13), the mass loss increases with SS-PCM content. This should be related to the high porosity of SS-PCM composites. After 14 days in 45 °C oven, the mortar samples were stabilized at 20 °C and 65% RH during 7 days (prior drying stage) [27]. They were then introduced into an accelerated carbonation chamber (3% CO_2, 20 °C and 65% RH).

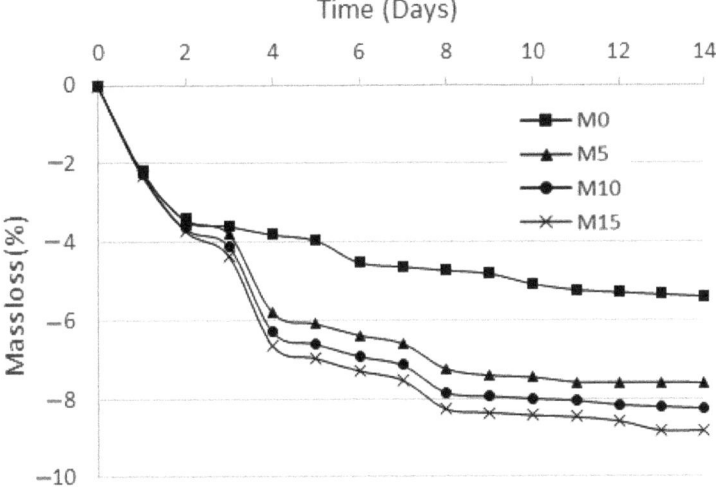

Figure 15. Mass loss of mortars over time in drying oven at 45 °C for 14 days.

The Figure 16 shows images of mortar samples after 14, 28, 48 and 360 days of accelerated carbonation. The phenolphthalein indicator was sprayed on the cutted surface to determine the carbonation depth. Phenolphthalein turns to purple coloration in area of pH higher than 9, so the purple coloration was observed in non-carbonated areas. The colorless outer area (carbonated area) thickened over time, i.e., the carbonation depth increased over time for all the tested mortars. At 360 days, the mortars with SS-PCM are fully colorless (i.e., complete carbonated).

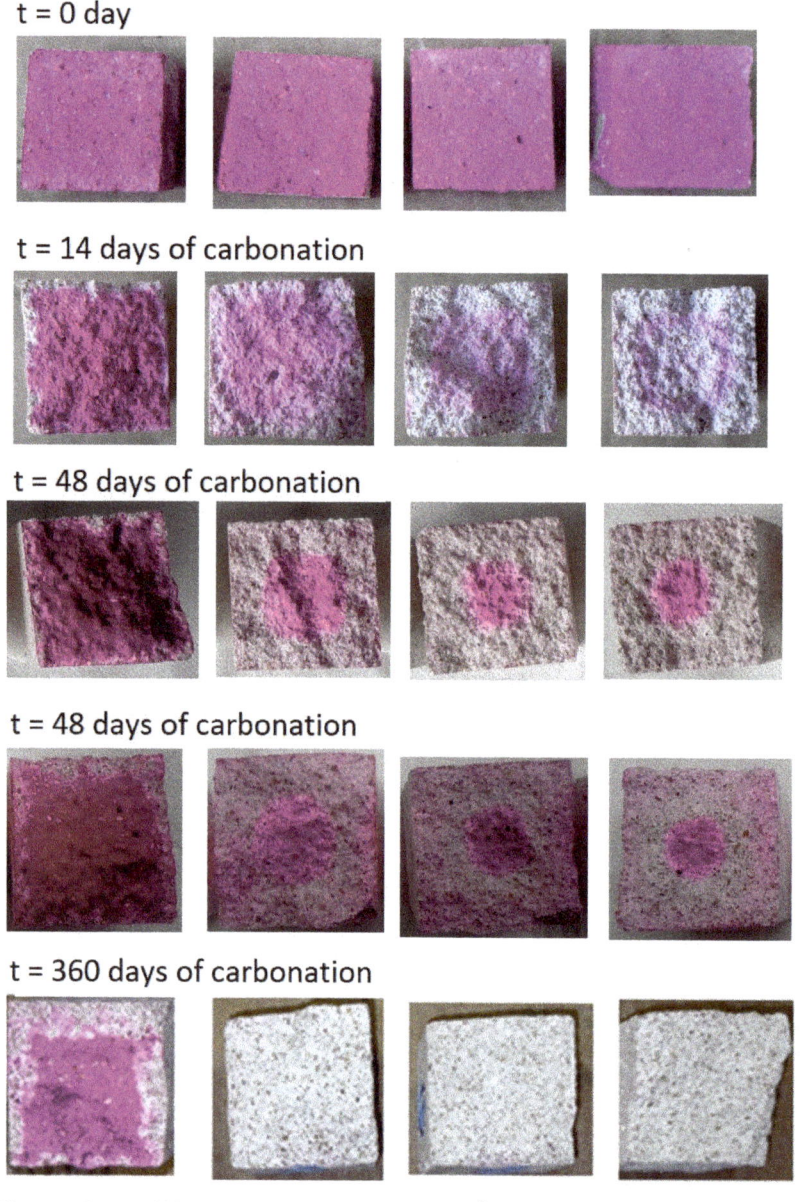

Figure 16. Image of SS-PCM composite mortars (4×4 cm^2) carbonated at 3% CO_2 chamber (20 °C, 65% RH).

Figure 17 shows the measurement of carbonation depth at 14, 28, 48 and 360 days for each mix-design mortar (M0, M5, M10 and M15). As seen in Figure 16, there was no carbonation at the beginning of accelerated carbonation (t = 0 day). Then, the carbonation depth at 14 days was observed and could be measured for all samples. The carbonation depth of all mix-design mortar increased over time. However, the carbonation kinetics of the reference mortar was slower compared to SS-PCM composite mortars. The carbonation kinetic increased with SS-PCM content. After 1 year of accelerated carbonation, the carbonation of mortars incorporating SS-PCM (M5, M10 and M15) was complete ($\geq$20 mm), while that of reference mortar reached 6.8 mm. The high kinetic of SS-PCM composite mortars in relation to the reference mortar was due to its higher and coarser porosity, promoting a high diffusion of CO_2 into the porous network.

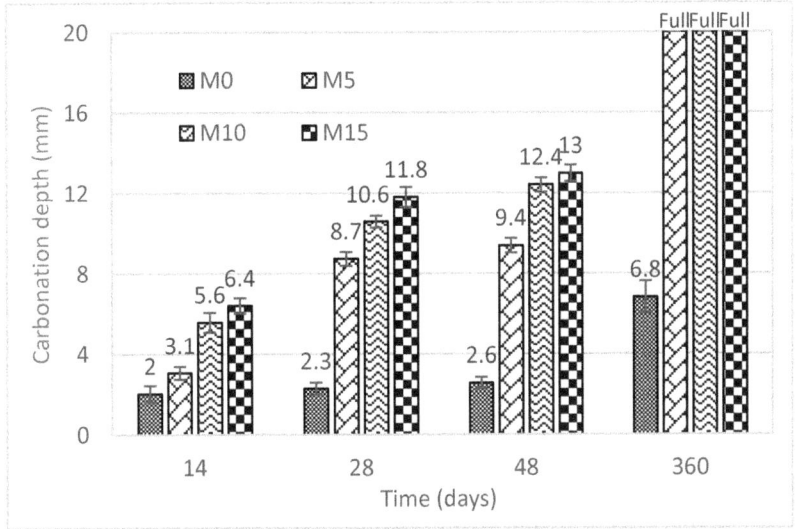

Figure 17. Carbonation depth of SS-PCM composite mortars over time.

It should be noted that the high carbonation depth of mortars incorporating SS-PCM is partially due to the severe carbonation conditions in the accelerated chamber (3%CO_2). In the real condition of use (building application), the carbonation kinetics in the atmospheric air was slower compared to the accelerated chamber. In fact, the Figure 18 shows the image of mortars carbonated for 360 days at atmospheric air chamber and accelerated carbonation chamber (3% CO_2). The SS-PCM composite mortars are fully carbonated in chamber at 3% CO_2 ($\geq$20 mm), unlike those in the air atmospheric chamber.

In the atmospheric carbonation, the SS-PCM composite mortars show similar carbonation depth between 5 and 6 mm, while the reference mortar achieved only 1.7 mm in carbonation depth (Figure 19).

This high carbonation rate of SS-PCM composite mortar could be an advantage in term of mechanical performance. Figure 20 shows an increase in mechanical strength with the carbonation rate. Such a behavior can be related to effects induced by the carbonation that leads to the calcium carbonate precipitation into the porous network reducing the porosity, and improving the compressive strength. However, the low pH of carbonated mortars could promote the corrosion of steels for reinforced concrete [41]. The SS-PCM mortar should be suitable for non-reinforced concrete: masonry coating, masonry bricks, roof coating, covering panels, etc.

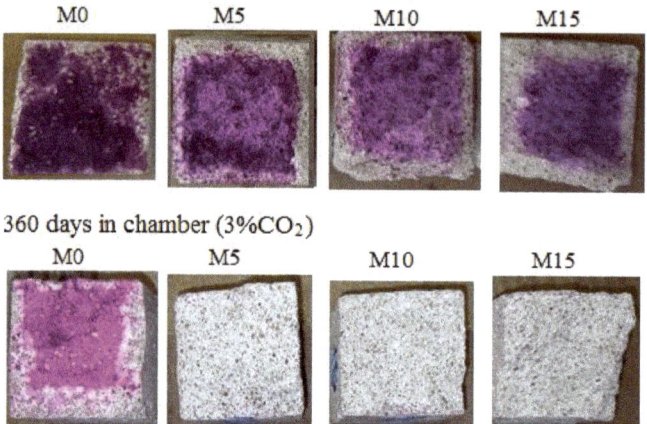

Figure 18. Images of carbonated mortars for one year in atmospheric air and in 3% CO$_2$ chamber.

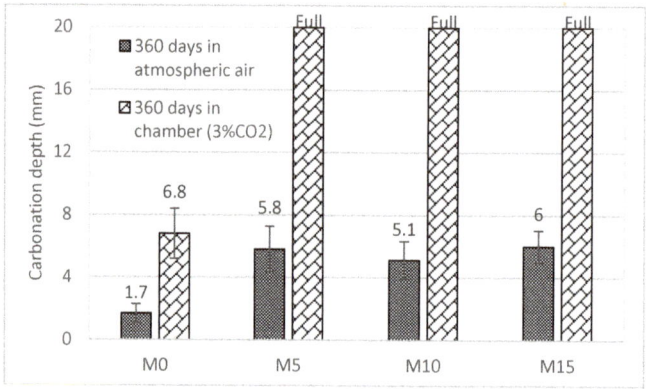

Figure 19. Carbonation depth of mortars for one year in atmospheric air and in 3% CO$_2$ chamber.

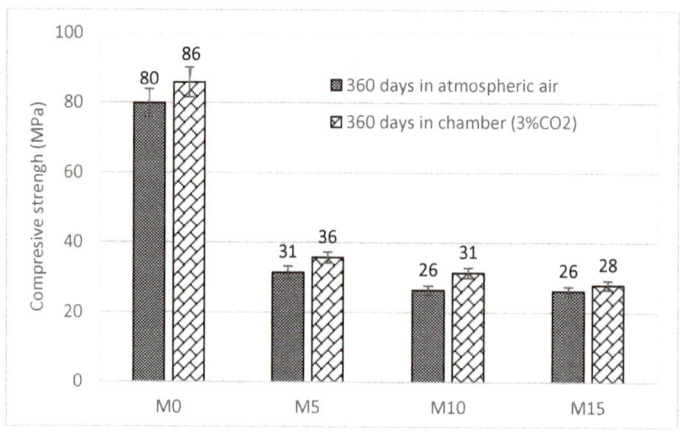

Figure 20. Compressive strength of carbonated mortars for one years.

3.2.3. Mechanical Stability: Reversibility over Thermal Cycles

It is relevant to check the stability of mortars incorporating SS-PCM over thermal cycles, or in other words its ability in terms of mechanical strength to withstand the SS-PCM phase transitions (heat storage) during the building lifespan. The heating at 40 °C and cooling at 15 °C led to charge and discharge heat into SS-PCM, respectively. The Figure 21 shows the results of the aging test in terms of compressive strength as a function of the number of phase transition cycles (0, 24, 48 and 100 cycles).

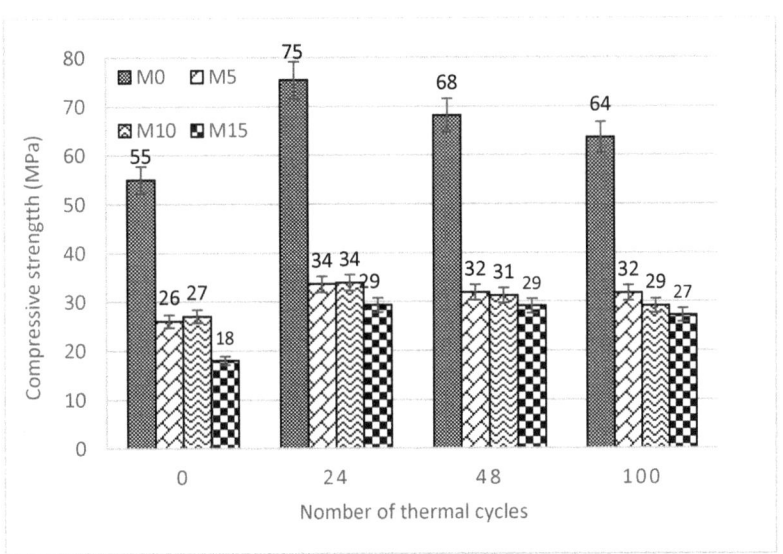

Figure 21. Compressive strength of mortars as a function of thermal aging cycles.

A significant increase in compressive strength of all mortar samples is noticed after 24 cycles, a slight decrease occurred between 48 and 100 cycles. This mechanical performance improvement after 24 cycles should be related to the thermal activation of hydration reactions, increasing the hydration kinetics of mortars. For mortars incorporating SS-PCM, there is a convergence towards the similar value of around 30 MPa after 100 cycles. Overall, the Figure 21 shows the mechanical stability of SS-PCM composite mortars over the heat storage cycles (until 100 cycles). This mechanical stability should be explained by the volume stability of SS-PCM during phase change. Indeed, unlike classic solid-liquid PCM (SL-PCM) as paraffin with solid-liquid transition phase (volume instability), the SS-PCM have the advantage of a stable solid-solid transition phase (volume stability). Hence, the volume stability of SS-PCM during the storage-distorage cycle allowed to avoid stress and cracking into the mortar skeleton. Furthermore, it should be noted that this study should be conducted over a larger number of cycles (200 cycles) to be more representative. For supporting structure applications in building, it is important to keep mechanical stability over heat storage cycles.

4. Conclusions

The influence of SS-PCM incorporations on properties and durability of mortars was examined. The rheological, structural, mechanical and thermal properties of SS-PCM mortars were measured and analyzed. Then, the durability study allowed assessing the volume (drying shrinkage), chemical (carbonation) and mechanical stabilities (over thermal storage cycles). The results showed that:

- The increase of SS-PCM content in the fresh mortar induces a better workability and increases the total porosity, hence a drop in mechanical strength and thermal

conductivity. However, the mechanical performance of SS-PCM composite mortars remains sufficient (27–18 MPa) for most building applications.
- The drying shrinkage amplitude decreases with its SS-PCM content. This result should be related to the compressibility of the SS-PCM amortizing the local capillary pressure, leading to less measurable bulk deformation.
- An increase in mechanical strength with the carbonation rate was noted. The mechanical stability of SS-PCM composite mortars over the heat storage cycles is confirmed. This should be explained by the volume stability of SS-PCM during phase change unlike classic SL-PCM.

The advantage of this innovative SS-PCM compared to the existing SL-PCM is its ability to store high thermal energy with a better durability thanks to its volume stability during solid-solid phase change. So, this storage material could be a solution for the durability problems inhibiting the use of phase change materials in building sector. The SS-PCM mortar was suitable for masonry coating, masonry bricks, roof coating, covering panels, etc. This allow to improve the thermal inertia of the building, thus avoiding sudden changes in the indoor temperature, all while maintaining the supporting structure capacity.

Author Contributions: Conceptualization: Y.R., K.N., L.P. and Z.E.A.T.; Methodology: Y.R., K.N., L.P., Z.E.A.T., A.C. and Y.M.; Formal analysis: Y.R., K.N., L.P., Z.E.A.T., A.C. and Y.M.; Investigation: Y.R. and K.N.; Writing—original draft: Y.R. and K.N.; Visualization: Y.R. and K.N.; Editing: Y.R. and K.N.; Writing—review: L.P., Z.E.A.T., A.C. and Y.M.; Supervision: Y.M. and K.N. All authors have read and agreed to the published version of the manuscript.

Funding: This research received no external funding.

Institutional Review Board Statement: Not applicable.

Informed Consent Statement: Not applicable.

Data Availability Statement: The data presented in this study are available on request from the corresponding author.

Acknowledgments: The SS-PCM are supplied by the "Laboratoire GEC" of Cergy Paris University (France). The SEM analyses were carried out in the "Microscopies & Analyses" imaging facility, Federation I-Mat (FR4122) of the CY Cergy Paris University. The mercury intrusion porosimetry (MIP) was performed in the "Laboratoire Matériaux et Durabilité des Constructions" of the Toulouse University (France).

Conflicts of Interest: The authors declare no conflict of interest.

References

1. Fallahi, A.; Guldentops, G.; Tao, M.; Granados-Focil, S.; Van Dessel, S. Review on solid-solid phase change materials for thermal energy storage: Molecular structure and thermal properties. *Appl. Therm. Eng.* **2017**, *127*, 1427–1441. [CrossRef]
2. Ndiaye, K.; Ginestet, S.; Cyr, M. Thermal energy storage based on cementitious materials: A review. *AIMS Energy* **2018**, *6*, 97–120. [CrossRef]
3. Šavija, B. Smart Crack Control in Concrete through Use of Phase Change Materials (PCMs): A Review. *Materials* **2018**, *11*, 654. [CrossRef] [PubMed]
4. Liu, L.; Li, J.; Deng, Y.; Yang, Z.; Huang, K.; Zhao, S. Optimal design of multi-layer structure composite containing inorganic hydrated salt phase change materials and cement: Lab-scale tests for buildings. *Constr. Build. Mater.* **2021**, *275*, 122125. [CrossRef]
5. Soares, N.; Costa, J.J.; Gaspar, A.R.; Santos, P. Review of passive PCM latent heat thermal energy storage systems towards buildings' energy efficiency. *Energy Build.* **2013**, *59*, 82–103. [CrossRef]
6. Baetens, R.; Jelle, B.P.; Gustavsen, A. Phase change materials for building applications: A state-of-the-art review. *Energy Build.* **2010**, *42*, 1361–1368. [CrossRef]
7. Jayalath, A.; Mendis, P.; Gammampila, R.; Aye, L. Applications of phase change materials in concrete for sustainable built environment: A Review. In Proceedings of the International Conference on Structural Engineering, Construction and Management, Kandy, Sri Lanka, 16–18 December 2011. Available online: http://dl.lib.mrt.ac.lk/handle/123/9349 (accessed on 8 March 2023).
8. Snoeck, D.; Priem, B.; Dubruel, P.; De Belie, N. Encapsulated Phase-Change Materials as additives in cementitious materials to promote thermal comfort in concrete constructions. *Mater. Struct.* **2014**, *49*, 225–239. [CrossRef]
9. Adesina, A.; Awoyera, P.; Sivakrishna, A.; Kumar, K.R.; Gobinath, R. Phase change materials in concrete: An overview of properties. *Mater. Today Proc.* **2020**, *27*, 391–395. [CrossRef]

10. Ren, M.; Wen, X.; Gao, X.; Liu, Y. Thermal and mechanical properties of ultra-high performance concrete incorporated with microencapsulated phase change material. *Constr. Build. Mater.* **2021**, *273*, 121714. [CrossRef]
11. Asimakopoulou, E.K.; Kolaitis, D.I.; Founti, M.A. Fire safety aspects of PCM-enhanced gypsum plasterboards: An experimental and numerical investigation. *Fire Saf. J.* **2015**, *72*, 50–58. [CrossRef]
12. Essid, N.; Loulizi, A.; Neji, J. Compressive strength and hygric properties of concretes incorporating microencapsulated phase change material. *Constr. Build. Mater.* **2019**, *222*, 254–262. [CrossRef]
13. Šavija, B.; Zhang, H.; Schlangen, E. Influence of Microencapsulated Phase Change Material (PCM) Addition on (Micro) Mechanical Properties of Cement Paste. *Materials* **2017**, *10*, 863. [CrossRef]
14. Halamickova, P.; Detwiler, R.J.; Bentz, D.P.; Garboczi, E.J. Water permeability and chloride ion diffusion in portland cement mortars: Relationship to sand content and critical pore diameter. *Cem. Concr. Res.* **1995**, *25*, 790–802. [CrossRef]
15. Dehdezi, P.K.; Hall, M.R.; Dawson, A.R.; Casey, S.P. Thermal, mechanical and microstructural analysis of concrete containing microencapsulated phase change materials. *Int. J. Pavement Eng.* **2013**, *14*, 449–462. [CrossRef]
16. Hunger, M.; Entrop, A.G.; Mandilaras, I.; Brouwers, H.J.H.; Founti, M. The direct incorporation of microencapsulated Phase Change Materials in the concrete mixing process. In Proceedings of the CMS 2009 Conference on Life Cycle Design of Building Systems and Materials, Enschede, The Netherlands, 12–15 June 2009. Available online: https://www.academia.edu/22180396 (accessed on 8 September 2023).
17. Wei, Z.; Falzone, G.; Wang, B.; Thiele, A.; Puerta-Falla, G.; Pilon, L.; Neithalath, N.; Sant, G. The durability of cementitious composites containing microencapsulated phase change materials. *Cem. Concr. Compos.* **2018**, *81*, 66–76. [CrossRef]
18. Drissi, S.; Ling, T.-C. Thermal and durability performances of mortar and concrete containing phase change materials. *IOP Conf. Ser. Mater. Sci. Eng.* **2018**, *431*, 062001. [CrossRef]
19. Jiang, Y.; Ding, E.; Li, G. Study on transition characteristics of PEG/CDA solid–solid phase change materials. *Polymer* **2002**, *43*, 117–122. [CrossRef]
20. Alkan, C.; Günther, E.; Hiebler, S.; Ensari, Ö.F.; Kahraman, D. Polyurethanes as solid-solid phase change materials for thermal energy storage. *Sol. Energy* **2012**, *86*, 1761–1769. [CrossRef]
21. Tang, B.; Yang, Z.; Zhang, S. Poly(polyethylene glycol methyl ether methacrylate) as Novel Solid-Solid Phase Change Material for Thermal Energy Storage. *J. Appl. Polym. Sci.* **2012**, *125*, 1377–1381. [CrossRef]
22. Yanshan, L.; Shujun, W.; Hongyan, L.; Fanbin, M.; Huanqing, M.; Wangang, Z. Preparation and characterization of melamine/formaldehyde/polyethylene glycol crosslinking copolymers as solid–solid phase change materials. *Sol. Energy Mater. Sol. Cells* **2014**, *127*, 92–97. [CrossRef]
23. Harlé, T.; Ledesert, B.; Nguyen, T.M.G.; Hebert, R.; Melinge, Y. Phase-Change Material for Storing Thermal Energy, Manufacturing Method and Uses of Such a Material. WO 2017198933 A1, 23 November 2017. Available online: https://patents.google.com/patent/WO2017198933A1 (accessed on 2 October 2023).
24. Plancher, L.; Pierre, A.; Nguyen, G.T.M.; Hébert, R.L.; Ledésert, B.A.; Di Martino, P.; Mélinge, Y. Rheological Behaviour of Cementitious Materials Incorporating Solid–Solid Phase Change Materials. *Materials* **2022**, *15*, 20. [CrossRef] [PubMed]
25. NF EN 196-1, Méthodes D'essais des Ciments, Partie 1: Détermination des Résistances Mécaniques 2016. Available online: https://www.boutique.afnor.org/norme/nf-en-196-1/methodes-d-essais-des-ciments-partie-1-determination-des-resistances/article/866862/fa184622 (accessed on 12 October 2023).
26. Benabed, B.; Kadri, E.-H.; Azzouz, L.; Kenai, S. Properties of self-compacting mortar made with various types of sand. *Cem. Concr. Compos.* **2012**, *34*, 1167–1173. [CrossRef]
27. The French National Research Project PERFDUB. Approche Performantielle de la Durabilité des Ouvrages en Béton. Available online: https://www.perfdub.fr/en/ (accessed on 14 October 2020).
28. Zhang, Y.; Yang, Z.; Ye, G. Dependence of unsaturated chloride diffusion on the pore structure in cementitious materials. *Cem. Concr. Res.* **2020**, *127*, 105919. [CrossRef]
29. Samson, G.; Phelipot-Mardelé, A.; Lanos, C.; Aogla, K.; Weekes, L.; Augusthus-Nelson, L.; Arora, S.; Singh, S.P.; Zhang, J.; Ma, H.; et al. A review of thermomechanical properties of lightweight concrete. *Mag. Concr. Res.* **2017**, *69*, 201–216. [CrossRef]
30. Narain, J.; Jin, W.; Ghandehari, M.; Wilke, E.; Shukla, N.; Berardi, U.; El-Korchi, T.; Van Dessel, S. Design and Application of Concrete Tiles Enhanced with Microencapsulated Phase-Change Material. *J. Arch. Eng.* **2016**, *22*, 05015003. [CrossRef]
31. Cao, V.D.; Pilehvar, S.; Salas-Bringas, C.; Szczotok, A.M.; Rodriguez, J.F.; Carmona, M.; Al-Manasir, N.; Kjøniksen, A.-L. Microencapsulated phase change materials for enhancing the thermal performance of Portland cement concrete and geopolymer concrete for passive building applications. *Energy Convers. Manag.* **2017**, *133*, 56–66. [CrossRef]
32. Lura, P.; Jensen, O.M.; van Breugel, K. Autogenous shrinkage in high-performance cement paste: An evaluation of basic mechanisms. *Cem. Concr. Res.* **2003**, *33*, 223–232. [CrossRef]
33. Grasley, Z.C.; Leung, C.K. Desiccation shrinkage of cementitious materials as an aging, poroviscoelastic response. *Cem. Concr. Res.* **2011**, *41*, 77–89. [CrossRef]
34. Di Bella, C.; Wyrzykowski, M.; Lura, P. Evaluation of the ultimate drying shrinkage of cement-based mortars with poroelastic models. *Mater. Struct.* **2017**, *50*, 52. [CrossRef]
35. Neville, A.M. *Properties of Concrete*, 4th ed.; Pearson Higher Education, Prentice Hall: Englewood Cliffs, NJ, USA, 1995. Available online: https://www.technicalbookspdf.com/properties-of-concrete-fifth-edition-a-m-neville/ (accessed on 28 November 2023).

36. Chuang, C.-W.; Chen, T.-A.; Huang, R. Effect of Finely Ground Coal Bottom Ash as Replacement for Portland Cement on the Properties of Ordinary Concrete. *Appl. Sci.* **2023**, *13*, 13212. [CrossRef]
37. Pacheco-Torgal, F.; Labrincha, J.; Leonelli, C.; Palomo, A.; Chindaprasirt, P. *Handbook of Alkali-Activated Cements, Mortars and Concretes*, 1st ed.; Abington Hall: Cambridge, UK, 2015. Available online: https://www.sciencedirect.com/book/9781782422761 (accessed on 5 October 2023).
38. Stefan, L.; Boulay, C.; Torrenti, J.-M.; Bissonnette, B.; Benboudjema, F. Influential factors in volume change measurements for cementitious materials at early ages and in isothermal conditions. *Cem. Concr. Compos.* **2018**, *85*, 105–121. [CrossRef]
39. Gilbert, R. *Time Effects in Concrete Structures*; Elsevier: Amsterdam, The Netherlands, 1988. Available online: https://trid.trb.org/view/311317 (accessed on 3 February 2024).
40. Rusch, H.; Jungwirth, D.; Hilsdorf, D.H. *Creep and Shrinkage, Their Effect on the Behaviour of Concrete Structures*; Springer: New York, NY, USA, 1983. Available online: https://www.springer.com/gp/book/9781461254263 (accessed on 7 September 2023).
41. Medvedev, V.; Pustovgar, A.; Adamtsevich, A.; Adamtsevich, L. Concrete Carbonation of Deep Burial Storage Constructions under Model Aging Conditions. *Buildings* **2023**, *14*, 8. [CrossRef]

Disclaimer/Publisher's Note: The statements, opinions and data contained in all publications are solely those of the individual author(s) and contributor(s) and not of MDPI and/or the editor(s). MDPI and/or the editor(s) disclaim responsibility for any injury to people or property resulting from any ideas, methods, instructions or products referred to in the content.

Article

Resistance of Concrete with Crystalline Hydrophilic Additives to Freeze–Thaw Cycles

Anita Gojević [1], Ivanka Netinger Grubeša [2,*], Sandra Juradin [3] and Ivana Banjad Pečur [4]

1. City of Osijek, Franje Kuhača 9, 31000 Osijek, Croatia; anitagojevic@gmail.com
2. Department of Construction, University North, 104. brigade 3, 42000 Varaždin, Croatia
3. Faculty of Civil Engineering, Architecture and Geodesy, University of Split, Matice hrvatske 15, 21000 Split, Croatia; sandra.juradin@gradst.hr
4. Faculty of Civil Engineering, University of Zagreb, Andrija Kačić Miošić Street 26, 10000 Zagreb, Croatia; ivana.banjad.pecur@grad.unizg.hr
* Correspondence: inetinger@unin.hr

Abstract: The study explores the hypothesis that crystalline hydrophilic additives (CA) can enhance concrete's resistance to freeze/thaw cycles, crucial for assessing building durability. Employing EU standards, the research evaluates concrete resistance through standardized European freeze/thaw procedures. Monitoring concrete slabs exposed to freezing in the presence of deionized water and in the presence of 3% sodium chloride solution, the study measures surface damage and relative dynamic modulus of elasticity. Additionally, it assesses internal damage through monitoring of relative dynamic modulus of elasticity on cubes and prisms submerged in water and exposed to freezing/thawing. The pore spacing factor measured here aids in predicting concrete behavior in freeze/thaw conditions. Results suggest that the standard air-entraining agent offers effective protection against surface and internal damage due to freeze/thaw cycles. However, the CA displays potential in enhancing resistance to freeze/thaw cycles, primarily in reducing internal damage at a 1% cement weight dosage. Notably, a 3% replacement of cement with CA adversely affects concrete resistance, leading to increased surface and internal damage. The findings contribute to understanding materials that can bolster concrete durability against freeze–thaw cycles, crucial for ensuring the longevity of buildings and infrastructure.

Keywords: concrete; durability; crystalline hydrophilic additives; freeze–thaw cycles; surface damage; internal damage; pore spacing factor

Citation: Gojević, A.; Netinger Grubeša, I.; Juradin, S.; Banjad Pečur, I. Resistance of Concrete with Crystalline Hydrophilic Additives to Freeze–Thaw Cycles. *Appl. Sci.* **2024**, *14*, 2303. https://doi.org/10.3390/app14062303

Academic Editors: Mouhamadou Amar and Nor Edine Abriak

Received: 21 February 2024
Revised: 5 March 2024
Accepted: 8 March 2024
Published: 9 March 2024

Copyright: © 2024 by the authors. Licensee MDPI, Basel, Switzerland. This article is an open access article distributed under the terms and conditions of the Creative Commons Attribution (CC BY) license (https://creativecommons.org/licenses/by/4.0/).

1. Introduction

The durability of buildings is mostly influenced by the durability of the materials used in their construction. A primary factor that undermines this durability is the freeze–thaw cycle [1]. When temperatures dip below zero, water within the material freezes and expands, exerting stress on the material's walls [2]. Through repeated freeze–thaw cycles, this stress leads to material damage, consequently diminishing its durability. In cement composites, such damage manifests as surface scaling or internal cracking [3].

A common approach to enhancing concrete's durability against freeze/thaw cycles involves incorporating air-entraining agents into the concrete mixture [4]. These agents introduce air bubbles during mixing, which disrupt the capillaries through which water could penetrate the concrete. By minimizing water content in the concrete, issues related to freeze–thaw cycles are mitigated. However, it is important to exercise caution with these agents, as they may adversely affect the compressive strength of the concrete [5]. The literature also suggests that concrete durability can be improved by incorporating mineral additives like slag [6], fly ash [7], and silica fume [8]. Furthermore, concrete durability can be enhanced by partially replacing aggregate with rubber [9–12], employing polymer binders [13,14], modifying [15–17] or impregnating concrete with polymers [18,19], using polycarbonate superplasticizers [20,21], employing biomimetic polymer additives [22], and utilizing polymer fibers and biofibers [23,24].

Crystalline admixtures (CA) are primarily commercially available products offered by various manufacturers (such as Xypex, Richmond, BC, Canada; Kryton, Vancouver, BC, Canada; Penetron, East Setauket, NY, USA; Harbin, China). They serve a dual function: reducing concrete permeability and repairing cracks [25]. The recommended dosage of CA in concrete typically ranges from 0.3% to 5% by the weight of the cement [26,27]. Several authors have studied the impact of CA on the durability properties of concrete by monitoring crack healing, with most concluding its effectiveness in this regard [28–32]. It was noted that the highest rate of healing was observed when samples containing CA were consistently immersed in water [28–30]. According to [29,33], calcium carbonate in the form of aragonite is formed in concrete cracks treated with CA, effectively sealing them.

Considering the confirmed effectiveness of CA in the concrete crack healing process and the fact that cracks occur in concrete during freeze–thaw cycles, it would be intriguing to precisely determine the effectiveness of CA in enhancing concrete resistance to freeze–thaw cycles. European legislation mandates testing the resistance of concrete to freeze–thaw cycles through procedures outlined in standards CEN/TS 12390-9 [34] and CEN/TR 15177 [35]. In the method outlined in CEN/TS 12390-9 [34], concrete samples saturated with deionized water or a 3% sodium chloride solution undergo freeze/thaw cycles (56 cycles), during which surface scaling and mass loss of concrete are measured. The procedure described in CEN/TR 15177 [35] can be employed to monitor damage to the internal structure. Additionally, EN 480-11 [36] is used to predict concrete behavior under freeze–thaw conditions, involving microscopic observation of hardened concrete samples, measurement of pore spacing, and calculation of the pore spacing factor which is defined as distance of any point in cement paste to the edge of the nearest air void. Cement-based materials are considered resistant to freeze–thaw cycles if the spacing factor is less than 0.2 mm.

Since the authors in [37] have already confirmed reduced water absorption by using CA in concrete as an indicator of concrete resistance to freeze–thaw cycles, and the authors in [32] have confirmed reduced water penetration in concrete with CA, the hypothesis arises that the application of CA could potentially improve concrete resistance to freeze–thaw cycles. Therefore, this study aims to examine the resistance of concrete to freeze–thaw cycles according to standardized procedures prescribed by EU standards.

2. Experimental Part

In the experimental part of the paper, four concrete mixtures were prepared; a reference mixture (M1), a mixture with an air entraining agent (M2), and mixtures with a crystalline hydrophilic admixture in two different amounts per cement weight (M3, M4).

2.1. Properties of Aggregates, Binders, and Additives to Concrete

In this research, dolomite was used as an aggregate in fractions 0–4 mm, 4–8 mm, 8–16 mm, and 16–31.5 mm, as well as a dolomite-type filler. The density of dolomite aggregate and filler determined according to EN 1097-6 standard [38] was 2780 kg/m^3. The specific surface area for filler determined using the BET method according to the standard ISO 9277 [39] was 2.32 m^2/g. Sieve curves for dolomite fractions are shown together with target and actual cumulative aggregate curve in Figure 1, where it should be noted that 5% of the 0–4 mm fraction was replaced with filler.

The cement used for making concrete mixtures was CEM I 52.5 N. In all mixtures, the superplasticiser ViscoCrete 5380, Sika Croatia, Zagreb, Croatia was used in the amount of 1% of the mass of binder. In mixture M2, the air-entraining agent LPS A 94 from Sika was used in the amount of 0.2% of the mass of cement. The crystalline hydrophilic admixture Penetron Admix from Penetra, Sesvete, Croatia was used in the amount of 1% of binder in mixtures M3 and 3% of the mass of binder in mixtures M4. The density of binders (cement and crystalline hydrophilic admixture) was determined according to the standard EN

1096-6 [40], and the specific surface area was determined using the BET method according to ISO 9277 [39]. The densities of the superplasticizer and air entraining agent are adopted from the additive producer. The densities of binders, superplasticizer, and air-entraining agent, as well as the specific surface areas of binders, are shown in Table 1.

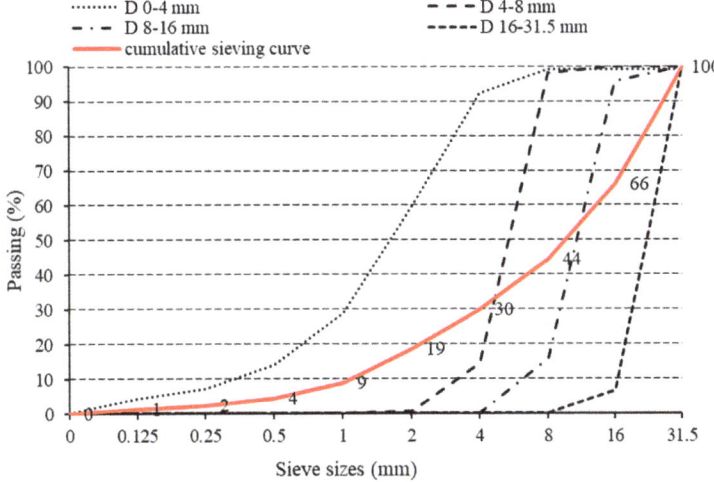

Figure 1. Fraction sieve curves, target, and cumulative sieving curves of aggregate.

Table 1. Densities of binders, superplasticizer, and air-entraining agent, and specific surface area of binders.

Components	Density, kg/m^3	Specific Surface Area, m^2/g
Cement, CEM I 52.5 N	2960	3.76
Superplasticiser, ViscoCrete 5380	1080	-
Air entraining agent, LPS A 94	1000	-
Crystalline hydrophilic admixture (CA), Penetron	2910	2.70

2.2. Composition of Concrete Mixtures

The composition of concrete mixtures is shown in Table 2. All mixtures have the same water/cement ratio of 0.35, the same amount of aggregate, and the same amount of binder (400 kg). In mixtures M1 and M2 it is cement, while in mixtures M3 and M4 it is the total amount of cement and crystalline hydrophilic additive.

Table 2. Composition of concrete mixtures for 1 m^3 of concrete.

Mixture/Components		M1	M2	M3	M4
	Cement (kg)	400	400	396	388
	Water (kg)	140	140	140	140
	Superplasticizer (kg)	4	4	4	4
	Air entraining agent (kg)	-	0.8	-	-
	Crystalline hydrophilic admixture (kg)	-	-	4	12
Aggregate	Dolomite 0–4 mm (kg)	576.6	576.6	576.6	576.6
	Dolomite 4–8 mm (kg)	195.6	195.6	195.6	195.6
	Dolomite 8–16 mm (kg)	469.8	469.8	469.8	469.8
	Dolomite 16–31.5 mm (kg)	685	685	685	685
	Filler (kg)	30.2	30.2	30.2	30.2

The aggregates used for preparing concrete were first saturated and then surface-dried. This was achieved in an artificial way by dipping the aggregates into a water tank for 24 h, taking them out, and then wiping excess water from their surface. First, coarse and fine aggregate was mixed for 1 min, then binder was added and the mixing was continued for an additional 2 min. In the end, water was added and the mixing was continued for an additional 2 min. Mixing the concrete in a pan mixer (DZ 100VS, Diemwerke, Hörbranz, Austria) took a total of 5 min.

2.3. Properties of Fresh and Hardened Concrete

The consistency of the concrete was determined according to EN 12350-2 [41], the density of fresh concrete according to EN 12350-6 [42], and the air content according to the standard EN 12350-7, with the pressure gauge method [43]. The obtained results are shown in Table 3.

Table 3. Properties of concrete mixtures in their fresh state.

Mixture	M1	M2	M3	M4
Consistency–slump (cm)	12	14	11	11
Density (kg/m^3)	2504	2439	2520	2489
Air content (%)	1.5	5	1.5	1.6

According to Table 3, all mixtures belong to consistency class S3 (10–15 cm) according to EN 206 [44]. The addition of crystalline hydrophilic admixture had no impact on workability, which is in accordance with [45]. In terms of density, all mixtures can be considered normal weight concrete. As expected, mixture M2 has the highest air content in fresh concrete, for which the air-entraining agent in the mixture is directly responsible. Crystalline hydrophilic admixture did not affect the air content in fresh concrete for both tested doses, but Shetiya et al. [46] tested mixtures with different concentrations of Penetron crystalline admixture (1% and 2.5% of the cement mass) and found that the mixture with 1% crystalline admixture had the highest air content of all tested concretes.

From each mixture, 14 cubes of dimensions 15 cm × 15 cm × 15 cm and 3 prisms of dimensions 10 cm × 10 cm × 40 cm were prepared. After casting, the concrete specimens were stored under cover for 24 h under laboratory conditions until demolding to prevent water evaporation. After demolding, the specimens were in the mist room at 20 ± 2 °C and RH ≥ 95% until the age of testing. On 3 out of 14 cubes, the compressive strength of 28-day-old specimens is determined according to EN 12390-3 [47], and the results and their corresponding standard deviations are shown in Figure 2.

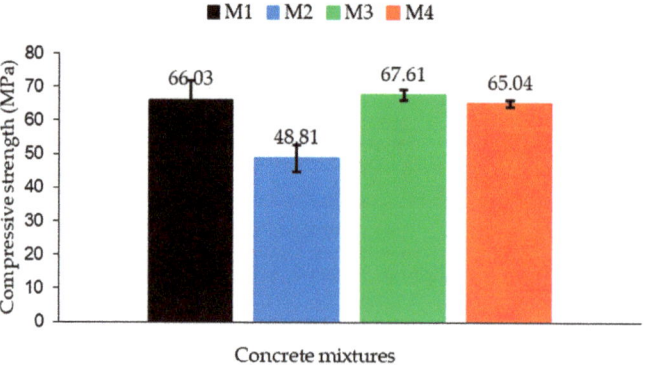

Figure 2. Compressive strength of concrete at the age of 28 days.

From Figure 2, it is evident that CA present in concrete mixtures M3 and M4 did not significantly affect the compressive strength of the concrete, which is in line with the conclusions presented in [32,48,49]. The presence of air entraining agent in mixture M2 significantly reduced the compressive strength of the concrete which is consistent with the well-known fact that air entraining agent negatively affects concrete strength [5].

Furthermore, from each of the eight cubes, one slab of dimensions 15 cm × 15 cm × 5 cm (total of eight slabs) was sawn out to monitor scaling due to freeze/thaw cycles according to CEN/TS 12390-9 [34], and the relative dynamic modulus of elasticity due to freeze/thaw cycles according to Clause 8 of CEN/TR 15177 standard [35] using an ultrasonic pulse transmission time device, and one slab of dimensions 10 cm × 15 cm × 4 cm to measure the spacing factor according to EN 480-11 standard [36]. Half of the slabs intended for scaling and relative dynamic modulus of elasticity monitoring were subjected to freeze/thaw attack in the presence of a 3 mm deep layer of deionized water, and the other half were subjected to freeze/thaw attack in the presence of a 3% sodium chloride solution. Figure 3 shows all the slab samples in the freezing and thawing chamber (producer: Schleibinger, Buchbach, Germany), and Figure 4 shows the monitoring of the amount of scaled material and dynamic modulus of elasticity during exposure to freeze/thaw cycles.

Figure 3. Slab samples in the freezing/thawing chamber.

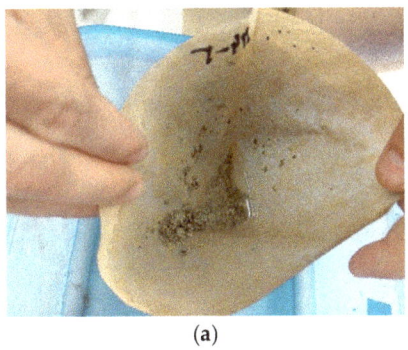

(a)   (b)

Figure 4. Measurements during exposure of panels to freeze/thaw cycles: (a) scaling; (b) monitoring of dynamic modulus of elasticity.

Figure 5 shows the slab prepared for measuring the spacing factor and the measuring device for the measuring. The remaining 3 out of a total of 14 cubes and 3 prisms of each mixture were subjected to freeze/thaw cycles in the presence of water in the Mis 600 chamber, LT, Slovenia at 28 days of their age and the relative dynamic modulus of elasticity was monitored during freeze/thaw cycles according to Clause 7 of CEN/TR 15177 standard [35]. Figure 6 shows specimens in the chamber immersed in water and measuring of the pulse transmission time on the cube and prism specimens.

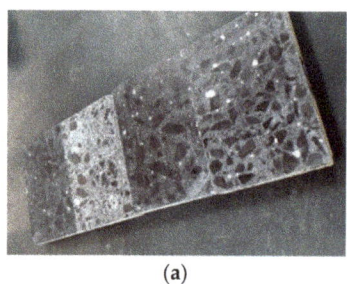
(a)
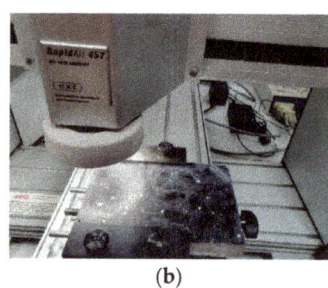
(b)

Figure 5. Spacing factor measuring: (a) concrete slab prepared for spacing factor measuring; (b) spacing factor measuring device.

(a)
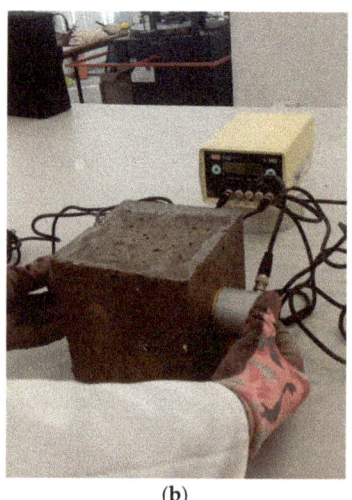
(b)

(c)

Figure 6. Measuring during exposure of cubes and prisms to freeze/thaw cycles in the presence of water: (a) Cubes and prisms immersed in water; (b) measuring of the pulse transmission time on the cubes; (c) measuring of the pulse transmission time on the prisms.

3. Test Results

The results of scaling tests due to freeze/thaw cycles according to CEN/TS 12390-9 [34] with corresponding standard deviations are shown in Figure 7, and the results of testing the relative dynamic modulus of elasticity due to freeze/thaw cycles according to Clause 8 of CEN/TR 15177 standard [35] are shown in Figure 8. Each point on the curves presented in Figure 7 represents the mean value of four measurements. While the standard deviation of results is expressed for absolute values (Figure 7), this was not possible for relative values (Figure 8). However, it should be noted that the relative values were calculated from the mean absolute values of four absolute measured values, with the exclusion of all values that deviated from the mean absolute value by more than 10%.

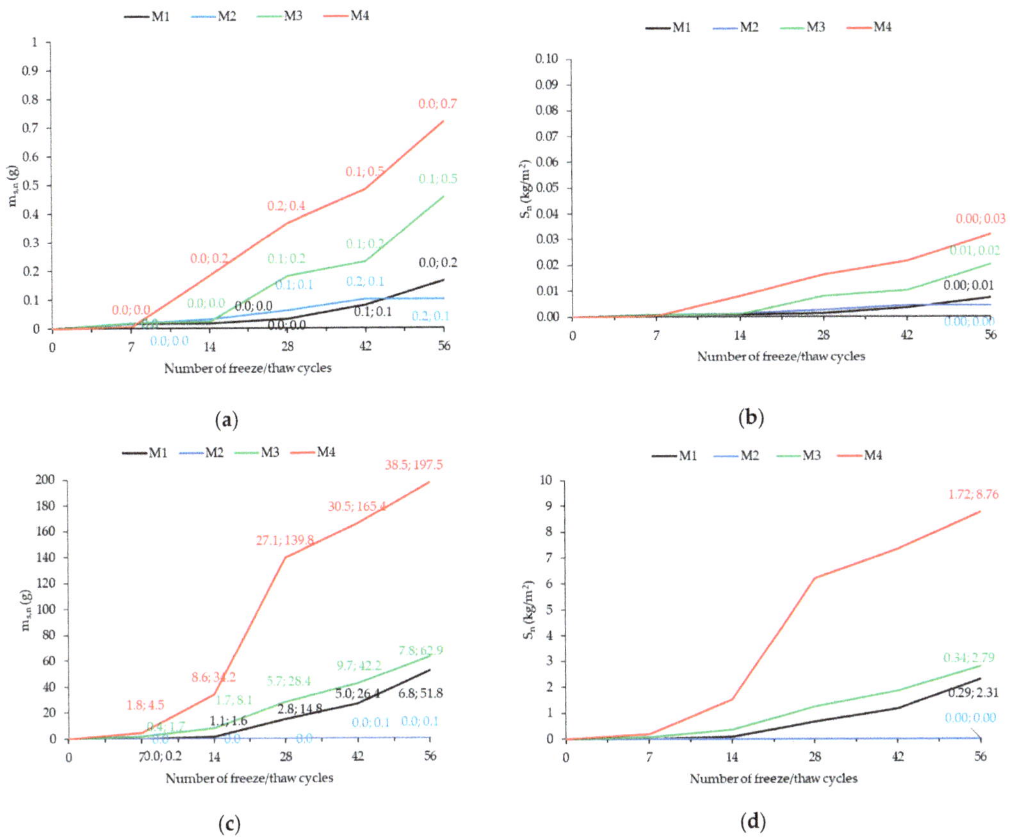

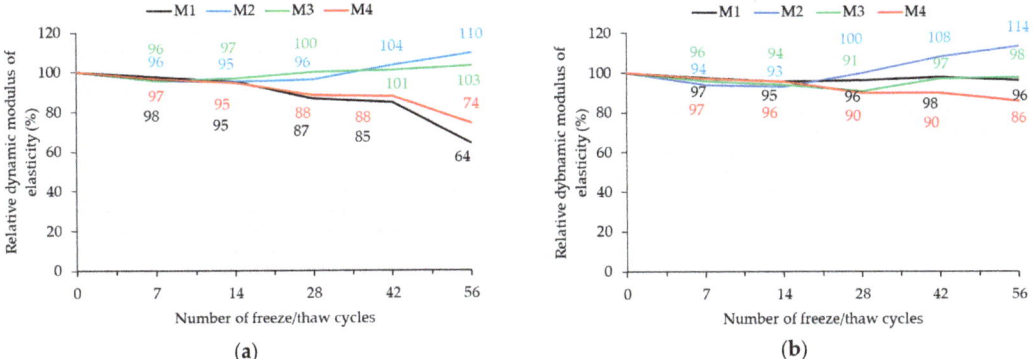

Figure 7. Scaling of concrete slabs: (**a**) scaled material mass due to freeze/thaw cycles in the presence of deionized water; (**b**) scaled material mass related to test surface due to freeze/thaw cycles in the presence of deionized water; (**c**) scaled material mass due to freeze/thaw cycles in the presence of 3% sodium chloride solution; (**d**) scaled material mass related to test surface due to freeze/thaw cycles in the presence of 3% sodium chloride solution.

Figure 8. Relative dynamic modulus of elasticity–measured on concrete slabs: (**a**) in the presence of deionized water; (**b**) in the presence of 3% sodium chloride solution.

The results of spacing factor measurements according to EN 480-11 [36] are shown in Figure 9.

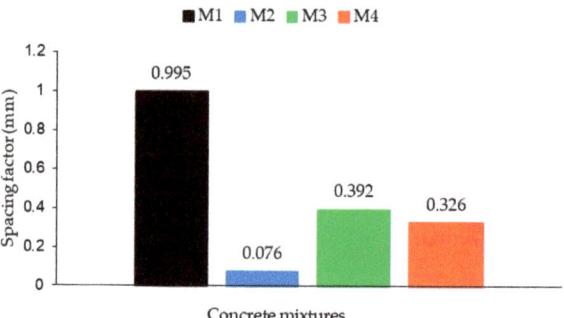

Figure 9. Spacing factor.

The results of testing the relative dynamic modulus of elasticity due to freeze–thaw cycles according to Clause7 of CEN/TR 15177 standard [35] are presented in Figure 10. The relative values were calculated from the mean absolute values of six measurements on cubes and three measurements on prisms, with the exclusion of all values that deviated from the mean absolute value by more than 10%.

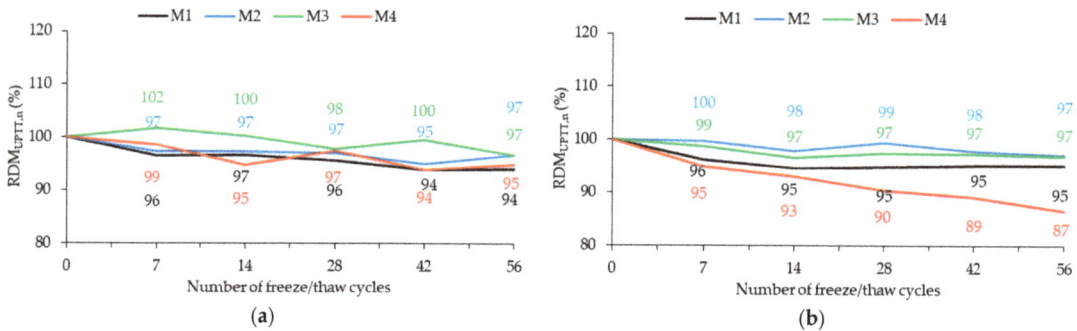

Figure 10. Relative dynamic modulus of elasticity: (**a**) measured on concrete cubes; (**b**) measured on concrete prisms.

4. Discussion

From Figure 7a,b, it is evident that in mixture M2 containing an air-entraining agent, there is a significant reduction in mass loss due to scaling after 56 freeze/thaw cycles compared to mixture M1, while the crystalline hydrophilic additive in mixtures M3 and M4 acted contrary to expectations, increasing the mass loss due to scaling, i.e., increasing the mass loss due to exposure of samples to deionized water. The mixture with a lower proportion of crystalline hydrophilic additive (M3) records a lower mass of scaled material compared to the mixture with a higher proportion of crystalline hydrophilic additive (M4). This is contrary to the observations in [50] where the mass loss ratio due to freeze/thaw cycles is significantly lower in mixtures with the addition of CA compared to the reference mixture. The mixture with the least amounts of scaled material, and therefore the best resistance to freeze/thaw cycles according to this method, and under conditions of exposure to deionized water, is mixture M2, followed by mixtures M1, M3, and M4 in sequence. From Figure 7c,d, it is noticeable that in mixture M2 containing an air-entraining agent, there is a drastic reduction in mass loss due to scaling after 56 freeze/thaw cycles compared to mixture M1, while the crystalline hydrophilic additive in mixtures M3 and M4 acted contrary to expectations, increasing the mass loss due to scaling, i.e., increasing the mass loss due to exposure of samples to a 3% sodium chloride solution. The mixture with a lower proportion of crystalline hydrophilic additive (M3) records significantly lower scaled material mass compared to the mixture with a higher proportion of crystalline hydrophilic additive (M4). The lowest scaled material mass, and thus the best resistance to freeze/thaw cycles according to this method and under conditions of exposure to a 3% sodium chloride solution, is recorded by mixture M2, followed by mixtures M1, M3, and M4 in sequence. When it comes to surface damage due to freeze/thaw cycles, the air-entraining agent is evidently the most effective additive for preventing damage, almost equally effective regardless of whether freezing/thawing occurs with or without salt presence, while the negative effect of the crystalline hydrophilic additive is significantly more pronounced during freezing/thawing in the presence of salt. Such research findings on the impact of the crystalline hydrophilic additive are even worse than the results presented in [51]. Specifically, Manhanga et al. [51] concluded in part of their study addressing the scaling of concrete exposed to a 3% sodium chloride solution that the crystalline hydrophilic additive (in amount of 0.8% per cement weight) does not affect this type of damage caused by freeze–thaw cycles.

From Figure 8a, it can be concluded that the drop in the dynamic modulus of elasticity as a measure of internal damage during exposure to freezing/thawing in the presence of deionized water is most pronounced in mixture M1. Mixture M4 has a smaller drop in the dynamic modulus of elasticity than mixture M1, while mixtures M3 and M2 recorded an increase in the dynamic modulus of elasticity. The increase in the dynamic modulus of elasticity during the freeze/thaw cycles is consistent with the increase in mass during freezing/thawing reported in [52]. The authors in [52] explain that freeze/thaw cycles promote the mobility of pore solution through osmosis. As a result, portlandite dissolved in pore water migrates, facilitating the reactions involved in the self-healing procedure. Additional ice formation in pores likely contributed to the reported mass increase, thus supporting the evolution of the self-healing process.

During exposure to freezing/thawing in the presence of a 3% sodium chloride solution (Figure 8b), the highest drop in the dynamic modulus of elasticity was recorded by mixture M4, while mixtures M1 and M3 recorded a somewhat smaller drop in the dynamic modulus of elasticity, and mixture M2 recorded an increase in the dynamic modulus of elasticity. In terms of internal damage, the air-entraining agent has shown the highest effectiveness in protecting concrete from damages caused by freeze/thaw cycles, but the crystalline hydrophilic additive at a 1% dosage (M3) has shown potential to improve concrete's resistance to freeze/thaw. This is in accordance with [53] where a positive effect on crack self-healing (monitored through the recovery of compressive strength of samples cured in water) of lower CA content has also been recorded, while a negative effect of higher CA content in the total binder quantity was noted.

Figure 9 shows that mixture M2 has the smallest pore-spacing factor, followed by mixtures M3 and M4, while mixture M1 has the largest pore-spacing factor. Considering that this testing method requires a pore spacing factor smaller than 0.2 mm for concrete to be considered resistant to freeze/thaw cycles, according to this method, only mixture M2 could be considered resistant to freeze/thaw cycles. The obtained value of the pore spacing factor of 0.076 mm for the M2 is in accordance with range from 0.07 mm to 0.16 mm for air-entrained concrete [54]. Compared with the mixture M1, the crystalline hydrophilic additive reduced the spacing factor more than 60%, but the obtained values of 0.392 mm (M3) and 0.326 mm (M4) are significantly higher than 0.2 mm requested for concrete to be considered resistant to freeze/thaw cycles. Figure 10a,b confirm the conclusions regarding Figure 8a,b, namely that regarding internal damage, the air-entraining agent has shown the highest effectiveness in protecting concrete from damages caused by freeze/thaw cycles, but the crystalline hydrophilic additive at a 1% dosage (M3) has shown potential to improve concrete's resistance to freeze/thaw cycles. Furthermore, regarding internal damages, the crystalline hydrophilic additive used in this study achieved better performance in enhancing the concrete's resistance to freeze/thaw cycles compared to the crystalline hydrophilic additive used in [51]. Specifically, Manhanga et al. [51] concluded, in part of their research focusing on the impact of the crystalline hydrophilic additive on the strength of cubic specimens exposed to freeze/thaw cycles in the presence of water, that the crystalline hydrophilic additive does not affect this type of damage caused by freeze–thaw cycles. On the other hand, Ferrara et al. [55] have indeed confirmed that the velocity of the ultrasonic wave passage is higher in concrete with a crystalline hydrophilic additive compared to reference concrete during the self-healing process of cracks in concrete, leading to the conclusion that the crystalline hydrophilic additive promotes crack healing. The results presented in this paper are in line with the results shown in Ferrara et al. [55] because cracks that occur as internal damage during freeze/thaw cycles are likely to be healed faster when concrete contains a 1% crystalline hydrophilic additive compared to reference concrete.

5. Conclusions

The paper investigates the effectiveness of the crystalline hydrophilic additive on concrete resistance to freeze/thaw cycles according to standardized EU methods. Scaling resulting from freeze/thaw cycles was observed on concrete slabs exposed to freezing/thawing under two conditions: in the presence of deionized water, and in the presence of a 3% sodium chloride solution. This served as a measure of surface damage. Additionally, the relative dynamic modulus of elasticity was assessed on concrete slabs subjected to freezing/thawing under the same conditions mentioned above. This measurement provided insight into internal damage. Furthermore, the relative dynamic modulus of elasticity was examined on concrete cubes and prisms submerged in water and exposed to freezing/thawing. This served as a measure of internal damage. Lastly, the paper explored the pore spacing factor. This factor is utilized more for predicting concrete behavior in freeze/thaw conditions rather than monitoring actual concrete behavior in such conditions. Based on the obtained results, it was concluded that the most effective protection against surface and internal damage to concrete is provided by the standardly used air-entraining agent, while the crystalline hydrophilic additive has the potential to improve concrete resistance to freeze/thaw cycles in the context of reducing internal damage only if used at a 1% cement weight dosage. A 3% replacement of cement with crystalline hydrophilic additive has shown a negative effect on concrete resistance to freeze/thaw cycles in terms of increased surface and internal damage.

Author Contributions: Conceptualization, A.G. and I.N.G.; methodology, A.G. and I.N.G.; investigation, A.G., I.N.G., S.J. and I.B.P.; resources, A.G., I.N.G., S.J. and I.B.P.; data curation, A.G. and I.N.G.; writing—original draft preparation, A.G., I.N.G. and S.J.; writing—review and editing, S.J. and I.B.P.; visualization, A.G.; funding acquisition, I.N.G. All authors have read and agreed to the published version of the manuscript.

Funding: This research received no external funding.

Institutional Review Board Statement: Not applicable.

Informed Consent Statement: Not applicable.

Data Availability Statement: The raw data supporting the conclusions of this article will be made available by the authors on request.

Acknowledgments: The authors are grateful for financial support within the project "Trajnost cementnih kompozita"-UNIN-TEH-23-1-7. This research is partially supported through projects: KK.01.1.1.02.0027, a project co-financed by the Croatian Government and the European Union through the European Regional Development Fund—the Competitiveness and Cohesion Operational Programme.

Conflicts of Interest: The authors declare no conflicts of interest.

References

1. Surej, K.R. Evaluation and Improvement of Frost Durability of Clay Bricks—A Thesis in The Centre for Building Studies, Ottawa, Canada. 1997. Available online: https://www.collectionscanada.gc.ca/obj/s4/f2/dsk3/ftp04/nq25926.pdf (accessed on 12 December 2023).
2. Pilehvar, S.; Szczotok, M.A.; Rodríguez, J.F.; Valentini, L.; Lanzón, M.; Pamies, R.; Kjøniksen, A.-L. Effect of freeze-thaw cycles on the mechanical behavior of geopolymer concrete and Portland cement concrete containing micro-encapsulated phase change materials. *Constr. Build. Mater.* **2019**, *200*, 94–103. [CrossRef]
3. Richardson, G.M. *Fundamentals of Durable Reinforced Concrete*, 1st ed.; CRC Press: Boca Raton, FL, USA, 2002; pp. 51, 77, 101, 133, 160–179, 194.
4. Qiu, Y.; Peng, H.; Zhao, H. Study on New Type of Concrete Air-Entraining Agent. In Proceedings of the International Conference on Artificial Intelligence and Electromechanical Automation (AIEA), Tianjin, China, 26–28 June 2020.
5. Nowak-Michta, A. Impact analysis of air-entraining and superplasticizing admixtures on concrete compressive strength. *Procedia Struct. Integr.* **2019**, *23*, 77–82. [CrossRef]
6. Nicula, L.M.; Corbu, O.; Iliescu, M. Influence of Blast Furnace Slag on the Durability Characteristic of Road Concrete Such as Freeze-Thaw Resistance. *Procedia Manuf.* **2020**, *46*, 194–201. [CrossRef]
7. Islam, M.M.; Alam, M.T.; Islam, M.S. Effect of fly ash on freeze–thaw durability of concrete in marine environment. *Aust. J. Struct. Eng.* **2018**, *19*, 146–161. [CrossRef]
8. Zang, P.; Li, Q.-F. Freezing–thawing durability of fly ash concrete composites containing silica fume and polypropylene fiber. *Proc. Inst. Mech. Eng. Part L J. Mater. Des. Appl.* **2013**, *228*, 241–246. [CrossRef]
9. Li, X.; Ling, T.C.; Mo, K.H. Functions and impacts of plastic/rubber wastes as eco-friendly aggregate in concrete—A review. *Constr. Build. Mater.* **2020**, *240*, 117869. [CrossRef]
10. Kumar, R.; Dev, N. Effect of acids and freeze-thaw on durability of modified rubberized concrete with optimum rubber crumb content. *J. Appl. Polym. Sci.* **2022**, *139*, 52191. [CrossRef]
11. He, Y.; Xu, F.; Wei, H. Effect of Particle Size on Properties of Concrete with Rubber Crumbs. *Am. J. Civ. Eng.* **2022**, *10*, 79–87. [CrossRef]
12. Pham, N.P.; Toumi, A.; Turatsinze, A. Effect of an enhanced rubber-cement matrix interface on freeze-thaw resistance of the cement-based composite. *Constr. Build. Mater.* **2019**, *207*, 528–534. [CrossRef]
13. Ribeiro, M.S.C.; Juvandes, L.F.P.; Rodrigues, J.D.; Ferreira, A.; Marques, A.T. Behaviour of Cement and Polymer Mortar Materials to Rapid Freeze-Thaw Cycling. *Mater. Sci. Forum* **2010**, *636–637*, 1329–1335. [CrossRef]
14. Khashayar, J.; Heidarnezhad, F.; Moammer, O.; Jarrah, M. Experimental investigation on freeze—Thaw durability of polymer concrete. *Front. Struct. Civ. Eng.* **2021**, *15*, 1038–1046. [CrossRef]
15. Qu, Z.; Guo, S.; Sproncken, C.C.M.; Surís-Valls, R.; Yu, Q.; Voets, I.K. Enhancing the Freeze–Thaw Durability of Concrete through Ice Recrystallization Inhibition by Poly(vinyl alcohol). *ACS Omega* **2020**, *5*, 12825–12831. [CrossRef] [PubMed]
16. Guo, Y.; Shen, A.; Sun, X. Exploring Polymer-Modified Concrete and Cementitious Coating with High-Durability for Roadside Structures in Xinjiang, China. *Adv. Mater. Sci. Eng.* **2017**, *2017*, 9425361. [CrossRef]
17. Hammodat, W.W. Investigate road performance using polymer modified concrete. *Mater. Today Proc.* **2021**, *42*, 2089–2094. [CrossRef]
18. Saeed, H. Properties of polymer impregnated concrete spacers. *Case Stud. Constr. Mater.* **2021**, *15*, e00772. [CrossRef]
19. Caiyun, W.; Li, W.; Zhang, C.; Jinpeng, F. Effect of Protective Coatings on Frost Resistance of Concrete Structures in Northeast Coastal Areas. *IOP Conf. Ser. Mater. Sci. Eng.* **2019**, *678*, 012108. [CrossRef]
20. Liu, T.; Zhang, C.; Zhou, K.; Tian, Y. Freeze-thaw cycling damage evolution of additive cement mortar. *Eur. J. Environ. Civ. Eng.* **2021**, *25*, 2089–2110. [CrossRef]
21. Setzer, M.J.; Fagerlund, G.; Janssen, D.J. CDF test—Test Method for the Freeze-Thaw Resistance of Concrete-Tests with Sodium Chloride Solution (CDF). *Mater. Struct.* **1996**, *29*, 523–528. [CrossRef]
22. Matar, M.G.; Aday, A.N.; Srubar III, W.V. Surfactant properties of a biomimetic antifreeze polymer admixture for improved freeze-thaw durability of concrete. *Constr. Build. Mater.* **2021**, *313*, 125423. [CrossRef]
23. Ji, Y.; Zou, Y.; Ma, Y.; Wang, H.; Li, W.; Xu, X. Frost Resistance Investigation of Fiber-Doped Cementitious Composites. *Materials* **2022**, *15*, 2226. [CrossRef]
24. Stefanidou, M.; Kamperidou, K.; Konstandinidis, A.; Koltsou, P.; Papadopoulos, S. Rheological properties of biofibers in cementitious composite matrix. In *Advances in Bio-Based Fibre*; Elsevier: Amsterdam, The Netherlands, 2022; pp. 553–573. [CrossRef]
25. Oliveira, A.S.; Toledo Filho, R.D.; Rego Fairbairn, E.M.; Cappa de Oliveira, L.F.; Martins Gomes, O.F. Microstructural characterization of self-healing products in cementitious systems containing crystalline admixture in the short- and long-term. *Cem. Concr. Compos.* **2022**, *126*, 104369. [CrossRef]
26. García Calvo, J.L.; Sánchez Moreno, M.; Carballosa, P.; Pedrosa, F.; Tavares, F. Improvement of the Concrete Permeability by Using Hydrophilic Blended Additive. *Materials* **2019**, *12*, 2384. [CrossRef] [PubMed]
27. Park, B.; Choi, Y.C. Effect of healing products on the self-healing performance of cementitious materials with crystalline admixtures. *Constr. Build. Mater.* **2021**, *270*, 121389. [CrossRef]

28. Roig-Flores, M.; Moscato, S.; Serna, P.; Ferrara, L. Self-healing capability of concrete with crystalline admixtures in different environments. *Constr. Build. Mater.* **2015**, *86*, 1–11. [CrossRef]
29. Escoffres, P.; Desmettre, C.; Charron, J.P. Effect of a crystalline admixture on the self-healing capability of high-performance fiber reinforced concretes in service conditions. *Constr. Build. Mater.* **2018**, *173*, 763–774. [CrossRef]
30. Reddy, T.C.S.; Ravitheja, A. Macro mechanical properties of self-healing concrete with crystalline admixture under different environments. *Ain Shams Eng. J.* **2019**, *10*, 23–32. [CrossRef]
31. Zhang, C.; Lu, R.; Li, Y.; Guan, X. Effect of crystalline admixtures on mechanical, self-healing and transport properties of engineered cementitious composite. *Cem. Concr. Compos.* **2021**, *124*, 104256. [CrossRef]
32. Gojević, A.; Ducman, V.; Netinger Grubeša, I.; Baričević, A.; Banjad Pečur, I. The Effect of Crystalline Waterproofing Admix-tures on the Self-Healing and Permeability of Concrete. *Materials* **2021**, *14*, 1860. [CrossRef]
33. Lauch, K.S.; Desmettre, C.; Charron, J.P. Self-healing of concrete containing different admixtures under laboratory and long-term real outdoor expositions based on water permeability test. *Constr. Build. Mater.* **2022**, *324*, 126700. [CrossRef]
34. *CEN/TS 12390-9:2016*; Testing Hardened Concrete—Part 9: Freeze-Thaw Resistance—Scaling. CEN: Brussels, Belgium, 2016.
35. *CEN/TR 15177:2006*; Testing the Freeze-Thaw Resistance of Concrete—Internal Structural Damage. CEN: Brussels, Belgium, 2006.
36. *EN 480-11:2005*; Admixtures for Concrete, Mortar and Grout—Test Methods—Part 11: Determination of Air Void Characteristics in Hardened Concrete (EN 480-11:2005). CEN: Brussels, Belgium, 2005.
37. Elsalamawy, M.; Mohamed, A.R.; Abdel-latif, E.A. Performance of crystalline forming additive materials in concrete. *Constr. Build. Mater.* **2020**, *230*, 117056. [CrossRef]
38. *EN 1097-6:2013*; Tests for Mechanical and Physical Properties of Aggregates—Part 6: Determination of Particle Density and Water Absorption. CEN: Brussels, Belgium, 2013.
39. *ISO 9277:2022*; Determination of the Specific Surface Area of Solids by Gas Adsorption BET Meth. International Organization for Standardization: Geneva, Switzerland, 2022.
40. *EN 196-6:2019*; Methods of Testing Cement—Part 6: Determination of Fineness. CEN: Brussels, Belgium, 2019.
41. *EN 12350-2:2019*; Testing Fresh Concrete—Part 2: Slump-Test. CEN: Brussels, Belgium, 2019.
42. *EN 12350-6:2019*; Testing Fresh Concrete—Part 6: Density. CEN: Brussels, Belgium, 2019.
43. *EN 12350-7:2019*; Testing Fresh Concrete—Part 7: Air Content—Pressure Methods. CEN: Brussels, Belgium, 2019.
44. *EN 206:2021*; Concrete—Specification, Performance, Production and Conformity (EN 206:2013+A2:2021). CEN: Brussels, Belgium, 2021.
45. Lin, X.; Li, W.; Castel, A.; Kim, T.; Huang, Y.; Wang, K. A comprehensive review on self-healing cementitious composites with crystalline admixtures: Design, performance and application. *Constr. Build. Mater.* **2023**, *409*, 134108. [CrossRef]
46. Shetiya, R.K.; Elhadad, S.; Salem, A.; Fülöp, A.; Orban, Z. Investigation into the Effects of Crystalline Admixtures and Coatings on the Properties of Self-Healing Concrete. *Materials* **2024**, *17*, 767. [CrossRef]
47. *EN 12390-3*; Testing Hardened Concrete—Part 3: Compressive Strength of Test Specimens. CEN: Brussels, Belgium, 2019.
48. Cappellesso, V.G.; Petry, N.D.S.; Molin, D.C.C.D.; Masuero, A.B. Use of crystalline waterproofing to reduce capillary porosity in concrete. *J. Build. Pathol. Rehabil.* **2016**, *1*, 9. [CrossRef]
49. García-Vera, V.E.; Tenza-Abril, A.J.; Saval, J.M.; Lanzón, M. Influence of Crystalline Admixtures on the Short-Term Behaviour of Mortars Exposed to Sulphuric Acid. *Materials* **2019**, *12*, 82. [CrossRef] [PubMed]
50. Zha, Y.; Yu, J.; Wang, R.; He, P.; Cao, Z. Effect of ion chelating agent on self-healing performance of Cement-based materials. *Constr. Build. Mater.* **2018**, *190*, 308–316. [CrossRef]
51. Manhanga, F.C.; Rudžionis, Ž.; Ivanauskas, E.; Augonis, A. The investigations on properties of self-healing concrete with crystalline admixture and recycled concrete waste. *MATEC Web Conf.* **2022**, *364*, 05002. [CrossRef]
52. Zhu, Y.; Yang, Y.; Yao, Y. Autogenous self-healing of engineered cementitious composites under freeze–thaw cycles. *Constr. Build. Mater.* **2012**, *34*, 522–530. [CrossRef]
53. Wu, H.; Zhao, Y.; Chen, X.; Li, S.; Zhao, Y. Effect of crystalline admixture and superabsorbent polymer on the self-healing and mechanical properties of basalt fibre mortars. *J. Asian Archit. Build. Eng.* **2023**. [CrossRef]
54. Wang, R.; Hu, Z.; Li, Y.; Wang, K.; Zhang, H. Review on the deterioration and approaches to enhance the durability of concrete in the freeze–thaw environment. *Constr. Build. Mater.* **2022**, *321*, 126371. [CrossRef]
55. Ferrara, L.; Krelani, V.; Carsana, M. A "fracture testing" based approach to assess crack healing of concrete with and without crystalline admixtures. *Constr. Build. Mater.* **2014**, *68*, 535–551. [CrossRef]

Disclaimer/Publisher's Note: The statements, opinions and data contained in all publications are solely those of the individual author(s) and contributor(s) and not of MDPI and/or the editor(s). MDPI and/or the editor(s) disclaim responsibility for any injury to people or property resulting from any ideas, methods, instructions or products referred to in the content.

Article

Long-Term Effects of External Sulfate Attack on Low-Carbon Cementitious Materials at Early Age

François El Inaty [1,2], Bugra Aydin [1], Maryam Houhou [1,3], Mario Marchetti [1], Marc Quiertant [2,4] and Othman Omikrine Metalssi [1,*]

1. Unité Mixte de Recherche-Matériaux pour une Construction Durable (UMR MCD), Université Gustave Eiffel, Cerema, F-77454 Marne-la-Vallée, France; francois.el-inaty@univ-eiffel.fr (F.E.I.); bugraadn@gmail.com (B.A.); mhouhou@cesi.fr (M.H.); mario.marchetti@univ-eiffel.fr (M.M.)
2. Institut de Recherche de l'ESTP, F-94230 Cachan, France; mquiertant@estp.fr
3. Laboratoire d'Innovation Numérique pour les Entreprises et les Apprentissages au Service de la Compétitivité des Territoires (LINEACT) Cesi EA, 7527 Saint-Nazaire, France
4. Expérimentation et Modélisation pour le Génie Civil et Urbain (EMGCU), Université Gustave Eiffel, F-77454 Marne-la-Vallée, France
* Correspondence: othman.omikrine-metalssi@univ-eiffel.fr

Citation: El Inaty, F.; Aydin, B.; Houhou, M.; Marchetti, M.; Quiertant, M.; Omikrine Metalssi, O. Long-Term Effects of External Sulfate Attack on Low-Carbon Cementitious Materials at Early Age. *Appl. Sci.* **2024**, *14*, 2831. https://doi.org/10.3390/app14072831

Academic Editors: Mouhamadou Amar and Nor Edine Abriak

Received: 13 February 2024
Revised: 14 March 2024
Accepted: 26 March 2024
Published: 27 March 2024

Copyright: © 2024 by the authors. Licensee MDPI, Basel, Switzerland. This article is an open access article distributed under the terms and conditions of the Creative Commons Attribution (CC BY) license (https:// creativecommons.org/licenses/by/ 4.0/).

Abstract: Placed in a sulfate-rich environment, concrete reacts with sulfate ions, influencing the long-term durability of reinforced concrete (RC) structures. This external sulfate attack (ESA) degrades the cement paste through complex and coupled physicochemical mechanisms that can lead to severe mechanical damage. In common practice, RC structures are generally exposed to sulfate at an early age. This early exposition can affect ESA mechanisms that are generally studied on pre-cured specimens. Moreover, current efforts for sustainable concrete construction focus on replacing clinker with supplementary cementitious materials, requiring a 90-day curing period, which contradicts real-life scenarios. Considering all these factors, the objective of this study is to explore ESA effects at an early age on cement-blended paste samples using various low-carbon formulations. The characterization techniques used demonstrated that the reference mix (100% CEM I) exhibits the weakest resistance to sulfate, leading to complete deterioration after 90 weeks of exposure. This is evident through the highest mass gain, expansion, cracking, formation of ettringite and gypsum, and sulfate consumption from the attacking solution. Conversely, the ternary mix, consisting of CEM I, slag, and metakaolin, demonstrates the highest resistance throughout the entire 120 weeks of exposure. All the blended pastes performed well in the sulfate environment despite being exposed at an early age. It can be recommended to substitute clinker with a limited quantity of metakaolin, along with blast furnace slag, as it is the most effective substitute for clinker, outperforming other combinations.

Keywords: external sulfate attack; early age; supplementary cementitious materials; physicochemical behavior; long term durability; low carbon cement

1. Introduction

In general, two types of sulfate attacks are distinguished. The first is known as internal sulfate attack (ISA, or ISR for internal sulfate reaction, or DEF for delayed ettringite formation), where sulfate originates from raw materials [1,2]. The second is external sulfate attack (ESA), and the source of sulfate, in this case, is the surrounding environment (groundwater, soil, seawater, etc.) [3,4]. When it comes to external sources, sulfate, a pervasive agent contributing to the deterioration of concrete structures, is inevitable due to its presence in various external sources where constructing RC structures is a necessity. Upon contact with RC structures, sulfate ions penetrate the cementitious matrix through the porous network, initiating chemical reactions with hydrated cementitious materials to produce ettringite and/or gypsum [5–9]. This process induces internal stresses, leading to expansion and cracking in the structure. Consequently, the structural integrity weakens, permeability increases, and a feedback loop ensues, facilitating further migration of sulfate ions. This phenomenon involves complex interactions among physical, chemical, and mechanical factors [10–14]. Despite ESA having historical roots dating back to 1818 [15], it remains a complex issue, influenced by many factors, such as cation type (calcium, sodium, or magnesium sulfate), concentration, pH of the surrounding environment, the type of cement used, and many other [16]. A study by Wu et al. [17] compared the degradation effects of blended cement under sodium sulfate and magnesium sulfate attack. Their findings revealed distinct mechanisms of attack, with magnesium sulfate causing significantly more damage compared to sodium sulfate.

Over the years, considerable efforts have been invested in developing methods for testing ESA resistance of all types of binder and concrete. However, each method comes with its strengths and limits, and many are criticized for their dissimilarity with respect to real-world field exposure conditions. Notably, various testing standards, such as ASTM C 1012, the Asian method, the Swiss method, and CSA A 3004-C8, employ high concentrations of sodium sulfate (greater than 30 g/L) [18] to generate an infinite source of sulfate and accelerate the test. A study conducted by Biczok [19] revealed that the preferential formation of ettringite occurs when a low concentration of sulfate is selected, while higher sulfate concentrations lead to an increased production of gypsum. These elevated concentrations, coupled with other testing conditions like extreme temperatures and harsh dry and wet cycles, are intended to accelerate the attack for concrete durability testing against sulfate. However, it is crucial to acknowledge that such conditions may alter the characteristics of the ESA, raising questions about the representativeness of these methods.

In addition, if not prefabricated, on-site cast concrete is directly exposed to the surrounding environment from an early age [20,21]. A recent investigation compared the resistance of prefabricated and cast-in-site concrete against ESA, revealing that prefabricated concrete demonstrates greater resistance to ESA [22]. Curing is recognized for enhancing the properties of structural concrete, including microstructure, mechanical strength, and durability, by promoting the hydration of cementitious materials, reducing creep and shrinkage, and providing protection against environmental aggressions [23]. However, real practices consist of exposing concrete to sulfate ions in certain environments at an early age. Most studies and standards focus on cured samples, typically after 28 or 90 days of curing before attacking the samples, leaving unanswered questions about the impact of ESA on RC structures during their early stages. A recent study [20] investigated the degradation mechanisms induced by ESA on both cured and uncured cement pastes. Surprisingly, pre-cured cement pastes exhibited rapid degradation, although sulfate ion ingress did not show significant differences when comparing pre-cured samples to the uncured ones. In contrast, Li et al. [24] conducted a study suggesting a minimum initial curing period of 14 days for cement-based materials exposed to sulfate. Their findings indicated that a longer initial curing period correlated with improved resistance to sulfate in the cement-based material.

Speaking of common practices while moving to the current trend to resolve environmental issues, one of the most effective and widely employed strategies today to jointly mitigate the substantial CO_2 emissions from the cement industry and to improve cementitious materials' durability is the partial substitution of clinker with supplementary cementitious materials (SCM) [25]. Due to their pozzolanic effects, these materials not only facilitate the formation of additional C-S-H and reduce the porosity size but also serve as a source of alumina, promoting the development of aluminum-containing phases [26]. The changes in cementitious materials composition significantly influence the mechanisms of ESA. In this context, research indicates that the incorporation of SCM enhances the durability of specimens under sulfate attack by reducing the C_3A content [27]. Additionally, numerous studies have demonstrated that the utilization of slag, metakaolin, silica fume, and fly ash improves durability when it comes to ESA [28–33]. In a recent investigation conducted by Miah et al. [34], specimens incorporating fly ash and slag exhibited enhanced mechanical properties after 90 days. Interestingly, these specimens demonstrated similar properties at

28 days compared to those made with ordinary Portland cement (OPC). Furthermore, these findings corresponded with observations of porosity and capillary water absorption. This suggests that the incorporation of SCM necessitates a 90-day curing period to exhibit their benefits. However, this leaves a gap regarding the mechanisms of ESA on materials cured in sulfate (exposed to sulfate at an early age) since previous studies focus on pre-cured samples when it comes to ESA.

Considering all the factors mentioned above, there is a gap in understanding the impact of ESA on samples exposed to sulfate ions at an early age over an extended period, utilizing an adjusted concentration of sulfate. Additionally, there is a need to investigate how ESA affects the resistance of low-carbon cementitious matrices, aligning with the developed low-carbon cement types. For this, this study aims to address two main objectives. Firstly, this study seeks to gain a comprehensive understanding of the long-term effects of ESA on cement paste samples exposed to sulfate ions at an early age. The second objective is to evaluate the resistance of SCM, namely fly ash, blast furnace slag, and metakaolin, under such attack conditions using low-carbon formulations. To achieve this, OPC pastes, binary, ternary, and quaternary blended cement pastes were utilized to address the objectives of the study. The samples underwent exposure to a sodium sulfate solution with a concentration of 15 g/L from an early age, for a duration of three years. This concentration was selected to both create an infinite source of sulfate and to shorten the duration of the test as much as possible while using a relatively low concentration of sulfate. Mass loss, expansion, and physicochemical changes were periodically monitored throughout the experiment, supplemented by visual inspections. The characterization techniques included thermogravimetric analysis, Fourier transform infrared spectroscopy, and the water-accessible porosity test. Furthermore, Raman spectroscopy was employed to monitor the sulfate content in the solutions in which cementitious material samples were immersed.

2. Materials and Methods

2.1. Materials and Mixes Design

For this study, six cementitious mixes were formulated, as detailed in Table 1. The selection of these formulations closely adhered to the specifications outlined in standards NF EN 197-1 and NF EN 197-5 [35,36], with a particular emphasis on chemical criteria. A water-to-binder (W/B) ratio of 0.55 was adopted across all mixes. This relatively high ratio was chosen to increase the porosity and, as a consequence, to accelerate the sulfate attack while mitigating potential segregation issues.

Table 1. Cementitious materials blend design.

	CEM I (%)	Fly Ash (%)	Blast Furnace Slag (%)	Metakaolin (%)
P1	100	-	-	-
S1	55	45	-	-
S2	55	-	45	-
T1	55	15	30	-
T2	55	-	35	10
Q1	55	15	20	10

The primary materials employed in these mixes were CEM I 52.5 N CE CP2 NF, manufactured by EQIOM, with clinker phases (calculated using the Bogue formula [37]) including C_3S at 56.5%, C_2S at 15.8%, C_3A at 5.0%, and C_4AF at 11.6%. Additionally, fly ash and blast furnace slag were incorporated, and their chemical compositions, as provided by the manufacturer, are detailed in Table 2, along with the chemical composition of the used CEM I. Furthermore, metakaolin from BASF MetaMax, described by the manufacturer as 100% calcined kaolin, was also included in the formulations. Those supplementary cementitious materials are the most used substitutes for clinker.

Table 2. Chemical composition of the used materials as given by the manufacturers.

Components	CEM I (w%) [1]	Fly Ash (w%) [1]	Blast Furnace Slag (w%) [1]
SiO_2	20.38	70.83	35.71
Al_2O_3	4.30	24.36	10.65
Fe_2O_3	3.80	2.24	0.45
TiO_2	0.24	1.48	0.73
MnO	0.08	0.05	0.23
CaO	62.79	0.06	43.32
MgO	1.25	0.23	3.97
SO_3	3.46	-	3.06
K_2O	0.73	0.64	0.45
Na_2O	0.35	0.1	0.16
P_2O_5	-	0.05	0.02
S^{2-}	Traces	-	-
Cl^-	0.05	-	-
Loss of ignition	2.54	-	-
Free lime	1.39	-	-

[1] Weight percent.

2.2. Mixing, Casting, and Exposure Procedures

The study was carried out on $40 \times 40 \times 160$ mm^3 prismatic specimens. The preparation protocol of specimens closely adheres to the NF EN 196-1 standard [38] and unfolds as follows: initially, water and cementitious materials are mixed at a low speed of 100 rpm for a duration of 60 s. Subsequently, the speed is increased to 300 rpm for an additional 30 s. Following this, the mixture is allowed to sit undisturbed for 90 s, with the edges of the mixer bowl scraped during the initial 20 s of this interval. The subsequent step involves a second round of high-speed (300 rpm) mixing lasting for 60 s. Various adjustments in mixing speed were used to guarantee the homogenization of the cementitious materials. Employing different mixing speeds helps minimize losses and promotes an even dispersion of components, thereby preventing uneven distribution. Such adjustments are crucial for maintaining the properties of the resulting paste. The resulting paste is poured in one layer into the molds without vibration, but rather, is subjected to four external shocks using a hammer. After a curing period of 24 h, demolding occurs, and the specimens are immediately immersed in sodium sulfate solutions with a concentration of 15 g/L (corresponding to approximately 10.14 g/L of sulfate ions). The water-to-solid volume ratio is maintained at 8 to simulate exposure to an almost infinite sulfate source. To ensure uniform exposure, the specimens are positioned on plastic supports in the attacking solution. The entire setup is preserved at a constant temperature of 20 °C and carefully monitored (expansion and mass gain for the samples and Raman spectroscopy to check the variation of sulfate concentration in the attacking solutions, refer to Section 2.3) and characterized throughout the study. The sodium sulfate solution is renewed weekly for the initial three months, bi-weekly for the subsequent six months, and then monthly, thereafter, to ensure exposure conditions equivalent to an infinite sulfate source, especially since the used concentration of sulfate is relatively low.

2.3. Characterizations Methods

In conjunction with the chemical and microstructural analyses detailed below, this study included mass and expansion monitoring, supplemented by visual observations, all performed at two-week intervals (for the mass and expansion variations) throughout the 120-week experimental phase. The expansion measurement assesses the axial expansion of the prismatic sample, with two axial bolts (or pins) fixed at the two extremities of the sample during casting. The measurements are recorded using an extensometer. The timing of the chemical and microstructural analyses (except for Raman spectroscopy) was determined based on the evolution of patterns of mass and expansion. Thermogravimetric analysis

(TGA) and Fourier transform infrared spectroscopy (FTIR) were conducted on powders extracted from the specimens at depths of 1 (surface), 2, 5, and 15 mm after drying them at a controlled temperature of 55 ± 5 °C for a period of 48 h. The extracted dry material was carried out utilizing the Accutom-100 microtome and a FRITSCH Pulverisette 6 planetary ball mill operating at a speed of 350 rpm for a duration of 2 min.

2.3.1. Thermogravimetric Analysis

TGA was employed as a chemical characterization technique, utilizing the NETZSCH STA 449 F1 apparatus (NETZSCH, Selb, Germany). The analysis spanned a temperature range of 25 to 1250 °C, with a heating rate of 10 °C per minute, all conducted under an inert nitrogen atmosphere. The derivative of the thermogravimetric analysis curve (DTG) was obtained to easily identify crucial mass losses occurring during the heating process.

2.3.2. Fourier Transform Infrared Spectroscopy

FTIR was employed as a chemical characterization technique to identify potential formations of ettringite and gypsum while making comparisons with the original chemical compounds present in the specimens. The Thermo Fisher Scientific Nicolet iS50 spectrometer (Waltham, MA, USA) was utilized for FTIR analysis, covering a spectral range between 400 and 4000 cm^{-1}. The tests were conducted using the integrated diamond ATR.

2.3.3. Raman Spectroscopy

Raman spectroscopy serves as a non-destructive chemical characterization method, including for civil engineering materials. In this study, it was employed to track sulfate consumption in the attacking solutions as an indicator of the extent of the ESA on each mix. Analyses were performed every 30 days (with every solution's renewal). The utilized equipment was an iRaman from BWTek (Newark, DE, USA), operating with a 50 mW laser at 532 nm, featuring a spectral range of 150–4000 cm^{-1} and a spectral resolution of 4 cm^{-1}, using a BAC101 immersion probe.

2.3.4. Water Porosity

The water porosity test, a microstructural assessment, was conducted in accordance with NF P18-459 standard [39]. Samples measuring $40 \times 40 \times 40$ mm^3 were cut from the original $40 \times 40 \times 160$ mm^3 prisms. These samples underwent oven drying at 105 °C until achieving a stable mass. Subsequently, they were placed into a desiccator and subjected to a vacuum for 4 h. The next step involved immersing the samples in water under a pressure of 25 mbar for 48 h. Throughout this process, mass measurements were recorded at various stages: the first being the mass of the dried samples, the second for the mass of the water-saturated samples, and the third capturing the mass of the water-saturated samples submerged in water. A precision scale, with an accuracy of 0.01 g, was employed, and the tests were also conducted at room temperature (20 °C $\pm$ 1 °C).

3. Results

3.1. Expansion

Expansion measurements were periodically recorded over the full 120-week duration of the study. The expansion evolution of the samples over time is presented in Figure 1. The provided values are derived from three measurements conducted on distinct samples.

Figure 1 clearly shows that the expansion behavior of the reference mix P1 differs significantly from that of the binary, ternary, and quaternary blended pastes, particularly after week 54. The expansion rate of P1 can be divided into five phases. These phases closely resemble those documented in the current literature [6]. In the initial phase, spanning the first 4 weeks, there is a substantial expansion, reaching 0.08%. The second phase involves a slower but continuous expansion, reaching 0.12% after 54 weeks of sulfate exposure. The third phase extends from week 54 to 72, characterized by a rapid and pronounced expansion, peaking at 0.29%. The fourth phase, between weeks 72 and 84, is characterized by stability, remaining around 0.3%. The fifth phase occurs after that period, reaching an expansion of 0.95% by week 90, at which point the samples were completely degraded. The behavior of the blended pastes (S1, S2, T1, T2, and Q1) exhibited two major phases globally (with three fluctuations in general). The first phase, similar to the reference mix P1, occurred during the initial 4 weeks, where the expansion was around 0.05%. Subsequently, insignificant changes were recorded until week 120, when the expansion reached 0.145%. Throughout this phase, three minor fluctuations were observed. The first fluctuation extended from week 4 to week 54, during which the expansion increased slowly. The second phase persisted until week 70, remaining almost stable before experiencing a subsequent increase at a higher rate.

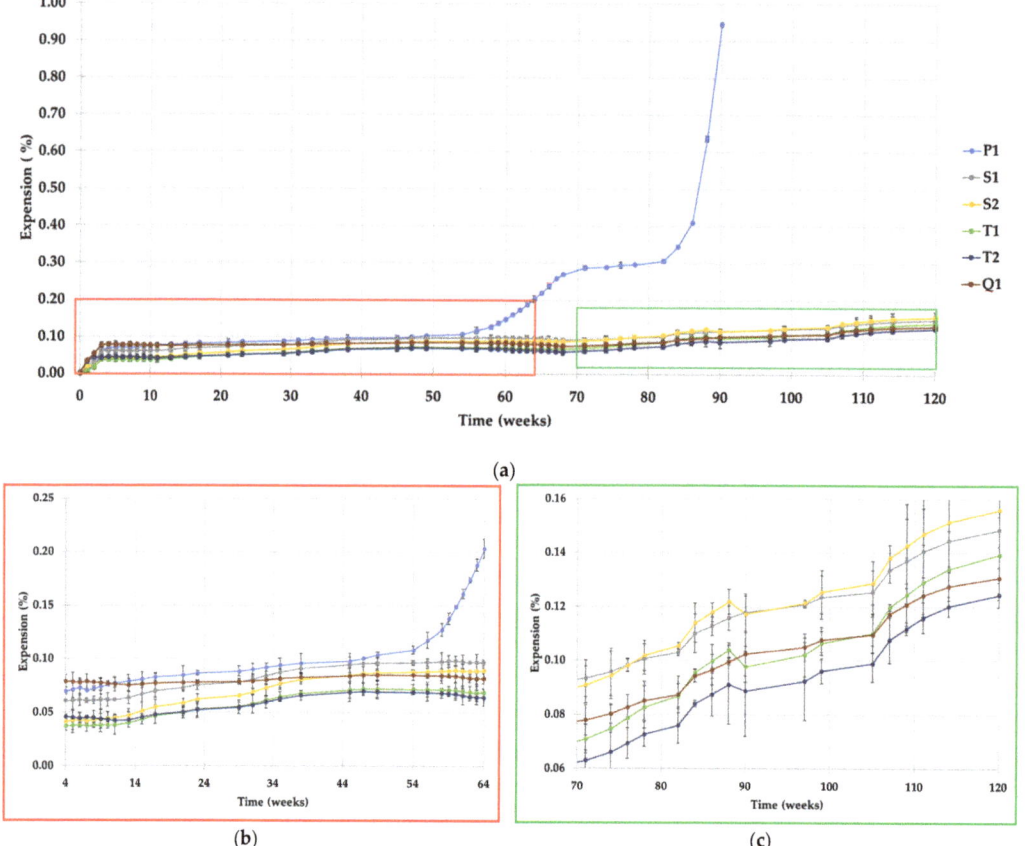

Figure 1. Longitudinal expansion of the six mixes over time during the ESA: (**a**) Results of the measurements over the total duration of the test; (**b**) Zoom of the results over the first 64 weeks; (**c**) Zoom of the results from week 70 to week 120.

Based on the description of the expansion rates provided earlier, a common pattern emerged between the expansion rates of the reference mix and the blended ones. After each increase in the expansion rates, a period of stability is observed, followed by a subsequent increase. This pattern is likely attributed to the formation of ettringite and/or gypsum in the pores, leading to internal stresses and causing cracking [20,40] (resulting in an increase in expansion). Subsequently, these products begin to fill the formed cracks (resulting in the stability of the expansion) until they recreate internal stresses and induce new cracks (leading to an increase in expansion). This phenomenon gains further confirmation when examining the continuous mass gain of the samples (refer to Section 3.2).

When comparing the mixes among themselves, the reference mix exhibits the least favorable behavior in terms of expansion and overall durability against ESA. While the other mixes maintained satisfactory expansion rates for 120 weeks, the reference mix P1 expanded by 0.95% after only 90 weeks and deteriorated thereafter. Regarding the blended pastes, as depicted in Figure 1 and considering the scale and low expansion rates, it is evident that they exhibit almost similar behavior. Upon closer examination, mix T2 (composed of CEM I, blast furnace slag, and metakaolin) stood out with the lowest expansion rates. These findings could suggest that the combination of blast furnace slag and a low quantity of metakaolin contributes to the enhanced durability of the overall mix (as previously shown in a study considering 10 to 15% of metakaolin replacement of cement [29]). This is particularly evident when comparing the behavior of other blended mixes that lack this specific combination of slag and metakaolin.

3.2. Mass Changes and Visual Inspection

The mass of all samples was regularly measured over 120 weeks. The mass evolution over time is illustrated in Figure 2 for the six different mixes. Each reported value is the average of three measurements from three different samples.

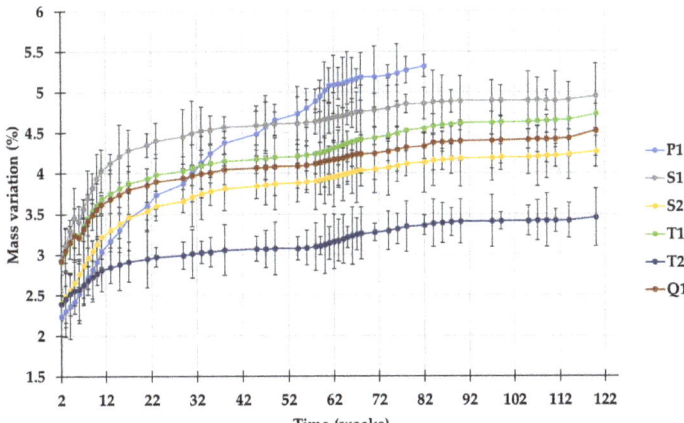

Figure 2. Mass of the six mixes during the ESA over time.

Figure 2 indicates that all samples experienced a mass increase throughout the entire attack period. This phenomenon is expected, as ESA leads to a mass gain, as shown by previous studies [41–43]. However, the rate of mass gain varied among the mixes. Binary (S1 and S2), ternary (T1 and T2), and quaternary (Q1) blended pastes exhibited similar rates of mass gain, distinct from the reference mix (P1) composed of 100% clinker. Blended pastes demonstrated a significant mass gain in the initial 35 weeks, stabilizing thereafter with a slight increase up to week 120. In contrast, the mass gain kinetics of the reference mix (P1) were different from the other formulations, showing a rather regular mass gain during the 62 weeks and a beginning of stabilization afterward. Mass measurements for P1 were terminated at week 82 due to the samples' full deterioration caused by ESA.

The initial 12 weeks witnessed substantial mass gain for all mixes, attributable to various factors. The samples, immersed in sulfate at an early age (24 h after casting), experienced both the hydration of cementitious pastes, leading to the formation of hydrated products, and water absorption [42,44]. Additionally, this gain can be associated with the chemical reaction between hydrated cementitious materials and sulfate, resulting in the formation of ettringite and gypsum within micro-pores and cracks, contributing to the observed mass gain [16,40,45]. Following this period, marking the completion of the hydration of cementitious materials (lasting 90 days, notably longer for blended pastes compared to 28 days for P1, as discussed by Miah et al. [34]), the blended mixes continued to experience mass gain but at reduced rates before almost stabilizing (having however minor gain), as discussed earlier. At this stage, with hydration nearly complete and the samples saturated, the observed mass gain could be attributed to the formation of gypsum and ettringite resulting from the chemical reaction between the hydrated products and sulfate. It is important to note that 90 days after the fabrication (upon completion of hydration), a significant reduction in porosity and capillary water absorption in the blended pastes is observed, according to previous studies [34]. This decrease in porosity contributes to stabilizing the mass gain. The above-explained results highlight that the reference mix (P1) exhibited the least favorable behavior in terms of mass gain and overall durability, as evidenced by its degradation after 90 weeks of exposure. In contrast, the other mixes, which incorporate SCM to partially substitute the clinker, demonstrated a more robust performance, maintaining their structural integrity even after 120 weeks of sulfate exposure.

Upon comparing the mass gains of the blended pastes, it becomes evident that the ternary mix (T2), composed of CEM I, blast furnace slag, and metakaolin, exhibited the lowest mass gain from the initiation of the sulfate attack. In contrast, the mix containing CEM I and fly ash (S1) displayed the highest mass gain among the blended mixes. This, coupled with the mass gain rates of the other blended mixes, could imply that the inclusion of blast furnace slag and metakaolin, within specified replacement percentages, enhanced the overall performance of the mixes.

In terms of visual inspection, the reference mix P1 exhibited the highest visual aging rate, as depicted in Figure 3. After 60 weeks of exposure, P1 underwent a significant color change compared to the minor changes observed in the other mixes, which exhibited similar behavior in terms of color alteration and crack development. Furthermore, after 95 weeks of exposure, mix P1 experienced complete degradation, while the rest of the mixes only displayed some cracking on the edges of the specimens. Additionally, even after 120 weeks, the blended mixes did not exhibit more than insignificant cracking, as illustrated with the mix Q1.

Figure 3. Visual inspection of P1 (reference mix) and Q1 after 4, 60, 95, and 120 weeks of exposure (the red circles highlight areas where cracks are observed in mix Q1).

In general, the expansion rates and the steps of increase aligned with the mass gain rates and visual observations. During the initial weeks, as the specimens underwent the curing process in a sulfate-enriched solution (initiated 24 h after casting), there was ongoing hydration, water absorption, and the formation of ettringite and gypsum due to sulfate (contributing to mass gain, internal stresses leading to expansion, and observable visual changes) [42]. Following this initial period (4 weeks for the reference mix P1 and approximately 12 weeks for the blended pastes), subsequent changes are primarily attributed to the reaction between sulfate and the hydrated products, leading the consequences of the ESA [46,47].

Based on those results, including expansion, mass variation, and visual observations, analyses such as TGA, FTIR, and water porosity were conducted at three key time points: 24 h after casting (Ti0), 8 weeks into the initiation of ESA (Ti1), and at 80 weeks (Ti2).

3.3. Thermogravimetric Analysis

TGA was conducted on the six mixes at two different time points: 8 (Ti1) and 80 (Ti2) weeks of immersion in the sulfate solution. The analysis included measurements on materials extracted at various depths within the $40 \times 40 \times 160$ cm^3 specimens, specifically at 1 (surface), 2, 5, and 15 mm.

According to some previous research [48,49], when examining the DTG curve, the initial peak, observed between 30 °C and 200 °C, signifies the decomposition of ettringite, C-S-H, and free water. Furthermore, the presence of monosulfoaluminates (AFm) was indicated by a peak at around 190 °C. The dihydroxylation process of portlandite (Ca(OH)$_2$) manifests as the third peak between 450 °C and 550 °C, while the decarbonation of calcite (CaCO$_3$) was responsible for the peak observed between 650 °C and 820 °C.

When investigating ESA, a careful analysis of portlandite consumption is imperative, as it provides valuable insights into the progression of this attack [16]. Additionally, AFm serves as a source of alumina during the sulfate attack, contributing to the formation of ettringite [50]. The results related to portlandite content are presented in Figure 4.

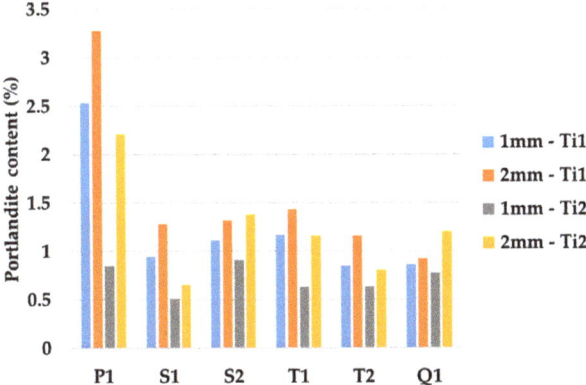

Figure 4. Portlandite content (%) of the six mix samples at 1 mm and 2 mm depths after being submerged in sulfate for 8 weeks (Ti1) and 80 weeks (Ti2).

Upon a general examination of portlandite content, carried out using the TGA test, it can be shown that the progression of the attack at 80 weeks overtakes that observed at 8 weeks, aligning with expectations. A more detailed analysis comparing depths at 1 mm and 2 mm reveals a heightened aggression of the attack at the initial 1 mm, in direct contact with sulfate. This is notably reflected in the portlandite content at these respective depths. Furthermore, the reference mix P1 exhibits a significantly higher overall portlandite content compared to the blended mixes. This is attributed to the absence of the pozzolanic effect of SCM, which typically favors the formation of C-S-H [34,51].

From the data presented in Figure 4, it becomes evident that the reference mix P1 exhibits weaker durability against sulfate attack. This conclusion is drawn from the higher portlandite loss observed in this mix, not only at different depths (1 and 2 mm) but also at different time points (Ti1 and Ti2). In the initial 8-week step, the portlandite content increased from 2.5% to 3.3% at depths of 1 and 2 mm, respectively. Subsequently, at Ti2, the portlandite content further increased from 0.85% at 1 mm to 2.2% at 2 mm. This consistent increase in portlandite content indicates a more pronounced degradation of the reference mix under sulfate attack conditions. However, when examining the blended mixes, their behavior in terms of portlandite loss appears to be quite similar. Considering the ternary mix T2, characterized by the best resistance based on the previously discussed results, its portlandite content, for instance, increased marginally from 0.63% at 1 mm depth to 0.8% at 2 mm depth at time Ti2, representing a minimal loss of 0.17%. This indicates a more robust performance of the blended mixes, particularly exemplified by mix T2, against sulfate attack conditions compared to the reference mix.

Analyzing the first peak of the DTG curve (refer to Figure 5), it is evident that the content of ettringite, C-S-H, and free water at all depths of the reference mix P1 is greater than in other mixes. Considering that all mixes underwent the same conditions and drying procedures and acknowledging that the C-S-H content is higher in the blended pastes [51], it can be concluded that the ettringite content in the reference mix is elevated, extending even to a depth of 15 mm. This indicates not only the poor resistance of P1 when it comes to ESA but also suggests that the attack has reached the core of the reference mix. In contrast, the attack remains superficial in the blended mixes even after 120 weeks. This conclusion is further supported by examining the AFm content (peak around 190 °C on the DTG curve), which does not exist in the reference mix, even at 15 mm, but is present in the other mixes at 5 and 15 mm depths. To validate these findings, an FTIR test is conducted and discussed in the following section.

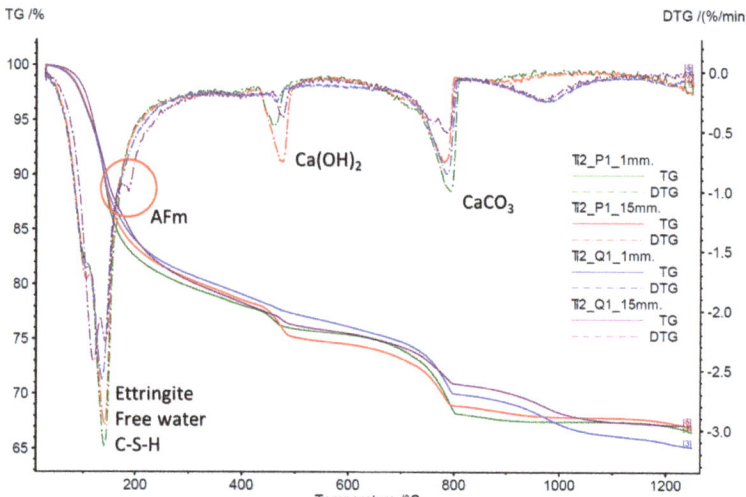

Figure 5. TG and DTG curves of the reference mix (P1) and of Q1 at 1 mm and 15 mm depths after 80 weeks of immersion in sulfate solution.

3.4. Fourier Transform Infrared Spectroscopy

FTIR analysis was conducted on the whole six mixes after 8 (Ti1) and 80 (Ti2) weeks of immersion in the sulfate solution at 1, 2, 5, and 15 mm depths.

In accordance with findings from previous studies [52–54], various chemical compounds and phases within the cementitious samples can be identified through FTIR. Portlandite, characterized by O-H stretching, is identified with a peak at 3642 cm^{-1}. The compounds AFm, ettringite, and gypsum, exhibiting S-O vibration, manifest by the peak of 1102 cm^{-1}. The presence of water is indicated by a broad band at the peak of 3373 cm^{-1} attributed to the stretch mode of the O-H bond. Ettringite, distinguished by S-O vibration, can be detected at the peak of 619 cm^{-1}, while gypsum shows up at the band of 1685 cm^{-1}. These distinctive peaks and bands provide valuable insights into the chemical composition and phases present in the cementitious samples, contributing to a comprehensive understanding of their behavior under ESA.

The analysis of the FTIR spectra presented in Figure 6a, which depicts the chemical composition of the reference mix P1 at 1, 2, and 5 mm after 80 weeks of exposure to sulfate, reveals a significant attack at 1 mm and a diminishing impact with increasing depth. This observation is distinguished by examining the intensity of peaks corresponding to gypsum (at 1685 cm^{-1}), ettringite, gypsum, and AFm (AFm's presence is questionable based on TGA results discussed in Section 3.3) at 1102 cm^{-1}, as well as ettringite at 619 cm^{-1}. Figure 6c, representing the same mix at 1 mm depth but comparing 8 weeks of exposure to 80 weeks, demonstrates that the attack has progressed and become more pronounced after 80 weeks of exposure. To assess the impact of SCM on enhancing durability against ESA, Figure 6d illustrates P1 and the quaternary mix Q1 at 1 mm depth after 80 weeks of exposure, highlighting a significant difference. Notably, the formation of ettringite and gypsum in the reference mix is substantial, showcasing the effectiveness of substituting clinker with SCM at specific percentages. This aligns with the overarching findings of the study. Further insights can be gained by examining Figure 6b, which compares the degree of formation of expansive products at different depths of the ternary mix T1 and the mix P1 shown in Figure 6a.

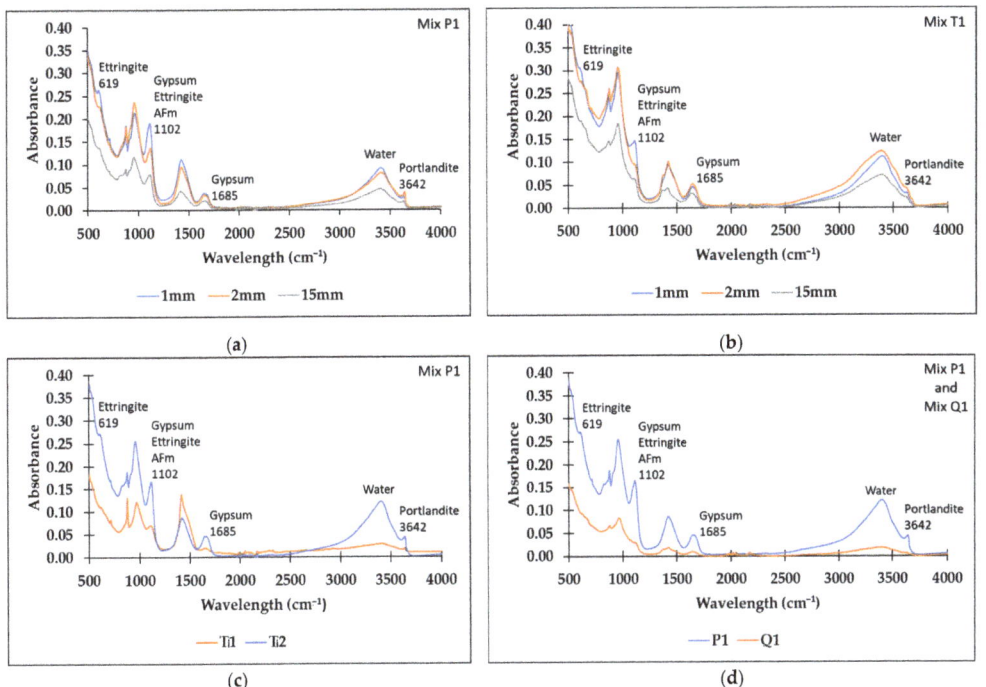

Figure 6. FTIR spectra, showcasing (**a**) the reference mix (P1) and (**b**) mix T1 at various depths after 80 weeks of exposure to sulfate. Additionally, it includes (**c**) mix P1 at 1 mm depth after 8 and 80 weeks of sulfate exposure, and (**d**) mix P1 and mix Q1 at 1 mm depth after 80 weeks of contact with sulfate.

3.5. Raman Spectroscopy

Raman spectroscopy was employed to analyze the solutions before each renewal (every 30 days), allowing for the monitoring of sulfate consumption.

According to previous studies [55,56], the presence of sulfate is identified by a peak found between 980 and 1160 cm^{-1}. Moreover, water is identified by the large peak found around 3400 cm^{-1} due to the vibration stretching mode of the O-H bond.

The results of the Raman spectroscopy, illustrated in Figure 7 and conducted at each renewal of the solutions, reveal the persistent presence of sulfate, indicated by its peak, which is never fully consumed by the samples. Interestingly, the intensity of the peak varies among the solutions and depends on the mix used to prepare the samples in contact with those solutions. The attacking solution in contact with the reference mix shows the lowest sulfate concentration after 30 days, suggesting that the reference mix consumes more available sulfate than the other mixes, up to the point that the sulfate consumption of the solution, considered as an infinite source of sulfate, become detectable. Moreover, the solution in contact with the ternary mix T2 exhibits the highest sulfate concentration after exposure, indicating that this mix has the highest resistance to sulfate. This aligns with the findings of the previous characterization tests discussed earlier. It also implies that the detection of ESA can be conducted by spectral monitoring of the sulfate solution without having to go through as many sample characterizations.

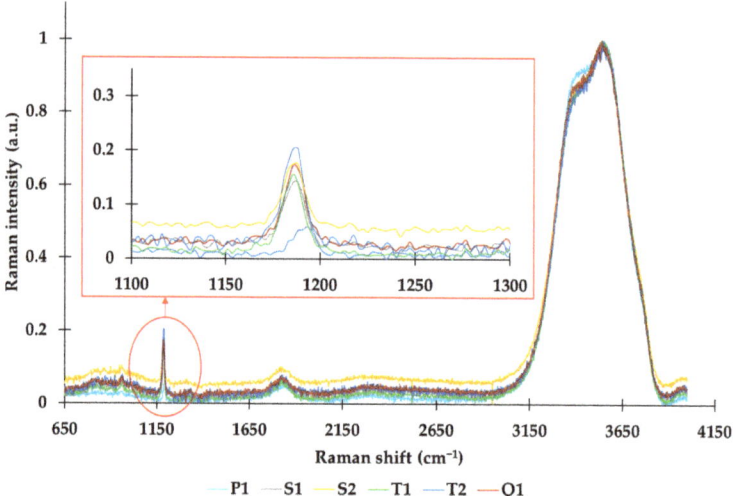

Figure 7. Raman specters of the sulfate solutions that are in contact with all the mixes after 80 weeks.

3.6. Water Porosity

A water porosity test was performed on all the mixes after 8 (Ti1) and 80 (Ti2) weeks of immersion in the sulfate solution in addition to 24 h after casting (Ti0).

As illustrated in Figure 8, the initial porosity of the samples was notably high, approximately around 51% at time Ti0, consistent with select a W/B ratio of 0.55. A previous study conducted by Nehdi et al. [57] demonstrated that increasing the W/B ratio leads to an increase in total porosity. The porosity then decreased by approximately 8% when cured in water (refer to Ti2 Ref) due to the ongoing hydration process. The significant initial porosity can be attributed to the fact that the samples were poured into the molds without vibration, as mentioned in Section 2.2. Additionally, it is evident that the porosity varies among the samples, and its evolution differs. This variability is attributed to differences in workability resulting from the substitution of clinker in the various mixes [58].

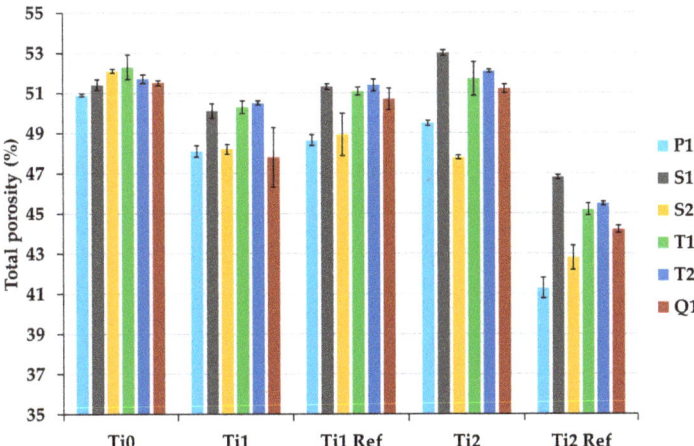

Figure 8. Total water porosity (%) of all the mixes 24 h after fabrication (Ti0), 8 (Ti1), and 80 (Ti2) weeks of being in contact with sulfate and in water.

Comparing water porosity results at 8 weeks between mixes exposed to sulfate (Ti1) and mixes cured in water (Ti1 Ref) reveals interesting trends. Notably, the reference mix, P1, showed slightly higher porosity in the water-cured environment (~49%) compared to its sulfate-exposed counterpart (~48%). A similar pattern was observed in other mixes, including S1, S2, T1, T2, and Q1, where slightly higher porosity values were seen underwater curing than with sulfate exposure. This is attributed to the formation of ettringite and gypsum due to sulfate exposure. The most significant increase in porosity occurred in S1, where the sulfate-exposed mix (~50%) surpassed the water-cured mix (~51%) by 1.2 percentage points. This aligns with the expansion rates and mass gain of mix S1, which exhibited the highest values among all mixes in the early stages of the attack. Similar findings have been documented by other studies, which have compared the porosity of mixes containing 100% CEMI with those incorporating supplementary cementitious materials [57].

However, examining water porosity results after 80 weeks of sulfate exposure (Ti2) and comparing them with mixes cured in water for the same duration (Ti2 Ref) reveals different trends when compared to Ti1 and Ti1 Ref. The reference mix (P1) demonstrated higher porosity in the sulfate-exposed environment (49.5%) compared to its water-exposed counterpart (~41%). This unexpected trend extended to other mixes, including S1, S2, T1, T2, and Q1, where sulfate-exposed mixes exhibited higher porosity values than their water-exposed counterparts. The most important increase in porosity was observed in P1, with the sulfate-exposed mix exceeding the water-cured mix by 8.2 percentage points. This is likely due to the formation of cracks due to ESA. These results are in accordance with all the previous findings of this study, including extensive mass and expansion gain, high formation of ettringite and gypsum, as well as visual deterioration of the reference mix. Moreover, the blended mixes exhibited similar behavior in the long-term exposure to sulfate, as previously demonstrated, with good values regarding porosity at this stage, with mix S2 leading. However, the difference between the blended mixes during the water porosity test is minor.

The water porosity results are valuable for comparing two time-points. However, this parameter cannot be considered a major durability indicator concerning ESA. This is because there can be significant fluctuations in the porosity. These fluctuations may be attributed to various phenomena, including the pore-filling effect by the formation of ettringite and/or gypsum, cracks opening due to the crystallization pressure of the expensive products coming from ESA, subsequent sealing of these cracks, the creation of new cracks, and so forth.

3.7. Results Overview Illustration

From the findings of this study, a visual representation was generated to describe the obtained results. The illustration provides a concise overview of the key outcomes and insights derived from the comprehensive analyses conducted in this study. It is presented in Figure 9.

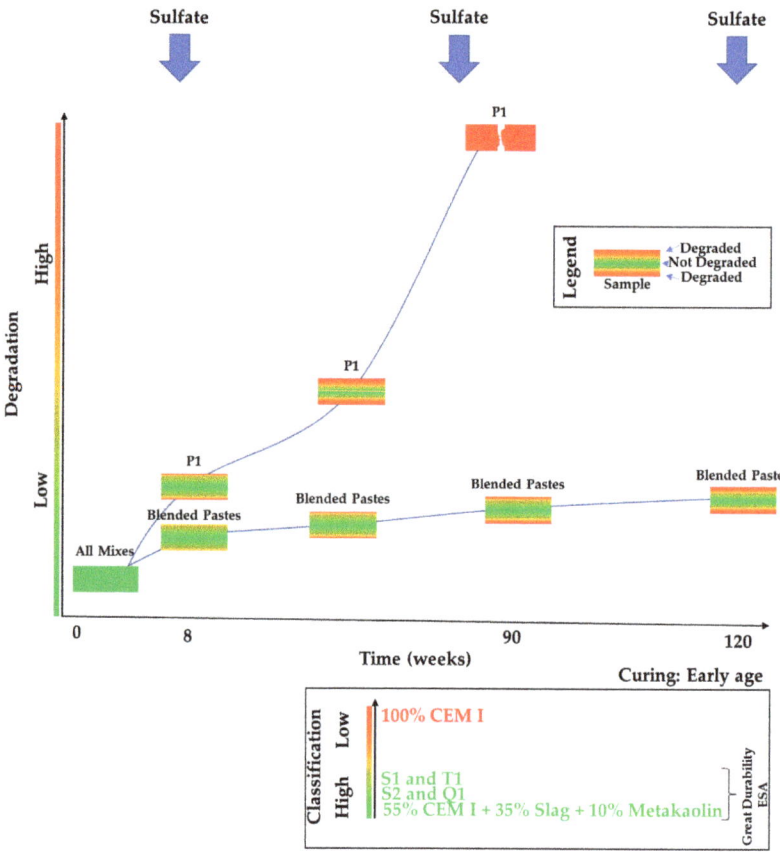

Figure 9. Illustrative overview of research results.

4. Conclusions

In this investigation, OPC pastes, as well as binary, ternary, and quaternary blended cement pastes, underwent long-term exposure to 15 g/L of sodium sulfate at an early age. Following comprehensive characterizations and continuous monitoring of the attacked samples, the study resulted in the following conclusions:

- CEM I exhibits the lowest resistance to ESA when compared to blended mixes. It shows the highest mass gain, expansion, formation of ettringite and gypsum, sulfate consumption from the solution, and microstructure alteration. Additionally, during the course of this experiment, it deteriorated only after 90 weeks, whereas the blended cements maintained their structural integrity even after 120 weeks;
- blended specimens demonstrated good durability, retaining structural integrity after 120 weeks of sulfate exposure from an early age;
- incorporating a low quantity (10%) of metakaolin along with blast furnace slag emerged as the most effective substitute for clinker, outperforming other combinations based on the observed behavior of various blended mixes in sulfate exposure;

- non-invasive Raman spectroscopy emerged as a reliable method for monitoring the ESA effect by quantifying the sulfate ions left in the attacking solutions.

Further investigations at the mortar and/or concrete scale would provide additional insights into the physical transport properties, in addition to the chemical aspects of ESA. This is particularly relevant as the current study focuses on cement paste materials that are highly porous. Moreover, further investigations are needed to explore the effects of adding metakaolin to slag as a substitute for clinker, aiming to develop highly resistant materials against ESA.

Author Contributions: Validation, conceptualization, methodology, writing—review, editing, and supervision O.O.M., M.M. and M.Q.; conceptualization, methodology, writing—original draft preparation, investigation, formal analysis, and editing, F.E.I.; validation, investigation, writing—review, and formal analysis, B.A. and M.H. All authors have read and agreed to the published version of the manuscript.

Funding: This research received no external funding.

Institutional Review Board Statement: Not applicable.

Informed Consent Statement: Not applicable.

Data Availability Statement: Data are contained within the article.

Conflicts of Interest: The authors declare no conflicts of interest.

References

1. Zhao, G.; Li, J.; Shao, W. Effect of mixed chlorides on the degradation and sulfate diffusion of cast-in-situ concrete due to sulfate attack. *Constr. Build. Mater.* **2018**, *181*, 49–58. [CrossRef]
2. Ouyang, C.; Nanni, A.; Chang, W.F. Internal and external sources of sulfate ions in portland cement mortar: Two types of chemical attack. *Cem. Concr. Res.* **1988**, *18*, 699–709. [CrossRef]
3. Geng, J.; Easterbrook, D.; Li, L.; Mo, L. The stability of bound chlorides in cement paste with sulfate attack. *Cem. Concr. Res.* **2015**, *68*, 211–222. [CrossRef]
4. Wongprachum, W.; Sappakittipakorn, M.; Sukontasukkul, P.; Chindaprasirt, P.; Banthia, N. Resistance to sulfate attack and underwater abrasion of fiber reinforced cement mortar. *Constr. Build. Mater.* **2018**, *189*, 686–694. [CrossRef]
5. Skaropoulou, A.; Sotiriadis, K.; Kakali, G.; Tsivilis, S. Use of mineral admixtures to improve the resistance of limestone cement concrete against thaumasite form of sulfate attack. *Cem. Concr. Compos.* **2013**, *37*, 267–275. [CrossRef]
6. Santhanam, M.; Cohen, M.D.; Olek, J. Mechanism of sulfate attack: A fresh look: Part 1: Summary of experimental results. *Cem. Concr. Res.* **2002**, *32*, 915–921. [CrossRef]
7. Yu, X.; Chen, D.; Feng, J.; Zhang, Y.; Liao, Y. Behavior of mortar exposed to different exposure conditions of sulfate attack. *Ocean Eng.* **2018**, *157*, 1–12. [CrossRef]
8. Ikumi, T.; Segura, I.; Cavalaro, S.H.P. Effects of biaxial confinement in mortars exposed to external sulfate attack. *Cem. Concr. Compos.* **2019**, *95*, 111–127. [CrossRef]
9. El Inaty, F.; Marchetti, M.; Quiertant, M.; Omikrine Metalssi, O. Chemical Mechanisms Involved in the Coupled Attack of Sulfate and Chloride Ions on Low-Carbon Cementitious Materials: An In-Depth Study. *Appl. Sci.* **2023**, *13*, 11729. [CrossRef]
10. Neville, A. The confused world of sulfate attack on concrete. *Cem. Concr. Res.* **2004**, *34*, 1275–1296. [CrossRef]
11. Yu, C.; Sun, W.; Scrivener, K. Mechanism of expansion of mortars immersed in sodium sulfate solutions. *Cem. Concr. Res.* **2013**, *43*, 105–111. [CrossRef]
12. Xiong, C.; Jiang, L.; Xu, Y.; Chu, H.; Jin, M.; Zhang, Y. Deterioration of pastes exposed to leaching, external sulfate attack and the dual actions. *Constr. Build. Mater.* **2016**, *116*, 52–62. [CrossRef]
13. Shao, W.; Li, Q.; Zhang, W.; Shi, D.; Li, H. Numerical modeling of chloride diffusion in cement-based materials considering calcium leaching and external sulfate attack. *Constr. Build. Mater.* **2023**, *401*, 132913. [CrossRef]
14. El Inaty, F.; Baz, B.; Aouad, G. Long-term durability assessment of 3D printed concrete. *J. Adhes. Sci. Technol.* **2022**, *37*, 1921–1936. [CrossRef]
15. Al-Amoudi, O.S.B. Sulfate attack and reinforcement corrosion in plain and blended cements exposed to sulfate environments. *Build. Environ.* **1998**, *33*, 53–61. [CrossRef]
16. Jabbour, M. Multi-Scales Study for the External Sulfatic Attack in Reinforced Concrete Structures. Ph.D. Thesis, Université Paris-Est, Paris, France, 2019. Available online: https://tel.archives-ouvertes.fr/tel-02956401 (accessed on 2 November 2023).
17. Wu, M.; Zhang, Y.; Ji, Y.; She, W.; Yang, L.; Liu, G. A comparable study on the deterioration of limestone powder blended cement under sodium sulfate and magnesium sulfate attack at a low temperature. *Constr. Build. Mater.* **2020**, *243*, 118279. [CrossRef]

18. Jabbour, M.; Metalssi, O.O.; Quiertant, M.; Baroghel-Bouny, V. A Critical Review of Existing Test-Methods for External Sulfate Attack. *Materials* **2022**, *15*, 7554. [CrossRef] [PubMed]
19. Imre BICZOK. *Concrete Corrosion and Concrete Protection*; Chemical Publishing Compan: New York, NY, USA, 1967. Available online: https://scholar.google.com/scholar_lookup?title=Concrete+Corrosion+and+Concrete+Protection&author=Biczok,+I.&publication_year=1967 (accessed on 20 November 2023).
20. Metalssi, O.O.; Ragoug, R.; Barberon, F.; d'Espinose de Lacaillerie, J.-B.; Roussel, N.; Divet, L.; Torrenti, J.-M. Effect of an Early-Age Exposure on the Degradation Mechanisms of Cement Paste under External Sulfate Attack. *Materials* **2023**, *16*, 6013. [CrossRef]
21. Li, Z.; Zhou, X.; Ma, H.; Hou, D. *Advanced Concrete Technology*; John Wiley & Sons: Hoboken, NJ, USA, 2022.
22. Zhao, G.; Shi, M.; Guo, M.; Fan, H. Degradation Mechanism of Concrete Subjected to External Sulfate Attack: Comparison of Different Curing Conditions. *Materials* **2020**, *13*, 3179. [CrossRef]
23. Pawar, Y.; Kate, S. Curing of Concrete: A Review. *Int. Res. J. Eng. Technol.* **2020**, *7*, 1820–1824. [CrossRef]
24. Li, X.; Yu, X.; Zhao, Y.; Yu, X.; Li, C.; Chen, D. Effect of initial curing period on the behavior of mortar under sulfate attack. *Constr. Build. Mater.* **2022**, *326*, 126852. [CrossRef]
25. Scrivener, K.; Martirena, F.; Bishnoi, S.; Maity, S. Calcined clay limestone cements (LC3). *Cem. Concr. Res.* **2018**, *114*, 49–56. [CrossRef]
26. Zou, Y.-X.; Zuo, X.-B.; Zhang, H.-L.; Wang, S.-Q. Influence of fly ash and chlorides on the behavior of sulfate attack in blended cement pastes. *Constr. Build. Mater.* **2023**, *394*, 132231. [CrossRef]
27. Elahi, M.M.A.; Shearer, C.R.; Naser Rashid Reza, A.; Saha, A.K.; Khan, M.N.N.; Hossain, M.M.; Sarker, P.K. Improving the sulfate attack resistance of concrete by using supplementary cementitious materials (SCMs): A review. *Constr. Build. Mater.* **2021**, *281*, 122628. [CrossRef]
28. Ramyar, K.; İnan, G. Sodium sulfate attack on plain and blended cements. *Build. Environ.* **2007**, *42*, 1368–1372. [CrossRef]
29. Al-Akhras, N.M. Durability of metakaolin concrete to sulfate attack. *Cem. Concr. Res.* **2006**, *36*, 1727–1734. [CrossRef]
30. Shi, Z.; Ferreiro, S.; Lothenbach, B.; Geiker, M.R.; Kunther, W.; Kaufmann, J.; Herfort, D.; Skibsted, J. Sulfate resistance of calcined clay–Limestone–Portland cements. *Cem. Concr. Res.* **2019**, *116*, 238–251. [CrossRef]
31. Al-Dulaijan, S.U.; Maslehuddin, M.; Al-Zahrani, M.M.; Sharif, A.M.; Shameem, M.; Ibrahim, M. Sulfate resistance of plain and blended cements exposed to varying concentrations of sodium sulfate. *Cem. Concr. Compos.* **2003**, *25*, 429–437. [CrossRef]
32. Lee, S.T.; Moon, H.Y.; Swamy, R.N. Sulfate attack and role of silica fume in resisting strength loss. *Cem. Concr. Compos.* **2005**, *27*, 65–76. [CrossRef]
33. Baghabra Al-Amoudi, O.S. Attack on plain and blended cements exposed to aggressive sulfate environments. *Cem. Concr. Compos.* **2002**, *24*, 305–316. [CrossRef]
34. Miah, M.J.; Huaping, R.; Paul, S.C.; Babafemi, A.J.; Li, Y. Long-term strength and durability performance of eco-friendly concrete with supplementary cementitious materials. *Innov. Infrastruct. Solut.* **2023**, *8*, 255. [CrossRef]
35. EN 197-1:2011; European Committee for Standardization. Cement—Part 1: Composition, Specifications and Conformity Criteria for Common Cements. iTeh, Inc.: Newark, DE, USA, 2011.
36. NF EN 197-5; European Committee for Standardization. Ciment—Partie 5: Ciment Portland Composé CEM II/C-M et Ciment composé CEM VI. Afnor EDITIONS. AFNOR: Saint-Denis, France, 2021. Available online: https://www.boutique.afnor.org/fr-fr/norme/nf-en-1975/ciment-partie-5-ciment-portland-compose-cem-ii-cm-et-ciment-compose-cem-vi/fa200094/264804 (accessed on 18 October 2023).
37. Bogue, R.H. *The Chemistry of Portland Cement*, 2nd ed.; LWW: Pennvania, PA, USA, 1955; Volume 79, p. 322.
38. NF 196-1; Méthodes d'Essais des Ciments. AFNOR: Saint-Denis, France, 2006.
39. NF P18-459; Béton-Essai pour Béton Durci—Essai de Porosité et de Masse Volumique. AFNOR: Saint-Denis, France, 2010. Available online: https://www.boutique.afnor.org/fr-fr/norme/nf-p18459/beton-essai-pour-beton-durci-essai-de-porosite-et-de-masse-volumique/fa160729/34961 (accessed on 9 January 2024).
40. Metalssi, O.O.; Touhami, R.R.; Barberon, F.; d'Espinose de Lacaillerie, J.-B.; Roussel, N.; Divet, L.; Torrenti, J.-M. Understanding the degradation mechanisms of cement-based systems in combined chloride-sulfate attack. *Cem. Concr. Res.* **2023**, *164*, 107065. [CrossRef]
41. Jiang, L.; Niu, D. Study of deterioration of concrete exposed to different types of sulfate solutions under drying-wetting cycles. *Constr. Build. Mater.* **2016**, *117*, 88–98. [CrossRef]
42. Zhang, Z.; Jin, X.; Luo, W. Long-term behaviors of concrete under low-concentration sulfate attack subjected to natural variation of environmental climate conditions. *Cem. Concr. Res.* **2019**, *116*, 217–230. [CrossRef]
43. Nosouhian, F.; Mostofinejad, D.; Hasheminejad, H. Concrete Durability Improvement in a Sulfate Environment Using Bacteria. *J. Mater. Civ. Eng.* **2016**, *28*, 04015064. [CrossRef]
44. Haufe, J.; Vollpracht, A. Tensile strength of concrete exposed to sulfate attack. *Cem. Concr. Res.* **2019**, *116*, 81–88. [CrossRef]
45. Santhanam, M.; Cohen, M.D.; Olek, J. Effects of gypsum formation on the performance of cement mortars during external sulfate attack. *Cem. Concr. Res.* **2003**, *33*, 325–332. [CrossRef]
46. Tian, B.; Cohen, M.D. Does gypsum formation during sulfate attack on concrete lead to expansion? *Cem. Concr. Res.* **2000**, *30*, 117–123. [CrossRef]
47. Ma, X.; Çopuroğlu, O.; Schlangen, E.; Han, N.; Xing, F. Expansion and degradation of cement paste in sodium sulfate solutions. *Constr. Build. Mater.* **2018**, *158*, 410–422. [CrossRef]

48. Gao, Y.; Cui, X.; Lu, N.; Hou, S.; He, Z.; Liang, C. Effect of recycled powders on the mechanical properties and durability of fully recycled fiber-reinforced mortar. *J. Build. Eng.* **2022**, *45*, 103574. [CrossRef]
49. Nochaiya, T.; Sekine, Y.; Choopun, S.; Chaipanich, A. Microstructure, characterizations, functionality and compressive strength of cement-based materials using zinc oxide nanoparticles as an additive. *J. Alloys Compd.* **2015**, *630*, 1–10. [CrossRef]
50. Irbe, L.; Beddoe, R.E.; Heinz, D. The role of aluminium in C-A-S-H during sulfate attack on concrete. *Cem. Concr. Res.* **2019**, *116*, 71–80. [CrossRef]
51. Fode, T.A.; Chande Jande, Y.A.; Kivevele, T. Effects of different supplementary cementitious materials on durability and mechanical properties of cement composite–Comprehensive review. *Heliyon* **2023**, *9*, e17924. [CrossRef]
52. Farcas, F.; Touzé, P. La spectrométrie infrarouge à transformée de Fourier (IRTF). *Bull. Lab. Ponts Chaussées* **2001**, *230*, 77–88.
53. Liu, P.; Chen, Y.; Wang, W.; Yu, Z. Effect of physical and chemical sulfate attack on performance degradation of concrete under different conditions. *Chem. Phys. Lett.* **2020**, *745*, 137254. [CrossRef]
54. Yue, Y.; Wang, J.J.; Basheer, P.A.M.; Bai, Y. Raman spectroscopic investigation of Friedel's salt. *Cem. Concr. Compos.* **2018**, *86*, 306–314. [CrossRef]
55. Tang, C.; Ling, T.-C.; Mo, K.H. Raman spectroscopy as a tool to understand the mechanism of concrete durability—A review. *Constr. Build. Mater.* **2021**, *268*, 121079. [CrossRef]
56. Water Molecule Vibrations with Raman Spectroscopy. PhysicsOpenLab. Available online: https://physicsopenlab.org/2022/01/08/water-molecule-vibrations-with-raman-spectroscopy/ (accessed on 19 October 2023).
57. Nehdi, M.L.; Suleiman, A.R.; Soliman, A.M. Investigation of concrete exposed to dual sulfate attack. *Cem. Concr. Res.* **2014**, *64*, 42–53. [CrossRef]
58. Toutanji, H.; Delatte, N.; Aggoun, S.; Duval, R.; Danson, A. Effect of supplementary cementitious materials on the compressive strength and durability of short-term cured concrete. *Cem. Concr. Res.* **2004**, *34*, 311–319. [CrossRef]

Disclaimer/Publisher's Note: The statements, opinions and data contained in all publications are solely those of the individual author(s) and contributor(s) and not of MDPI and/or the editor(s). MDPI and/or the editor(s) disclaim responsibility for any injury to people or property resulting from any ideas, methods, instructions or products referred to in the content.

Article

Exploring the Effect of Moisture on CO_2 Diffusion and Particle Cementation in Carbonated Steel Slag

Shenqiu Lin [1], Ping Chen [1,2], Weiheng Xiang [2,3,*], Cheng Hu [2,4,*], Fangbin Li [2], Jun Liu [2] and Yu Ding [1]

[1] School of Materials Science and Engineering, Guilin University of Technology, Guilin 541004, China; linshenqiu6@163.com (S.L.); chenping8383@188.com (P.C.)
[2] College of Civil and Architectural Engineering, Guilin University of Technology, Guilin 541004, China
[3] Guangxi Engineering and Technology Center for Utilization of Industrial Waste Residue in Building Materials, Guilin University of Technology, Guilin 541004, China
[4] Collaborative Innovation Center for Exploration of Nonferrous Metal Deposits and Efficient Utilization of Resources, Guilin University of Technology, Guilin 541004, China
* Correspondence: 2021072@glut.edu.cn (W.X.); hucheng42@glut.edu.cn (C.H.)

Abstract: The study of the mechanisms affecting the preparation parameters of carbonated steel slag is of great significance for the development of carbon sequestration materials. In order to elucidate the mechanism of the influence of moisture on CO_2 diffusion and particle cementation in steel slag, the effects of different water–solid ratios and water contents on the mechanical properties, carbonation products, and pore structure of steel slag after carbonation were investigated. The results show that increasing the water–solid ratio of steel slag can control the larger initial porosity and improve the carbon sequestration capacity of steel slag, but it will reduce the mechanical properties. The carbonation process relies on pores for CO_2 diffusion and also requires a certain level of moisture for Ca^{2+} dissolution and diffusion. Increasing the water content enhances particle cementation and carbonation capacity in steel slag specimens; however, excessive water hinders CO_2 diffusion. Reducing the water content can increase the carbonation depth but may compromise gelling and carbon sequestration ability. Therefore, achieving a balance is crucial in controlling the water content. The compressive strength of the steel slag with suitable moisture and initial porosity can reach 118.7 MPa, and 217.2 kg CO_2 eq./t steel slag can be sequestered.

Keywords: steel slag; carbon sequestration; water content; carbonation diffusion; pore

Citation: Lin, S.; Chen, P.; Xiang, W.; Hu, C.; Li, F.; Liu, J.; Ding, Y. Exploring the Effect of Moisture on CO_2 Diffusion and Particle Cementation in Carbonated Steel Slag. *Appl. Sci.* **2024**, *14*, 3631. https://doi.org/10.3390/app14093631

Academic Editors: Mouhamadou Amar and Nor Edine Abriak

Received: 20 March 2024
Revised: 2 April 2024
Accepted: 4 April 2024
Published: 25 April 2024

Copyright: © 2024 by the authors. Licensee MDPI, Basel, Switzerland. This article is an open access article distributed under the terms and conditions of the Creative Commons Attribution (CC BY) license (https://creativecommons.org/licenses/by/4.0/).

1. Introduction

Steel slag is a prominent by-product generated during the smelting process in the iron and steel industry, constituting approximately 8–15% of the total crude steel production [1–3]. In 2021, global crude steel production reached around 1.952 billion tons, resulting in an annual emission of steel slag exceeding 250 million tons [4]. Among these figures, China's crude steel production accounted for about 1.03 billion tons with a corresponding output of over 120 million tons of steel slag [5]. However, steel slag possesses drawbacks such as high free CaO content and low hydration activity [6], leading to its combined utilization rate remaining below 30% [7]. Consequently, it is primarily stockpiled or landfilled, posing significant environmental pollution concerns. Therefore, there is an urgent need to develop an efficient and environmentally friendly approach to the disposal of steel slag.

Meanwhile, the steel production process is accompanied by a substantial amount of carbon emissions. The carbon emissions of China's iron and steel sector account for approximately 15% of the national total [8]. Carbon Capture, Utilization and Storage (CCUS) is widely regarded as an optimal approach to achieve significant carbon emission reduction in the iron and steel industry [9,10]. By harnessing CO_2, it becomes possible to convert the high calcium content present in steel slag into highly stable $CaCO_3$, thereby facilitating long-term carbon sequestration [11]. The CO_2 carbonation of steel slag has been acknowledged as the most favorable method for disposing of this by-product.

Steel slag can be classified into basic oxygen furnace slag (BOFS), electrical arc furnace slag (EAFS), argon oxygen decarburization slag (AODS), and ladle-refining slag (LFS) based on the production process [4,10]. Among these, LFS is a by-product generated during secondary or alkaline steelmaking [12], and it exhibits the highest calcium content among all types of slags [13]. The cooling process of the LFS results in the generation of a substantial quantity of the self-powdered γ-Ca_2SiO_4 (γ-C_2S) material phase, thereby conferring a powdery nature upon the LFS. The γ-C_2S enables a rapid reaction with CO_2 to form calcium carbonates ($CaCO_3$) and calcium silicate hydrates (C-S-H) [14,15]. Consequently, LFS holds great potential for application in CCUS technology.

The carbonation process of steel slag building materials typically involves mixing water with steel slag powder to form cubic blocks, which are then subjected to a direct gas–solid reaction by passing CO_2 gas through a closed stainless steel reactor [16,17]. Some researchers have also utilized liquid–solid reactions [18]. During the reaction between CO_2 and steel slag, CO_3^{2-} formed from dissolved CO_2 reacts with Ca^{2+} precipitated from the slag to produce stable $CaCO_3$ that provides high mechanical strength and permanently fixes CO_2 in the material [6]. Although the theoretical carbon fixation rate of steel slag ranges between 25% and 50%, various factors during carbonation limit its actual rate to about 15% [19]. Each ton of cement produced will emit 0.94 tons of CO_2 [20], so when carbonation of steel slag is applied in cementitious materials or aggregates to replace cement, the reduction in cement production and the amount of CO_2 absorbed by the steel slag makes it possible to achieve a carbon reduction of approximately 1.09 t of CO_2 for each ton of steel slag. This would be expected to achieve the carbon-negative production of building materials.

Gas–solid carbonation is mainly restricted by reaction kinetics and CO_2 diffusion [4]; hence, researchers strive to study different conditions in this process for improved efficiency. Currently, there is a substantial body of mature research on the carbonation reaction of steel slag, which has revealed the effects of various conditions such as particle size, temperature, CO_2 pressure, and carbonation time on the carbonation results [21–25]. For instance, Ukwattage et al. improved the carbonation efficiency of steel slag by optimizing the three parameters of CO_2 pressure, temperature, and water–solid ratio [26]. Zhang et al. increased the carbon sequestration efficiency of steel slag to 16.65% by increasing the CO_2 pressure [27]. The study conducted by Tian et al. [28] concluded that temperature is the primary determinant influencing the direct gas–solid carbonation reaction of steel slag. However, it appears that the researchers overlooked the significance of moisture control in specimen preparation prior to carbonation. Steel slag not only needs water for Ca^{2+} dissolution and particle cementation during carbonation, but the carbonation reaction process also consumes a certain amount of water.

Furthermore, numerous unresolved issues persist in the investigation of the carbonation reaction mechanism of steel slag. For instance, there is a lack of synergistic studies on the water content and pore structure of specimens prior to carbonation. The alteration in pore structure during carbonation significantly impacts the compressive strength of specimens [14,29]. Secondly, existing research primarily focuses on pressed steel slag and lacks investigation into pouring molding commonly employed in cement-molding processes. Pouring molding can yield higher initial porosity compared to pressing molding, which theoretically results in higher carbon sequestration. The initial porosity of cast-molded specimens can be significantly controlled by manipulating the water–solid ratio, while the pore structure and porosity play a pivotal role in facilitating CO_2 diffusion. Thirdly, some researchers have observed that the center of the specimen has lower levels of carbonation than the surface part [30,31], but the mechanism underlying the microscopic effect remains elusive.

Considering the aforementioned issues, this study aims to utilize small poured cubic blocks of steel slag and manipulate different water–solid ratios to control their initial porosity. The water content is simultaneously regulated through air-drying of the steel slag. Initially, the impact of various water–solid ratios and water contents on the properties of steel slag is investigated, while microscopic analysis is employed to examine the carbonation products and pore structure of these. Consequently, a mechanism involving CO_2 diffusion and particle cementation is proposed by integrating the analytical findings with previous research studies. Notably, this research introduces a novel approach by employing three-dimensional pore structure analysis to explore the relationship between CO_2 diffusion and pore structure.

2. Materials and Methods

The ladle-refining slag (LFS) utilized in this study was obtained from a steel mill located in Guangxi, China. It underwent ball milling and sieving through a 100-mesh sieve to obtain a powder with an average particle size of 1.34 µm, as depicted in Figure 1. Particle size analysis was performed using a Mastersizer 3000 laser particle sizer from Malvern Panalytical, EA Almelo, The Netherlands. LFS exhibits a significant composition of CaO (~71.23%) and SiO_2 (~21.31%) as presented in Table 1. The mineral composition of LFS is illustrated in Figure 2, where the primary mineral phases include calcio-olivine, akermanite, and fluorite, among others.

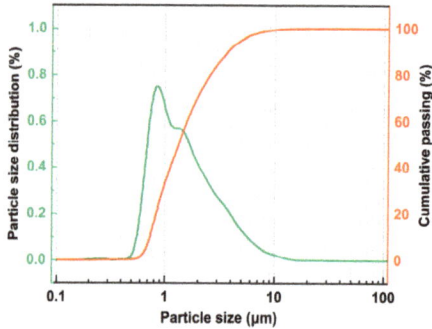

Figure 1. Particle size distribution of LFS powder.

Table 1. Chemical compositions of LFS (wt%).

Oxides	CaO	SiO_2	Cr_2O_3	MgO	TiO_2	Fe_2O_3	Others
LFS	71.23	21.31	2.37	1.96	1.72	1.27	0.12

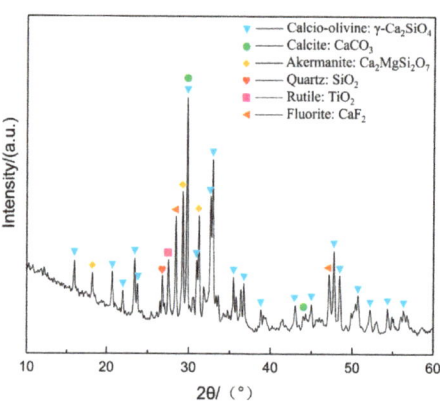

Figure 2. X-ray diffraction pattern of LFS powder.

The steel slag specimens were prepared by mixing water and LFS powder at different water–solid ratios of 0.25, 0.3, and 0.35. Subsequently, the resulting slurry was poured into a 20 × 20 × 20 mm² steel film mold and allowed to cure through natural air-drying for a duration of 24 h at room temperature. Afterward, the cured steel slag specimens were divided into two groups, and uniform quality was ensured across all specimens. The first group of steel slag specimens underwent oven-drying until they reached a constant weight in order to determine their dry basis weight (M_0). Meanwhile, the second group of steel slag specimens was subjected to oven-drying at a temperature of 40 °C for t time to obtain their wet basis weight (M_t). The water content of each specimen was calculated using Equation (1). Throughout the weighing process and water content control procedures, efforts were made to maintain an error margin within ±0.1%. The control of water content can be facilitated by deriving an empirical formula through the analysis of the water content change curve in relation to drying time.

$$w\% = (M_t - M_0)/M_0 \times 100\% \tag{1}$$

After controlling the water content of the steel slag specimens, the sample was promptly introduced into the reaction vessel as depicted in Figure 3. This carbonation reactor is a single-layer stainless steel reactor with a volume of 20 L in a constant-temperature (25 °C) environment. The CO_2 pressure of the carbonation reactor can be controlled by adjusting the valve of the CO_2 gas tank. The reaction vessel was evacuated to −0.08 MPa and then carbonated at room temperature (25 °C) and a CO_2 pressure of 0.4 MPa for a duration of 24 h. The purity of CO_2 employed for this carbonation process was determined to be 99%. A schematic diagram for the preparation step is illustrated in Figure 4.

Figure 3. Carbonation reaction vessel.

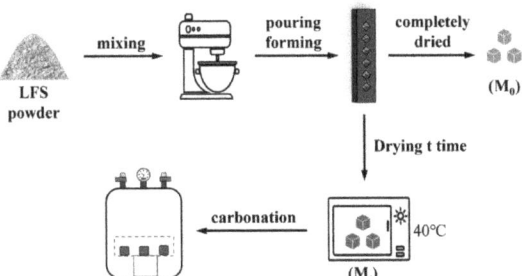

Figure 4. Schematic diagram for the preparation step.

The compressive strength analysis was performed using a UTM5105 microcomputer control electronic universal material testing machine from Suns Technology Stock Co., Ltd., Shenzhen, China; the testing process was carried out in accordance with the Chinese standard GB/T 17671-2021 [32]. Thermogravimetric analysis was performed using the

STA 449 F5 thermal analyzer from NETZSCH, CO., Germany. The carbon sequestration rate of the specimens was calculated from the thermogravimetric analysis results using Equations (2) and (3) [15]. X-ray diffraction (XRD) analysis was performed using the X'Pert PRO X-ray diffractometer from Malvern Panalytical Ltd., The Netherlands. SEM analysis was performed using the S-4800 scanning electron microscope from Carl Zeiss AG, Germany. Three-Dimensional Computerized Tomography (3D CT) analysis was performed using a Xradia 510 Versa high-resolution 3D X-ray microscope from Carl Zeiss AG, Germany, equipped with a voltage of 100 kV, a pixel size of 3 µm, and an exposure time of 2 s.

$$CO_2(wt.\%) = \frac{\Delta m_{CO2}}{m_{105°C}} \times 100 \tag{2}$$

$$CO_2 uptake(wt.\%) = \frac{CO_{2carbonated}\ (wt.\%) - CO_{2initial}\ (wt.\%)}{100 - CO_{2carbonated}\ (wt.\%)} \times 100 \tag{3}$$

where $CO_2(wt.\%)$ represents the rate of carbon dioxide loss; Δm_{CO2} denotes the rate of mass loss resulting from calcium carbonate decomposition in the specimen at elevated temperatures; $m_{105°C}$ indicates the weight of the dried specimen; $CO_2 uptake(wt.\%)$ signifies the rate of carbon sequestration; $CO_{2carbonated}(wt.\%)$ refers to the rate of carbon dioxide loss in the specimen after carbonation; $CO_{2initial}(wt.\%)$ represents the rate of carbon dioxide loss in the uncarbonated specimen.

3. Results

3.1. Regulation of Water Content Regulation

The variation rule of water content in specimens under different water–solid ratios at a 40 °C drying temperature was initially investigated, and the corresponding results are presented in Figure 5. Moisture loss from the specimen is evaporative moisture loss driven by the external environment, which is related to the relative humidity of the environment [33]. The higher the water–solid ratio, the greater the initial water content of the specimen observed. Additionally, a higher water–solid ratio leads to an accelerated overall drying rate due to decreased specimen density and increased porosity, facilitating easier drying. Initially, the drying rate was faster but gradually slowed down over time, possibly attributed to the difficulty in drying deeper internal water compared to surface and pore water. By adjusting different drying times, it was possible to control the water content of specimens according to fitted curves.

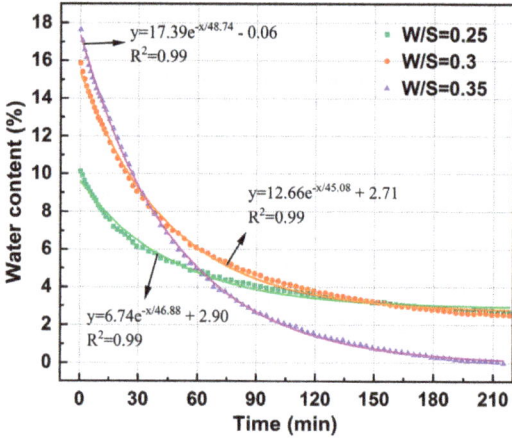

Figure 5. Variation in the water content of specimens with time at different water–solid ratios.

3.2. Compressive Strength

The steel slag specimens were prepared with varying water–solid ratios ranging from 0.25 to 0.35 and water contents ranging from 4% to 9%. These specimens were then carbonated using the aforementioned method for controlling water content. The resulting compressive strengths for each specimen are depicted in Figure 6. Notably, the compressive strength of the specimens exhibited a decreasing trend as the water–solid ratio increased. This was attributed to the fact that specimens with lower water–solid ratios tend to be denser and less porous, leading to an optimal compressive strength of 118.7 MPa observed at a water–solid ratio of 0.25 and a water content of 6%. The strength obtained by this specimen in 1 d exceeded the 28 d compressive strength value of the researcher's specimen [17]. However, a higher water–solid ratio offers improved pourability and moldability.

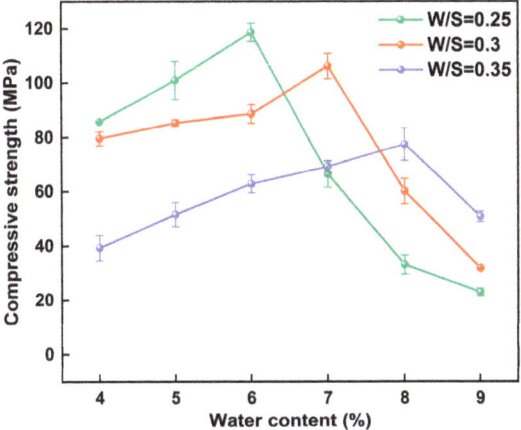

Figure 6. The effects of different water–solid ratios and water content on the compressive strength of the samples.

The initial compressive strength of the specimens demonstrated an increase, followed by a decline as water content increased. This phenomenon is attributed to the fact that γ-C_2S requires a certain amount of water to dissolve Ca^{2+} during the carbonation reaction (see Equation (4)) [34]. An increase in water content provides more water for the carbonation reaction, which makes γ-C_2S and the other reactive constituents more susceptible to carbonation, thus increasing the strength. Excess water may obstruct the pores of the steel slag specimens, resulting in the hindered diffusion of carbon dioxide [35].

$$Ca_2SiO_4 + 4H_2O \rightarrow 2Ca^{2+} + H_4SiO_4 + 4OH^- \qquad (4)$$

The higher the water–cement ratio, the greater the optimal water content required for achieving maximum compressive strength in the specimen. This is attributed to the fact that specimens with higher water–cement ratios possess a more porous structure, thereby rendering it more challenging for water to hinder the carbonation process compared to specimens with lower water–cement ratios. This was consistent with the discussion in Wang et al.'s study [36].

3.3. Carbon Sequestration

The thermogravimetry (TG) and DTG results of carbonated LFS with varying water–solid ratios and water contents are presented in Figure 7. The DTG results reveal that the mass loss peak between 400 and 800 °C represents the decomposition peak of $CaCO_3$. By employing Equations (2) and (3), the rate of mass loss over the temperature range can be used to calculate the carbon sequestration rate, as demonstrated in Figure 7a. It is observed that the peak carbon sequestration rate of carbonated LFS increases with the water–solid ratio, in contrast to the result of compressive strength presented in Figure 7b. Although a higher water–solid ratio results in increased porosity, leading to lower densification and strength, these pores facilitate greater penetration of CO_2 during the curing process. The carbon sequestration rate of LFS samples, characterized by a 0.35 water–solid ratio and 8% water content, amounts to 21.72%, indicating that each ton of LFS can effectively absorb and fix approximately 217.2 kg of CO_2. This is much higher than the average carbon sequestration rate value of 15% proposed by researchers [19]. From the TG plot, it can be inferred that the carbonated LFS with a higher water–solid ratio is less influenced by the water content, suggesting that an increase in pore space mitigates the adverse effects of water on CO_2 diffusion. The carbonated LFS exhibits an increasing then decreasing trend of the carbon sequestration rate with respect to water content, which is consistent with the compressive strength results presented in Figure 6. The current findings imply that maintaining an adequate water supply is crucial to ensure proper dissolution and solidification of CO_2, yet an excessive increase in water content beyond a certain threshold potentially leads to adverse effects on CO_2.

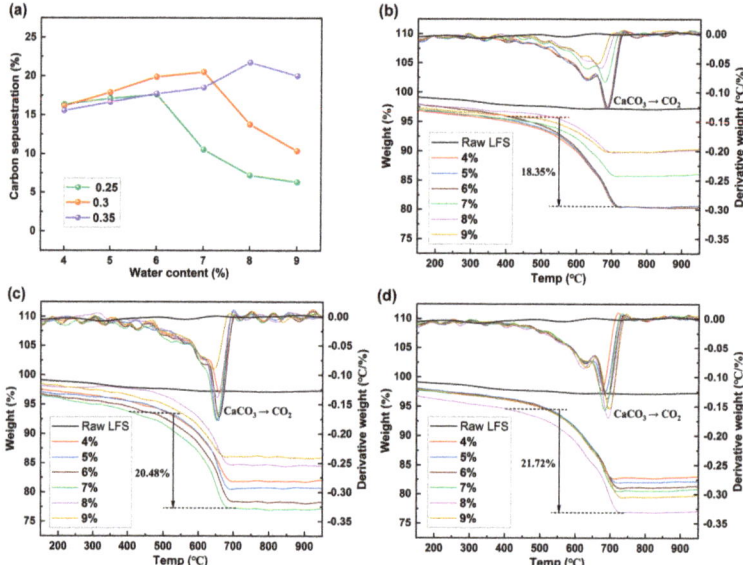

Figure 7. (**a**) Carbon sequestration rate of carbonated LFS with different water–solid ratios and different water contents; (**b**) TG and DTG results of carbonated LFS with 0.25 water–solid ratio; (**c**) TG and DTG results of carbonated LFS with 0.3 water–solid ratio; (**d**) TG and DTG results of carbonated LFS with 0.35 water–solid ratio.

In order to observe the distribution of water in samples with different water contents, the center of samples with a water–solid ratio of 0.3 and different water contents was cut to obtain a profile, and phenolphthalein powder was evenly sprayed on the profile. The LFS, a highly alkaline substance, would show color in the presence of water, as shown in the upper part of Figure 8. The sample with 0% water content did not show color due to the absence of water, which proved the feasibility of the test method. The increase in water content was observed to be accompanied by a progressively more pronounced coloration of the specimen's cut surface.

Figure 8. Sample center profiles of LFS before and after carbonation.

After carbonation, the sample was uniformly covered with standard phenolphthalein solution on the central profile to indicate the carbonation depth, as shown in the bottom part of Figure 8. The sample with 16% water content has a darker color before carbonation and almost covers the whole sample. After carbonation, it was found that the sample also showed a darker color in the phenolphthalein solution, indicating that the degree of carbonation of the sample was small. This was because the excessive water content led to the hindered diffusion of CO_2 during the carbonation process. With the decrease in water content, the color of the sample before carbonation became lighter and lighter. This proves that the moisture inside and outside the sample was reduced, and the resistance of CO_2 diffusion was reduced. So when the water content was lower, the depth of carbonation would increase. Samples with 0% to 7% water content were fully carbonated to the inside. When the water content was 7%, the obstruction effect of water on CO_2 diffusion became very small, and there was enough water to dissolve Ca^{2+} and CO_2. After carbonation, phenolphthalein drops in the sample did not show color, the carbonation degree was high, and tightly packed cementation products could be clearly observed.

3.4. Mineralogical Compositions

The X-ray diffraction patterns and semi-quantitative mineral-phase analyses of samples with 6–8% water content and varying water–solid ratios are presented in Figure 9. The semi-quantitative analytical procedure exclusively considers calcium oxides and represents a corrected outcome. The predominant constituents in the LFS feedstock were calcio-olivine and akermanite, denoted by the chemical formulas γ-C_2S and $Ca_2Mg(Si_2O_7)$, respectively. $Ca_2Mg(Si_2O_7)$ is a mineral-phase variant of $3CaO \cdot 2SiO_2(C_3S_2)$, primarily formed due to the incorporation of Mg^{2+} ions into the C_3S_2 lattice, replacing some of the Ca^{2+} ions during the high-temperature melting process of LFS. It shares similar properties to C_3S_2. Previous studies have demonstrated the significant carbonation activity of γ-C_2S and C_3S_2 [37]. Therefore, the primary sources of carbonation activity in the LFS were calcium silicate and akermanite. In contrast, fluorite does not react with CO_2 and was included in the analysis solely for comparative purposes with other phases. Additionally, the presence of other trace substances in the LFS was not analyzed; thus, the results of this semi-quantitative analysis should be considered as a reference only. The primary product of the reaction between steel slag and CO_2 is $CaCO_3$ [38], and the three primary crystalline forms of laboratory-generated $CaCO_3$ are calcite, aragonite, and vaterite [39]. γ-C_2S carbonated at low temperatures produces $CaCO_3$, primarily in the form of calcite [33]. Moreover, it can be discerned from Figure 9a that a prominent calcite diffraction peak appeared subsequent to carbonation, while no diffraction peaks of aragonite or vaterite were detected.

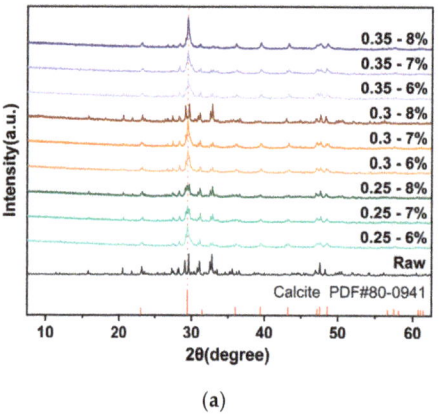 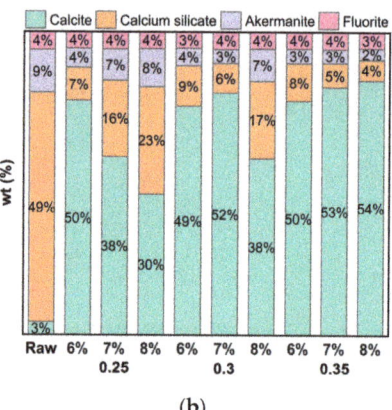

(a) (b)

Figure 9. LFS with different water–solid ratios and water contents: (**a**) X-ray diffraction pattern; (**b**) semi-quantitative mineral-phase analysis.

From Figure 9b, it can be discerned that an increase in the water–solid ratio results in a corresponding elevation of calcite generation following carbonation. The generation of calcite demonstrates an initial rise followed by a decline as water content increases, in line with the findings of carbon sequestration and thermogravimetry presented in Figure 7. Upon carbonation, the contents of calcium silicate and akermanite decrease as the degree of carbonation intensifies, thereby indicating that γ-C_2S and $Ca_2Mg(Si_2O_7)$ serve as the primary active agents in the carbonation of LFS, which are transformed into stable calcite upon carbonation, concurrently fixing CO_2 within the calcite. However, calcium silicate and akermanite do not experience complete carbonation, likely due to the challenges associated with CO_2 penetration within the particles of LFS.

3.5. Morphology

The microscopic morphology of LFS after carbonation with a water–solid ratio of 0.3 and a water content of 6–8% is shown in Figure 10, magnified by 20,000 times. It can be observed that as the water content increases, the size of carbonation products initially increases and then decreases. Additionally, specimens with higher degrees of carbonation exhibit larger and denser calcium carbonate crystals. The formation of calcite crystals on the particle surface can be observed in specimens with a water content of 6%. These fine particles, approximately 0.3 µm in size, are cemented to the larger matrix as depicted in Figure 10a. At a water content of 7%, calcite exhibits enhanced crystallinity and grows in rock-like layers, while some small clusters of calcium carbonate crystals fill pores and increase compactness (Figure 10b). This may explain why specimens with a water content of 7% possess higher strength compared to others. Figure 10c reveals that small calcium carbonate crystals measuring around 0.2 µm are cemented together to form larger particles measuring approximately 1.2 µm; however, these cemented particles fail to continue growing due to certain limitations, resulting in low crystallinity levels. This phenomenon could be attributed to an abundance of surface moisture hindering CO_2 diffusion on high-water-content specimen particles. Pores associated with high-water-content samples are smaller than those found in low-water-content samples but occur at a greater frequency.

Figure 10. SEM images of carbonated LFS with (**a**) 6% water content; (**b**) 7% water content; (**c**) 8% water content.

Figure 11 shows the 5000×-magnification electron image, and the EDS surface scan results are shown in Figure 10b. The dense structure of the 7%-water-content specimen can be observed in the electron image. A large number of overlapping regions of Ca, Mg, C, and O elements can be found, which further proves that the carbonation products of the LFS are calcium/magnesium carbonates and predominantly $CaCO_3$, which is consistent with the results in Figure 9. The areas exhibited smaller crystalline structures and high brightness corresponding to $CaCO_3$ products, whereas the larger particles with darker regions represent uncarbonated calcium silicate. This distinction arose from the fact that calcium carbonate could solely precipitate on the surface of calcium silicate particles, leading to a higher number of small calcium carbonate crystals formed on larger calcium silicate particles. Carbonation was more challenging for larger calcium silicate particles compared to their smaller counterparts.

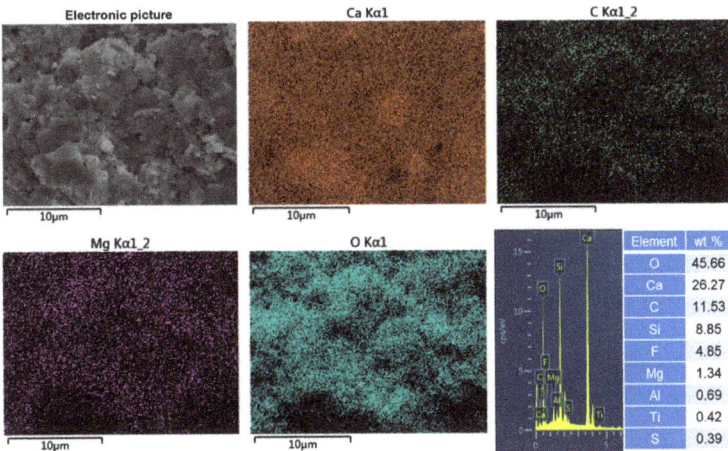

Figure 11. EDS result of carbonated LFS with 7% water content.

3.6. Pore Structure

In order to investigate the distribution of pore structures within carbonated LFS samples with a water–solid ratio of 0.3 and varying water contents, a high-resolution three-dimensional X-ray microscope was employed to analyze the central region of the samples. Pore identification is achieved through grayscale threshold segmentation, wherein the lower density of water compared to solids enables direct pore recognition during analysis. The porosity can be determined by reconstructing the test results in three dimensions. Figure 12 illustrates the porosity distribution along the Z-axis direction for two-dimensional cross-sections of samples with different water contents. It is evident that most curves exhibit overall flatness, with only a few positions displaying significant fluctuations attributed to large pore size defects present in the original samples. However, these defects remain within acceptable limits and do not undermine the value of our findings. Furthermore, it was observed that all samples exhibited reduced porosity after carbonation compared to their pre-carbonation state. This reduction in porosity correlates with changes in compressive strength, carbon sequestration rate, and physical phase content as discussed elsewhere in this paper. Notably, as the water content approaches 7%, higher degrees of carbonation lead to increased generation of calcium carbonate, which fills up pores and contributes to enhanced strength.

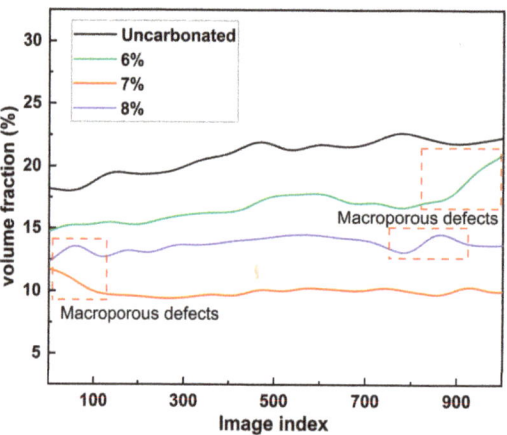

Figure 12. Axial pore size distribution of carbonated LFS with different water contents.

The test results were reconstructed in three dimensions, classifying holes based on connectivity into open and closed categories. The rendered results are presented in Figure 13. The upper section of the figure illustrates cut sections of the samples with varying water contents, while the lower section showcases the three-dimensional rendering. In the lower half of the 3D rendering, only the original shape is displayed, whereas in the upper half, both open and closed pores are visible with a priority given to displaying closed pores. It can be observed that micropores are distributed throughout the specimen, while closed pores are predominantly concentrated in its central region. Following carbonation treatment, there was a significant increase in closed pores within the specimen. This phenomenon may be attributed to calcium carbonate production blocking connecting pores and leading to an increased formation of closed pores.

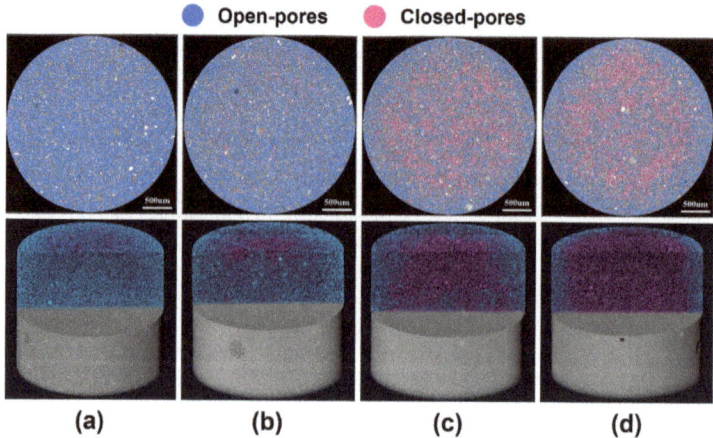

Figure 13. Cutaway and 3D rendering of 3D CT of specimens with (**a**) uncarbonated; (**b**) 6% water content; (**c**) 7% water content; (**d**) 8% water content.

The results of total porosity, open porosity, and closed porosity obtained through calculation are presented in Table 2. In total, 98.71% of the pores in the uncarbonated specimen are classified as open pores (Figure 13a), which facilitates favorable conditions for CO_2 diffusion during the carbonation process. The variation in water content shows a similar trend between open porosity and total porosity, indicating that carbonation generates calcium carbonate that fills most of the open pores, enhancing overall compactness but reducing pore connectivity. Closed pores consistently increase with increasing water content. Notably, specimens with 8% water content exhibit higher levels of closed pores compared to those with 7% water content; however, they show lower levels of carbonation. This observation suggests that higher-water-content specimens possess more bridging water between particles and result in calcium carbonate deposition within particle gaps, leading to increased formation of closed pores. Consequently, this impedes CO_2 diffusion pathways and affects later-stage carbonation progress. At a water content of 7%, although more closed pores are formed, there are still interconnected pores within the specimen allowing CO_2 penetration (Figure 13c). Beyond a water content threshold above 7%, closed pores occupy almost all internal spaces within the specimen (Figure 13d), resulting in incomplete carbonation and subsequent reductions in strength and carbon sequestration rate.

Table 2. Porosity of carbonated LFS with different water contents (%).

Samples	Water Content	Total Porosity	Open Porosity	Closed Porosity
a	uncarbonated	20.44%	20.18%	0.26%
b	6%	17.69%	16.90%	0.79%
c	7%	10.60%	7.14%	3.46%
d	8%	13.61%	8.52%	5.09%

4. Discussion

Previous studies have demonstrated that γ-C_2S undergoes low-temperature carbonation to produce calcite and silica, with the silica encapsulating uncarbonated γ-C_2S particles and the calcite filling the pores [40,41]. Similar structural characteristics are observed in steel slag carbonation as well [42]. Figure 14 illustrates the cementation mechanism of LFS, primarily composed of γ-C_2S, under a carbonation environment based on the findings presented in this paper. Distinguishing itself from previous studies, this mechanism provides a comprehensive understanding of the carbonation reaction behavior by focusing on pore and particle cementation, particularly in relation to both open and closed pores. For clarity purposes, γ-C_2S particles are used instead of LFS particles.

This study describes the differentiation between <7% and >7% water content during carbonation behavior. At <7% water content, γ-C_2S particles exhibit reduced formation of bridging water, facilitating easy diffusion of CO_2 between and within the particles through interconnected pores (see Figure 14a). The enhanced contact area between CO_2 and the particles enables a greater carbonation extent for γ-C_2S particles. This leads to easier interior carbonation of specimens with lower water content, as demonstrated in Figure 8. However, complete carbonation of steel slag due to inherent particle size constraints leads to the presence of a porous carbonate layer in its products [43].

At >7% water content, there is sufficient bridging water to cement the particles; however, the presence of this bridging water hinders CO_2 diffusion into the interior. Calcium ions are released into the bridging water and react with CO_2, resulting in precipitation of calcium carbonate. This process leads to the solidification of the bridging water and closure of gaps between particles, transforming original open pores into closed ones after carbonation (see Figure 14b). Consequently, this impedes subsequent particle gaps and internal carbonation.

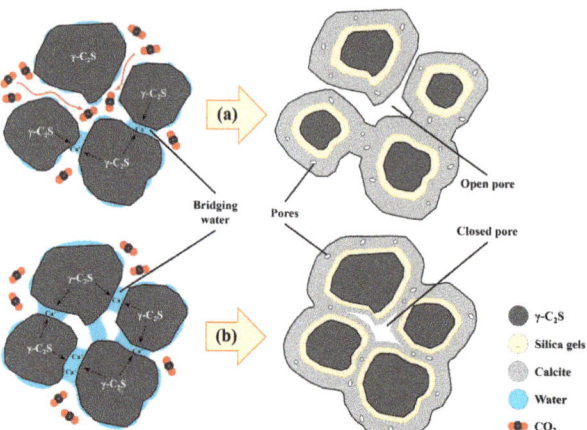

Figure 14. Carbonation cementation mechanism of LFS particles with (**a**) <7% water content; (**b**) >7% water content.

In summary, at <7% water content, the specimens were susceptible to interior carbonation due to the prolonged presence of open pores. At >7% water content, the specimens formed closed pores as a result of bridging water hardening, thereby hindering the carbonation process. At the optimum water content, although closed pores were also formed alongside open holes in the carbonated specimen, they did not impede carbonation from reaching the interior. Furthermore, Figure 14b illustrates a significantly more cemented structure with an appropriate amount of enhancing strength.

5. Limitations and Future Directions

This paper specifically addresses the significant impact of moisture on the carbonation reaction, without establishing correlations with other influencing factors. Future investigations should aim to comprehensively explore the carbonation mechanism of steel slag by concurrently studying moisture, temperature, CO_2 pressure, and other relevant variables. The objective of this paper is to emphasize the importance of moisture in the carbonation mechanism through controlled manipulation of its content; however, implementing such control measures in practical engineering scenarios may prove challenging based on the methods employed herein. Therefore, it is imperative for researchers to develop a suitable approach for regulating water content.

6. Conclusions

1. A method was developed to control the water content of steel slag specimens, and the significant influence of moisture and pore space on the carbonation reaction of steel slag specimens was innovatively identified. The carbonation process relies on pores for CO_2 diffusion and also requires a certain level of moisture for Ca^{2+} dissolution and diffusion. Increasing the water content enhances particle cementation and carbonation capacity in steel slag specimens; however, excessive water hinders CO_2 diffusion. Reducing the water content can increase the carbonation depth but may compromise gelling and carbon sequestration ability. Therefore, achieving a balance is crucial in controlling the water content;
2. The pour-molding process was employed to enhance the porosity of steel slag specimens. The increase in pores reduces the density of the steel slag specimens but enhances their CO_2 diffusion capacity. This is different from the common compression-molding process, achieving the absorption and fixation of approximately 217.2 kg of CO_2 for each ton of LFS at a 0.35 water–solid ratio and optimum water content. This carbon sequestration capacity is at an industry-leading level;

3. An innovative three-dimensional pore structure analysis method was used to analyze the connected and closed holes, and it was found that steel slag particles with low water content have a minimal impact on pore connectivity during cementation, allowing most of the pores to remain open. Conversely, steel slag particles with higher water content result in the formation of more closed pores that hinder CO_2 diffusion into the interior of the steel slag specimens, leading to a lower degree of carbonation compared to their surfaces. This study reveals the key mechanisms by which pore space and moisture affect carbonation reactions.

Author Contributions: Conceptualization, P.C.; methodology, P.C.; software, S.L.; validation, J.L.; formal analysis, W.X.; investigation, C.H.; resources, W.X.; data curation, S.L.; writing—original draft preparation, S.L.; writing—review and editing, C.H.; visualization, Y.D.; supervision, F.L.; project administration, P.C.; funding acquisition, P.C., W.X. and C.H. All authors have read and agreed to the published version of the manuscript.

Funding: This research was funded by the Guangxi Science Base and Talents Special Program (GuikeAD22035126), the Guangxi Natural Science Foundation project (2023GXNSFBA026130), and the Natural Science Foundation of China (No. 52368029 and No. 52062009).

Data Availability Statement: The raw data supporting the conclusions of this article will be made available by the authors on request.

Conflicts of Interest: The authors declare no conflicts of interest.

References

1. Xu, Y.; Lv, Y.; Qian, C. Comprehensive multiphase visualization of steel slag and related research in cement: Detection technology and application. *Constr. Build. Mater.* **2023**, *386*, 131572. [CrossRef]
2. Yu, X.; Chu, J.; Wu, S.; Wang, K. Production of biocement using steel slag. *Constr. Build. Mater.* **2023**, *383*, 131365. [CrossRef]
3. Xu, M.; Zhang, Y.; Yang, S.; Mo, L.; Liu, P. Effects of internal CO_2 curing provided by biochar on the carbonation and properties of steel slag-based artificial lightweight aggregates (SALAs). *Cem. Concr. Compos.* **2023**, *142*, 105197. [CrossRef]
4. Zhang, Y.; Yu, L.; Cui, K.; Wang, H.; Fu, T. Carbon capture and storage technology by steel-making slags: Recent progress and future challenges. *Chem. Eng. J.* **2023**, *455*, 140552. [CrossRef]
5. Gao, W.; Zhou, W.; Lyu, X.; Liu, X.; Su, H.; Li, C.; Wang, H. Comprehensive utilization of steel slag: A review. *Powder Technol.* **2023**, *422*, 118449. [CrossRef]
6. Song, Q.; Guo, M.Z.; Wang, L.; Ling, T.C. Use of steel slag as sustainable construction materials: A review of accelerated carbonation treatment. *Resour. Conserv. Recycl.* **2021**, *173*, 105740. [CrossRef]
7. Rui, Y.; Qian, C. CO_2-fixing steel slag on hydration characteristics of cement-based materials. *Constr. Build. Mater.* **2022**, *354*, 129193. [CrossRef]
8. Shen, J.; Zhang, Q.; Xu, L.; Tian, S.; Wang, P. Future CO_2 emission trends and radical decarbonization path of iron and steel industry in China. *J. Clean Prod.* **2021**, *326*, 129354. [CrossRef]
9. Yang, Y.; Xu, W.; Wang, Y.; Shen, J.; Wang, Y.; Geng, Z.; Wang, Q.; Zhu, T. Progress of CCUS technology in the iron and steel industry and the suggestion of the integrated application schemes for China. *Chem. Eng. J.* **2022**, *450*, 138438. [CrossRef]
10. Pan, S.Y.; Adhikari, R.; Chen, Y.H.; Li, P.; Chiang, P.C. Integrated and innovative steel slag utilization for iron reclamation, green material production and CO_2 fixation via accelerated carbonation. *J. Clean Prod.* **2016**, *137*, 617–631. [CrossRef]
11. Yadav, S.; Mehra, A. Experimental study of dissolution of minerals and CO_2 sequestration in steel slag. *Waste Manage.* **2017**, *64*, 348–357. [CrossRef] [PubMed]
12. Espinosa, A.B.; López-Ausín, V.; Fiol, F.; Serrano-López, R.; Ortega-López, V. Analysis of the deformational behavior of a clayey foundation soil stabilized with ladle furnace slag (LFS) using a finite element software. *Mater. Today Proc.* **2023**, *in press.* [CrossRef]
13. Pan, S.Y.; Chang, E.E.; Chiang, P.C. CO_2 capture by accelerated carbonation of alkaline wastes: A review on its principles and applications. *Aerosol Air Qual. Res.* **2012**, *12*, 770–791. [CrossRef]
14. Mo, L.; Zhang, F.; Deng, M. Mechanical performance and microstructure of the calcium carbonate binders produced by carbonating steel slag paste under CO_2 curing. *Cem. Concr. Res.* **2016**, *88*, 217–226. [CrossRef]
15. Li, L.; Jiang, Y.; Pan, S.Y.; Ling, T. Comparative life cycle assessment to maximize CO_2 sequestration of steel slag products. *Constr. Build. Mater.* **2021**, *298*, 123876. [CrossRef]
16. Humbert, P.S.; Castro-Gomes, J.P.; Savastano, H., Jr. Clinker-free CO_2 cured steel slag based binder: Optimal conditions and potential applications. *Constr. Build. Mater.* **2019**, *210*, 413–421. [CrossRef]
17. Ghouleh, Z.; Guthrie, R.I.; Shao, Y. High-strength KOBM steel slag binder activated by carbonation. *Constr. Build. Mater.* **2015**, *99*, 175–183. [CrossRef]

18. Zhu, F.; Cui, L.; Liu, Y.; Zou, L.; Hou, J.; Li, C.; Wu, G.; Xu, R.; Jiang, B.; Wang, Z. Experimental Investigation and Mechanism Analysis of Direct Aqueous Mineral Carbonation Using Steel Slag. *Sustainability* **2023**, *16*, 81. [CrossRef]
19. Wang, Y.; Liu, J.; Hu, X.; Chang, J.; Zhang, T.; Shi, C. Utilization of accelerated carbonation to enhance the application of steel slag: A review. *J. Sustain. Cen.-Based Mater.* **2023**, *12*, 471–486. [CrossRef]
20. Liu, J.; Liu, G.; Zhang, W.; Li, Z.; Jin, H.; Xing, F. A new approach to CO_2 capture and sequestration: A novel carbon capture artificial aggregates made from biochar and municipal waste incineration bottom ash. *Constr. Build. Mater.* **2023**, *398*, 132472. [CrossRef]
21. Zhong, X.; Li, L.; Jiang, Y.; Ling, T.C. Elucidating the dominant and interaction effects of temperature, CO_2 pressure and carbonation time in carbonating steel slag blocks. *Constr. Build. Mater.* **2021**, *302*, 124158. [CrossRef]
22. Luo, Z.; Wang, Y.; Yang, G.; Ye, J.; Zhang, W.; Liu, Z.; Mu, Y. Effect of curing temperature on carbonation behavior of steel slag compacts. *Constr. Build. Mater.* **2021**, *291*, 123369. [CrossRef]
23. Zhang, S.; Ghouleh, Z.; Mucci, A.; Bahn, O.; Provençal, R.; Shao, Y. Production of cleaner high-strength cementing material using steel slag under elevated-temperature carbonation. *J. Clean Prod.* **2022**, *342*, 130948. [CrossRef]
24. Zhang, S.; Ghouleh, Z.; Liu, J.; Shao, Y. Converting ladle slag into high-strength cementing material by flue gas carbonation at different temperatures. *Resour. Conserv. Recycl.* **2021**, *174*, 105819. [CrossRef]
25. Li, J.; Ni, W.; Wang, X.; Zhu, S.; Wei, X.; Jiang, F.; Zeng, H.; Hitch, M. Mechanical activation of medium basicity steel slag under dry condition for carbonation curing. *J. Build. Eng.* **2022**, *50*, 104123. [CrossRef]
26. Ukwattage, N.L.; Ranjith, P.G.; Li, X. Steel-making slag for mineral sequestration of carbon dioxide by accelerated carbonation. *Measurement* **2017**, *97*, 15–22. [CrossRef]
27. Zhang, X.; Su, Y.; Liu, S. The effect of carbonation pressure on microbial assisted steel slag-based carbon sequestration material. *J. Build. Eng.* **2024**, *86*, 108974. [CrossRef]
28. Tian, S.; Jiang, J.; Chen, X.; Yan, F.; Li, K. Direct Gas-Solid Carbonation Kinetics of Steel Slag and the Contribution to Insitu Sequestration of Flue Gas CO_2 in Steel-Making Plants. *ChemSusChem* **2013**, *6*, 2348–2355. [CrossRef]
29. Nielsen, P.; Boone, M.A.; Horckmans, L.; Snellings, R.; Quaghebeur, M. Accelerated carbonation of steel slag monoliths at low CO_2 pressure–microstructure and strength development. *J. CO_2 Util.* **2020**, *36*, 124–134. [CrossRef]
30. Tang, P.; Xuan, D.; Cheng, H.W.; Poon, C.S.; Tsang, D.C. Use of CO_2 curing to enhance the properties of cold bonded lightweight aggregates (CBLAs) produced with concrete slurry waste (CSW) and fine incineration bottom ash (IBA). *J. Hazard. Mater.* **2020**, *381*, 120951. [CrossRef]
31. Jiang, Y.; Ling, T.C. Production of artificial aggregates from steel-making slag: Influences of accelerated carbonation during granulation and/or post-curing. *J. CO_2 Util.* **2020**, *36*, 135–144. [CrossRef]
32. GB/T 17671-2021; Test Method of Cement Mortar Strength (ISO Method). State Administration of Market Supervision and Standardization Administration: Beijing, China, 2021. (In Chinese)
33. McCarter, W.J.; Chrisp, T.M.; Starrs, G.; Owens, E.H. Setting, hardening and moisture-loss within a cement-based backfill grout under simulated repository environments. *Measurement* **2012**, *45*, 235–242. [CrossRef]
34. Tan, Y.; Liu, Z.; Wang, F. Effect of temperature on the carbonation behavior of γ-C_2S compacts. *Cem. Concr. Compos.* **2022**, *133*, 104652. [CrossRef]
35. Yu, C.; Cui, C.; Zhao, J.; Liu, F.; Fu, S.; Li, G. Enhancing mechanical properties of dredged sludge through carbonation stabilization employing steel slag: An eco-friendly and cost-effective approach. *Constr. Build. Mater.* **2024**, *412*, 134748. [CrossRef]
36. Wang, D.; Xiao, J.; Duan, Z. Strategies to accelerate CO_2 sequestration of cement-based materials and their application prospects. *Constr. Build. Mater.* **2022**, *314*, 125646. [CrossRef]
37. Mu, Y.; Liu, Z.; Wang, F. Comparative study on the carbonation-activated calcium silicates as sustainable binders: Reactivity, mechanical performance, and microstructure. *ACS Sustain. Chem. Eng.* **2019**, *7*, 7058–7070. [CrossRef]
38. Humbert, P.S.; Castro-Gomes, J. CO_2 activated steel slag-based materials: A review. *J. Clean Prod.* **2019**, *208*, 448–457. [CrossRef]
39. Wray, J.L.; Daniels, F. Precipitation of calcite and aragonite. *J. Am. Chem. Soc.* **1957**, *79*, 2031–2034. [CrossRef]
40. Guan, X.; Liu, S.; Feng, C.; Qiu, M. The hardening behavior of γ-C_2S binder using accelerated carbonation. *Constr. Build. Mater.* **2016**, *114*, 204–207. [CrossRef]
41. Moon, E.J.; Choi, Y.C. Development of carbon-capture binder using stainless steel argon oxygen decarburization slag activated by carbonation. *J. Clean Prod.* **2018**, *180*, 642–654. [CrossRef]
42. Ren, E.; Tang, S.; Liu, C.; Yue, H.; Li, C.; Liang, B. Carbon dioxide mineralization for the disposition of blast-furnace slag: Reaction intensification using NaCl solutions. *Greenh. Gases* **2020**, *10*, 436–448. [CrossRef]
43. Kim, J.; Azimi, G. The CO_2 sequestration by supercritical carbonation of electric arc furnace slag. *J. CO_2 Util.* **2021**, *52*, 101667. [CrossRef]

Disclaimer/Publisher's Note: The statements, opinions and data contained in all publications are solely those of the individual author(s) and contributor(s) and not of MDPI and/or the editor(s). MDPI and/or the editor(s) disclaim responsibility for any injury to people or property resulting from any ideas, methods, instructions or products referred to in the content.

Article

Shear Performance of the Interface of Sandwich Specimens with Fabric-Reinforced Cementitious Matrix Vegetal Fabric Skins

Lluís Gil *, Luis Mercedes, Virginia Mendizabal and Ernest Bernat-Maso

Department of Strength of Materials and Structural Engineering, School of Industrial, Aerospace and Audiovisual of Terrassa, Universitat Politècnica de Catalunya, C/Colom 11, TR45, 08222 Terrassa, Spain; luis.enrique.mercedes@upc.edu (L.M.); virginia.dolores.mendizabal@upc.edu (V.M.); ernest.bernat@upc.edu (E.B.-M.)
* Correspondence: lluis.gil@upc.edu

Featured Application: The design and manufacturing of sandwich solutions using FRCM vegetal fabric skins improve sustainability because they provide solutions with a lower global carbon footprint.

Abstract: The utilization of the vegetal fabric-reinforced cementitious matrix (FRCM) represents an innovative approach to composite materials, offering distinct sustainable advantages when compared to traditional steel-reinforced concrete and conventional FRCM composites employing synthetic fibers. This article introduces a design for sandwich solutions based on a core of extruded polystyrene and composite skins combining mortar as a matrix and diverse vegetal fabrics as fabrics such as hemp and sisal. The structural behavior of the resulting sandwich panel is predominantly driven by the interaction between materials (mortar and polyurethane) and the influence of shear connectors penetrating the insulation layer. This study encompasses an experimental campaign involving double-shear tests, accompanied by heuristic bond-slip models for the potential design of sandwich solutions. The analysis extends to the examination of various connector types, including hemp, sisal, and steel, and their impact on the shear performance of the sandwich specimens. The results obtained emphasize the competitiveness of vegetal fabrics in achieving an effective composite strength comparable to other synthetic fabrics like glass fiber. Nevertheless, this study reveals that the stiffness of steel connectors outperforms vegetal connectors, contributing to an enhanced improvement in both stiffness and shear strength of the sandwich solutions.

Keywords: sandwich panels; FRCM; cementitious matrix; vegetal fibers; shear test

Citation: Gil, L.; Mercedes, L.; Mendizabal, V.; Bernat-Maso, E. Shear Performance of the Interface of Sandwich Specimens with Fabric-Reinforced Cementitious Matrix Vegetal Fabric Skins. *Appl. Sci.* 2024, 14, 883. https://doi.org/10.3390/app14020883

Academic Editors: Mouhamadou Amar and Nor Edine Abriak

Received: 22 December 2023
Revised: 13 January 2024
Accepted: 17 January 2024
Published: 19 January 2024

Copyright: © 2024 by the authors. Licensee MDPI, Basel, Switzerland. This article is an open access article distributed under the terms and conditions of the Creative Commons Attribution (CC BY) license (https://creativecommons.org/licenses/by/4.0/).

1. Introduction

Sandwich panels crafted with concrete skins and insulating cores are a competitive solution for building structural components with energy-efficient added value (see a review in Oliveira et al.) [1]. The concern about climate change and sustainable solutions drives research towards greener and more bio-based engineering sandwich technologies according to Oliveira et al. [2]. In the present work, a solution based on the vegetal FRCM and polystyrene core serves as a lightweight construction solution with noteworthy insulating properties for building enclosures. The mechanical properties of these panels depend significantly on the composite action between materials. An adequate material connection is crucial, as insufficient bonding may result in a problematic stress distribution within the panel, potentially causing detachment failure or a substantial decrease in mechanical strength.

In this order, different authors like Cox et al. [3] developed a composite shear connector system of a glass-fiber-reinforced polymer (GFRP) used to transfer interface shear forces in a precast concrete sandwich panel. This study developed push-off, pullout, and flexural tests to evaluate the structural performance of the shear connectors, and the effect of the bond between concrete and insulation was evaluated with push-off tests. The results showed satisfactory performance with a lower bound of 90% of composite action for specimens with a 100 mm thick insulation wythe and full composite action for most panels with 50 mm thick insulation. Pantano et al. [4] studied a numerical model by evaluating zigzag functions by enforcing the continuity of transverse shear stresses at layer interfaces, being able to predict accurately the distribution of stresses and displacements in laminated plates and sandwich panels. Another study by Portal et al. [5] presented experimental and numerical methods to analyze the structural behavior related to a sandwich panel with a glass-fiber-reinforced polymer (GFRP) plate connection system, where double-shear tests were conducted on sandwich specimens to characterize the available shear capacity provided by the connectors and panel configuration. The authors conclude that for well-balanced composite action, it is necessary to use the least amount of material in the plate connectors and an increased amount of bending capacity of the outer panel to avoid a significant drop in the load after the peak load. At the same line, Hulin et al. [6] presents an experimental campaign at elevated temperatures for panels stiffened by structural ribs, insulation layers, and steel shear connectors. The results highlighted insulation shear failure from differential thermal expansion at the interface with concrete, where the shear connectors induced stress concentrations, leading to local failure. A study presented by Tomlinson et al. [7] carried out push-through tests on a precast-concrete-insulated sandwich panel using combined angled and horizontal connectors, where basalt-fiber-reinforced polymer and steel connectors were used. This study evaluated various connector inclination angles and diameters, diagonal connector orientations relative to the loading, and panels with or without an active foam-to-concrete bond. The results show that steel connectors failed by yielding in tension and inelastic buckling in compression. In the case of larger-diameter basalt-FRP connectors, they pulled out under tension and crushed in compression, and smaller-diameter basalt-FRP connectors ruptured in tension and buckled in compression. Also, it is demonstrated that the strength and stiffness improved with the connector angle and diameter. Lou et al. [8] performed 24 double-shear tests on precast-concrete-insulated sandwich panels using stainless-steel plate connectors. The authors analyzed parameters like connector directions, insulation effects, cavity widths, and connector heights. The authors developed a consistent campaign with in-plane and out-of-plane shear tests and concluded that the cavity size and the presence of insulation significantly contributed to shear transfer. Choi et al. [9] analyzed some precast concrete sandwich panels used for exterior cladding. Specimens were experimentally tested with a push-out test, with and without corrugated shear connectors. The investigation of the in-plane shear performance showed a relevant impact of the core material in the structural response. And later, Choi et al. [10] extended the study of the shear flow to the type of connectors. There are other relevant contributions about the shear performance of sandwich insulation panels using other types of tests like Sylaj et al. [11], Hou at al [12], Meng et al. [13], or Wang et al. [14].

To advance towards the utilization of more sustainable materials compared to synthetic fibers and steel, the present study concentrates on the creation of sandwich panels comprising a vegetal FRCM and expanded polystyrene as insulation. Authors have previously presented other complementary studies about FRCM vegetal fibers (see Mercedes et al. [15,16]) and some bending tests for sandwich FRCM solutions (see Mercedes et al. [16]). For the present work, the innovation lies in the use of vegetal FRCM skins and the introduction of flexible connectors made from vegetal fibers. The connection between layers is a must to have the necessary mechanical properties to develop composite materials that are competent with those commonly used in the construction industry. In this study, sandwich specimens were created using different fabrics (hemp, sisal, and glass) and connectors (hemp, sisal, and steel). These specimens underwent a double-shear test to examine how these fabrics and connectors influence the panel's strength against shear.

Additionally, a simplified connector slip-load model was developed and compared to the experimental results.

2. Materials and Methods

The experimental campaign included the manufacturing of specimens of FRCM bonded to an extruded polystyrene core and shear tests. These specimens were produced using the following specific procedures and materials.

2.1. Mortar

To manufacture the FRCM component, a thixotropic commercial mortar was used. This mortar is a single-component mixture comprising cement, synthetic resins, and polyamide fibers, with the addition of silica fume. The choice of this mortar was based on its proven effectiveness in previous studies such as Mercedes et al. [15]. The average results of the compression and flexion tests using norm EN1015-11:2000 [17] have been previously presented in the cited work with values of 39.25 MPa and 6.56 MPa, respectively.

2.2. Fabrics

Two types of vegetal fabrics and another type of synthetic fiber were used to manufacture the FRCM part: hemp, sisal, and glass (contrast material).

Vegetal fabrics were crafted using hemp and sisal yarns (both with a diameter of 2.5 mm). This arbitrary choice was justified by the notable tension levels achieved by hemp and sisal FRCM specimens in a prior study by Mercedes et al. [18]. In that study, the fabrics and yarns were coated with epoxy resin. This was performed to prevent fiber degradation produced by the environment of cementitious matrix composites (high alkalinity and humidity cycles) (see Ardanuy et al.) [19].

The size of the free cells in the grids of vegetal fabrics was determined by referencing the geometry of a commercial glass fabric (see Figure 1). In the case of vegetal fabrics, it was necessary to craft meshes with a greater volume of material to achieve the load capacity of glass fabrics, producing thicker meshes than synthetic ones. Two yarns were utilized for each tuft, underscoring the fact that the tensile strength and effectiveness were comparable to synthetic fiber meshes just by simply increasing the volume of vegetal fibers.

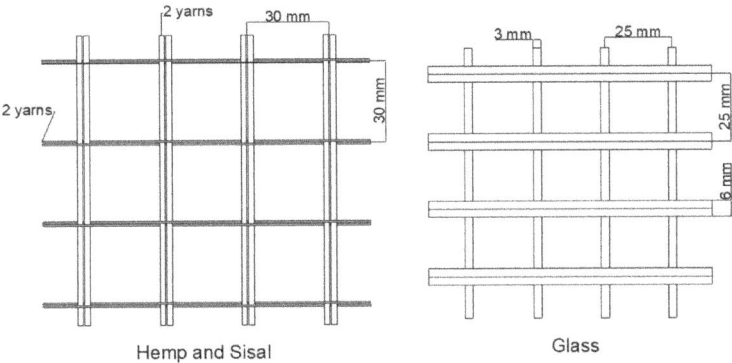

Figure 1. Reinforcing fabrics geometry: hemp and sisal (**left**); glass (**right**).

Weft yarns of hemp and sisal fabric were made of hemp yarns of 0.5 mm in order to reduce the thickness of the weft and wrap crossing point, and because the load capacity in the weft direction was not relevant for the shear test setup used in this study.

The fabrics were woven (Figure 2) with the same procedures used in Donnini et al. [20] and D'Antino et al. [21]. After one day of curing, the meshes were cut into pieces with dimensions of 45 mm × 35 mm.

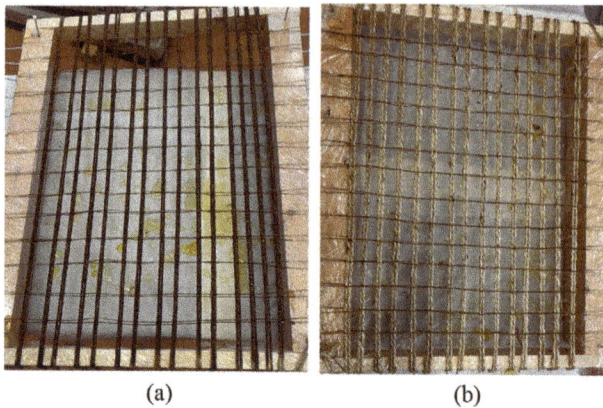

Figure 2. Vegetal fabric preparation: (**a**) hemp fabric and (**b**) sisal fabric.

The mechanical properties of the tuft (two yarns in the vegetal fabrics case) are shown in Table 1. The coated tuft data were obtained experimentally in this study using the tensile test procedure using norm EN ISO 13934-1/2 [22].

Table 1. Tuft properties ((%) = coefficient of variation, A_f = fiber area F_{fu} = maximum load mean, σ_{fu} = tensile strength mean, E_f = Young's modulus mean, ε_{fpick} = deformation peak mean).

Fibres	Number of Test	A_f (mm^2)	F_{fu} (N)	σ_{fu} (MPa)		E_f (GPa)		ε_{fpick} (%)	
Hemp	5	9.81	1701.00	173.35	(3%)	8.59	(11%)	1.45	(16%)
Sisal	5	9.81	1467.00	137.25	(16%)	4.87	(36%)	2.31	(14%)
Glass	5	1.05	708.00	668.50	(8%)	61.25	(2%)	1.32	(6%)

2.3. Connectors

To assess the impact of connectors on the shear behavior of FRCM bonded to an extruded polystyrene core, connectors of hemp, sisal, and steel were used. Hemp and sisal were selected as they are the vegetal fibers used in this study for crafting vegetal fabrics, while steel was chosen as it is the most commonly commercial material used for connectors in such types of sandwich solution.

Vegetal fiber connectors were crafted hook-shaped and impregnated with epoxy resin, featuring an equivalent area of 29.43 mm^2 (6 yarns). Steel connectors were in the shape of a pin or bolt with a cross-sectional area of 0.79 mm^2, accompanied by a nut at one end to enhance the anchoring effect with the mortar (see Figure 3).

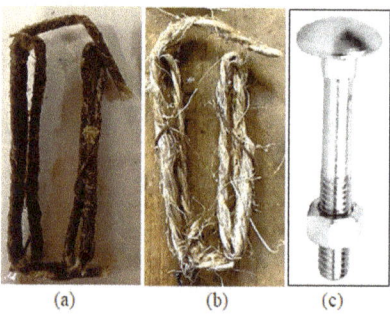

Figure 3. Types of connectors: (**a**) hemp, (**b**) sisal, (**c**) steel.

2.4. Extruded Polystyrene

Rigid extruded polystyrene foam boards with a thickness of 40 mm were used as the insulating core in the sandwich sample configuration.

2.5. Specimens

The experimental program included 40 specimens. The specimens consisted of a sandwich panel combining FRCM and extruded polystyrene, following the geometry in Figure 4.

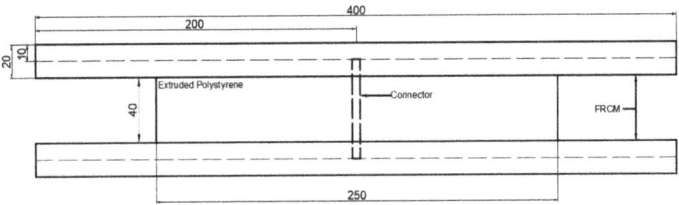

Figure 4. Specimen geometry in mm.

The dimensions of the FRCM were 50 × 400 mm with a thickness of 20 mm. These 40 samples included 3 different connectors (steel, hemp, and sisal) and 3 different reinforcement fabrics (hemp, sisal, and glass).

The mold to manufacture the FRCM specimens was prepared with a grid of wooden strips with 50 mm × 400 mm gaps (see Figure 4). These strips had a height of 20 mm. The manufacturing procedure was as follows:

- Prepare the mold base with a demolding agent (Figure 5a).
- Mix the mortar and pour it to a depth of approximately 15 mm.
- Place the fabric so that it slightly penetrates the first mortar layer (Figure 5b).
- Cover the fabric with a second layer of mortar to reach a thickness of 20 mm for the bottom FRCM layers.
- Place the extruded polystyrene boards. In the case of specimens with connectors, it has a hole in the middle.
- Place the mold of the other wooden strips (without a base) in the same location as the first mold.
- Add a third layer of mortar to reach the final thickness.
- Place the second fabric so that it slightly penetrates the first layer of mortar.
- For panels with connectors, place the connectors so that the top is above the fabric.
- Cover the second fabric (and connectors) with a fourth layer of mortar to reach a thickness of 20 mm for the top FRCM layers.
- Demold and leave samples to cure in laboratory conditions for 28 days. After this period, the specimens are ready for testing (Figure 5c).

The nomenclature used to identify the specimens is provided in Table 2.

Table 2. Nomenclature of sandwich panels.

Specimen	Fibres	Connectors	Numbers of Samples
SH-N	Hemp	Without	5
SH-H	Hemp	Hemp	5
SH-S	Hemp	Sisal	5
SH-St	Hemp	Steel	5
SS-N	Sisal	Without	5
SS-S	Sisal	Sisal	5
SG-N	Glass	Without	5
SG-St	Glass	Steel	5

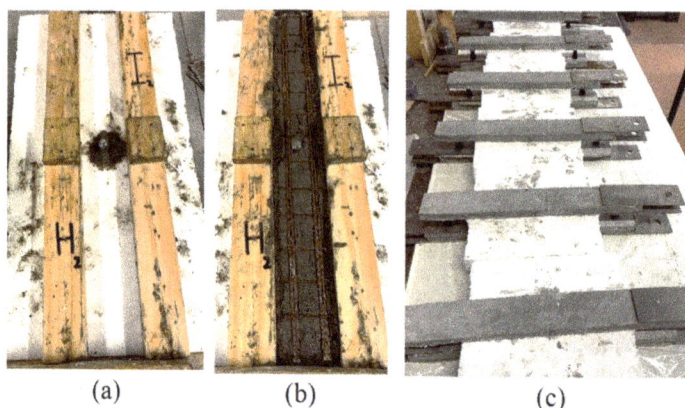

Figure 5. Manufacturing of sandwich panels: (**a**) mold, (**b**) fabric with steel connector and first layer of mortar, (**c**) final specimen.

3. Experimental Campaigns

3.1. Test Setup and Instrumentation

The specimens were subjected to a double-shear test (see Figure 6). In this test, metal plates (similar to those used in tension tests) were bonded to one end of the FRCM on each side of the sandwich specimen. This shear test is an adaptation inspired by the tension test with the clevis system according to AC434-0213-R1 [23]. In this configuration, auxiliary plates of aluminum were attached externally on opposite sides of the load application, simply to prevent the turning of the specimens during the test. An electromechanical press (MTS Insight 10 kN range) was used to perform the tests with a test rate of 2 mm/min.

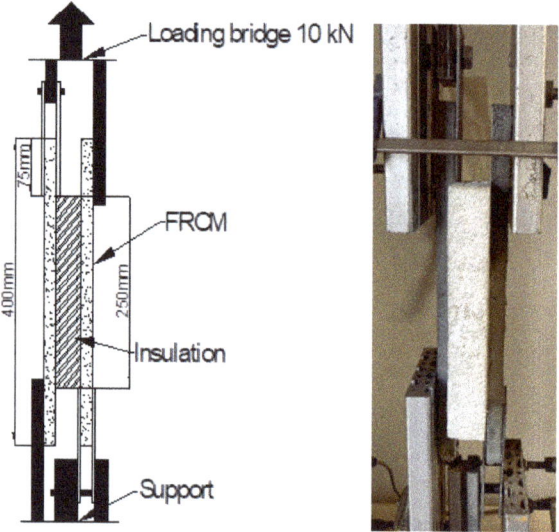

Figure 6. Double-shear test setup.

3.2. Type of Failures

In general, all the specimens had peeling failure because polystyrene is a low-strength material. Nevertheless, in the case of the samples with connectors, there was also a slippage of the connectors accompanied by the detachment of the mortar in the connector area, in some cases. Consequently, specimens with connectors displayed more cracking and ductile failure compared to the sudden and brittle failure observed in specimens without connectors (Figure 7).

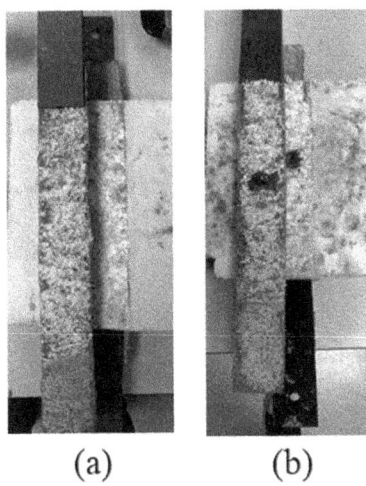

Figure 7. Type of failures: (**a**) peeling failures, (**b**) peeling failures and connector slipping.

3.3. Experimental Results

Table 3 shows the experimental results of the maximum load and shear stress (Fmax and τmax, respectively), the elastic stiffness (Ke), and the shear modulus (Ge) obtained from the linear stage in the load–displacement diagrams. Also, the following table presents the displacement (δmax) and angular distortion (δmax/te) at the maximum load, where te is the distance (60 mm) between the fabric embedded in the FRCM skins. To calculate the shear stress and shear modulus, the shear value from the dethatched FRCM skin (50 × 250 mm) was used.

Table 3. Experimental results.

Specimen	F_{max} (N)	τ_{max} (MPa)	C.V	K_e (N/mm)	G_e (MPa)	C.V	δ_{max} (mm)	δ_{max}/t_e (%)	C.V
SH-N	1409.20	0.11	(12%)	467.94	1.87	(10%)	4.85	9.70	(23%)
SH-H	1694.60	0.14	(6%)	601.05	2.40	(19%)	5.49	10.98	(21%)
SH-S	1684.40	0.13	(14%)	537.22	2.15	(15%)	4.50	9.01	(31%)
SH-St	2151.20	0.17	(8%)	653.73	2.61	(13%)	7.04	14.08	(22%)
SS-N	1369.75	0.11	(9%)	432.29	1.73	(7%)	4.89	9.79	(19%)
SS-S	1543.40	0.12	(9%)	479.70	1.92	(12%)	7.21	14.41	(24%)
SG-N	1340.40	0.11	(10%)	487.95	1.95	(7%)	5.60	11.19	(47%)
SG-St	1333.25	0.11	(12%)	608.29	2.43	(18%)	4.41	8.81	(40%)

The results in Table 3 show coefficients of variation ranging between 2 and 14% for maximum load and shear strength, respectively, indicating the good repeatability of the experiments, following Donnini and Corinaldesi [20]. It is noteworthy that specimens without connectors presented a similar shear strength. However, for stiffness and displacements, the variation was higher, ranging between 7 and 47%. This variability represents the expected scattering of data for elements composed of cementitious materials with a high non-linear behavior. The results presented in Table 3 are better appreciated in Figures 8–11.

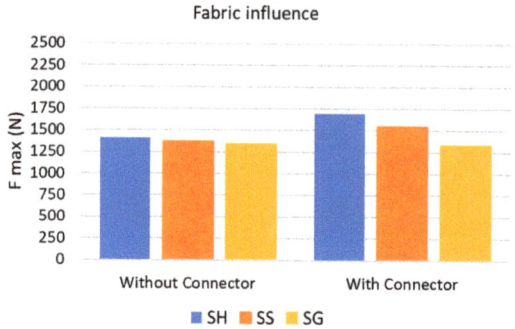

Figure 8. Fabric influence: maximum load.

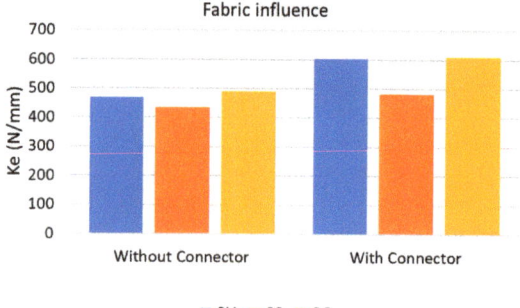

Figure 9. Fabric influence: elastic stiffness.

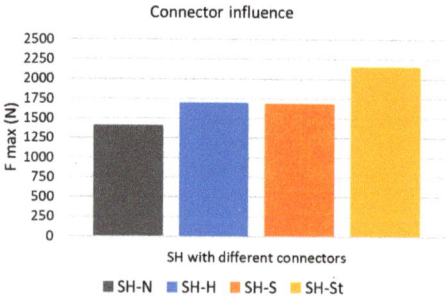

Figure 10. Connector influence: maximum load.

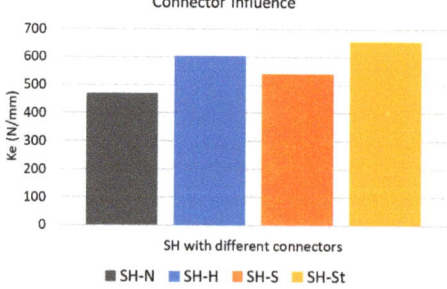

Figure 11. Connector influence: elastic stiffness.

Figures 8 and 9 show the impact of the fabric on the maximum load and elastic stiffness of the specimens. The performance of different fabric types remains consistent for both maximum load and stiffness values, regardless of the presence of connectors. Concerning maximum load, the inclusion of connectors significantly enhances the capacity for vegetal specimens, ranging from 13% to 53%. However, there is no notable change for glass fabric. This suggests that the FRCM skin may not reach the cracking strength required to activate the glass fabric, unlike what occurs with vegetal fabrics. Nevertheless, connectors play a beneficial role in maintaining cohesion between materials, activating vegetal fabrics at the achieved level of strain until significantly producing a higher ultimate load.

In terms of elastic stiffness, the response of the FRCM sandwich is directly tied to the Young's modulus of the fabric—a stiffer fabric correlates with higher specimen stiffness. Despite glass fabric being seven times stiffer than hemp fabric, this stiffness difference is not prominently reflected in the specimens. This is due to the initiation of non-linear behavior in the core deformation for low values of FRCM strain, minimizing the activation of fabric capacity in the composite and resulting in negligible stiffness differences.

Connectors prove efficient in ensuring strain compatibility among components, leading to an increase in stiffness values ranging from 11% to 40%. Although the influence of connectors during the elastic phase is minimal compared to the effect over the ultimate load, they play a crucial role in maintaining overall specimen compatibility.

Figures 10 and 11 illustrate the impact of the type of connector on the maximum load and on elastic stiffness for the hemp fabric specimens. The presence of connectors, independently of their material, increases both the maximum load and the elastic stiffness. The steel connector reaches the highest load and the highest stiffness. Therefore, the presence of stiff connectors maintains the strain field and the compatibility between layers in a more efficient manner than flexible vegetal connectors. The difference in stiffness between the fabric and the steel connector seems to not be a disadvantage, even though some local damage takes place in the mortar because of the concentration effects of the steel bolt.

Figures 12–16 show the load–displacement plots of the tested panels. It can be seen that specimens without connectors exhibit a quasi-brittle failure with a limited range of deformation (dashed lines) compared to specimens with connectors (continuous line). The presence of connectors enhances the activation of vegetal fabrics, effectively tightening the interfaces between materials and contributing to an increased strength of the sandwich structure. In the case of glass, the levels of strain are low in the FRCM and the fabric is not activated; therefore, there is no large difference in the load–displacement plots.

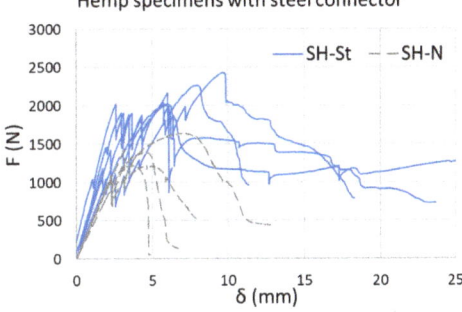

Figure 12. Load–displacement curves of SH-St compared with SH-N.

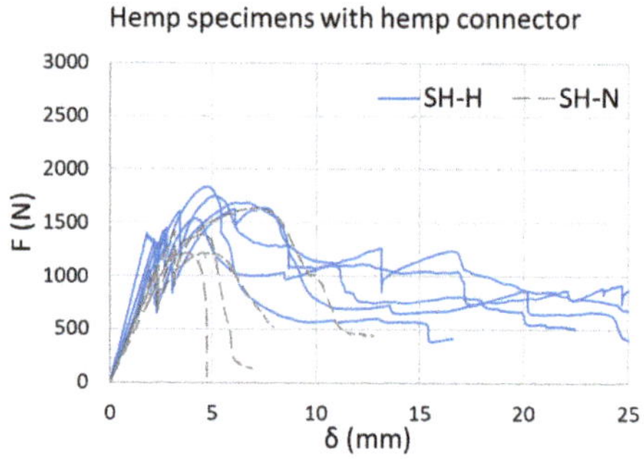

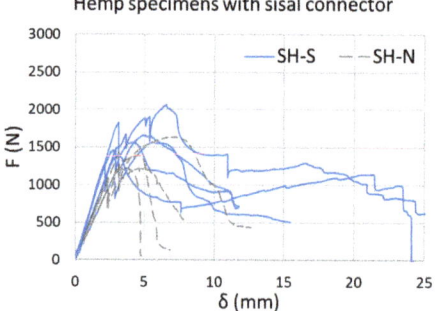

Figure 13. Load–displacement curves of SH-H compared with SH-N.

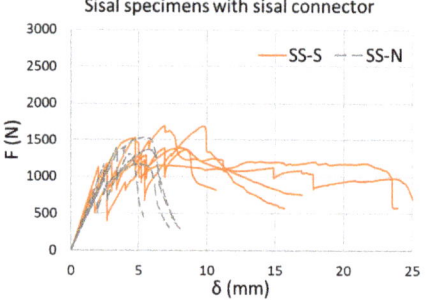

Figure 14. Load–displacement curves of SH-S compared with SH-N.

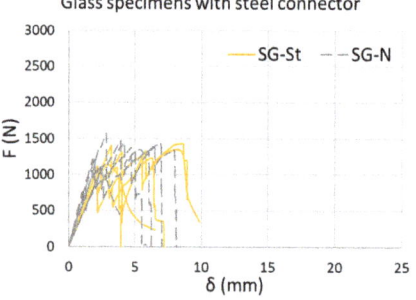

Figure 15. Load–displacement curves of SS-S compared with SS-N.

Figure 16. Load–displacement curves of SG-St compared to SG-N.

4. Connectors' Interlock-Slip Simplified Model

According to the experimental results, it is worth building a simple model to easily generate pre-dimension solutions for FRCM sandwich panels. Therefore, it is necessary to estimate the contribution of the connectors to the response of the structure. The problem is very complex, involving the nonlinear behavior of FRCM skins, interface interactions among materials, and debonding and slipping failure. A real model is far from the scope of the contribution. Nevertheless, a rough approach might take advantage of the comparison between the response of specimens without connectors and those with connectors. Therefore, in a simple manner:

$$F_{max} = F_{max\_none\_connector} + F_{max\_connector}$$

From Table 3 and the plots of Figures 12–14, it is feasible to estimate the contribution of the interface of the FRCM and polystyrene. To study the effect of the connector, only hemp FRCM specimens (SH) were used because they contain all type of samples.

The connectors interact, exhibiting a bi-linear behavior. Each one interacts, increasing its contribution until the maximum load is reached, while after, they are capable of maintaining it without a significant reduction, due to its stiffness.

As stated in Figure 11, steel connectors showed the highest stiffness, followed by hemp and finally sisal connectors. This property explains the reason why steel connectors are those that contribute more significantly to the shear strength of the specimen, providing a contribution 165% more than the vegetal connectors. Hemp and sisal show a similar contribution, due to their similar mechanical properties seen in Table 1.

Therefore, Figure 17 estimates the connector contribution.

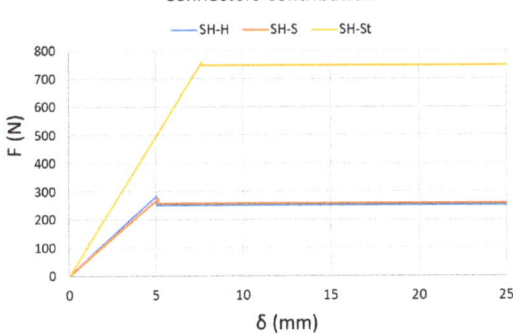

Figure 17. Connectors' contribution.

5. Conclusions

In this work, an experimental and analytical study was conducted to investigate the shear performance of a sandwich specimen with a vegetal-FRCM and polystyrene. According to the achieved results:

- All specimens experienced a peeling failure. However, specimens with connectors exhibited additional slippage of the connectors, resulting in more cracking and ductile failure compared to the fragile failure observed in specimens without connectors.
- The results showed that there was no significant influence of the kind of fabric in the maximum load of the specimen without the connector. This took place because the FRCM skin did not reach the cracking level required for the fabric to be activated and effectively contribute to the strength.
- In the case of the specimens with connectors, the levels of maximum loads and elastic stiffness were both increased. Vegetal fabrics were effectively activated by the cracking

while glass was very little activated. Therefore, the comparative performance produced a more ductile response in vegetal fabrics, due to their elongation capacity.
- All the types of connectors increased the maximum load and elastic stiffness of the sandwich specimens. The steel connector reached the highest maximum load and elastic stiffness. Hence, stiff connectors produced a tightening effect between the layers of materials, and the higher the stiffness in connectors, the higher the sandwich response.
- An interlock-slip model based on experimental evidence showed the potential to design FRCM solutions for sandwich applications with connectors. It showed that the connector contributes significantly.

The key findings indicate that FRCM vegetal fibers demonstrated mechanical competitiveness when compared to glass fiber. Sandwich solutions require connectors to enhance the mechanical capacity and, in this specific instance, steel connectors exhibit a more efficient performance than vegetal connectors. Therefore, future improvements could focus on designing vegetal connectors with increased stiffness and conduct additional tests to develop a more suitable interlock-slip model.

Author Contributions: Conceptualization, L.M.; methodology, L.M.; software, E.B.-M.; validation, L.G. and E.B.-M.; formal analysis, L.M.; investigation, L.M., E.B.-M., V.M. and L.G.; resources, L.G.; data curation, L.G.; writing—original draft preparation, L.M.; writing—review and editing, L.G., E.B.-M. and V.M.; visualization, L.M. and V.M.; supervision, L.G. and E.B.-M.; project administration, L.G.; funding acquisition, L.G. All authors have read and agreed to the published version of the manuscript.

Funding: This research was funded by the research project "Investigación para la impresión 3D de compuestos aislantes de FRCM para cerramientos inteligentes en edificación" COMPI3D, PID2022-137156OB-I00. It is funded by Agencia Estatal de Investigación (AEI), Ministerio de Ciencia e Innovación.

Institutional Review Board Statement: Not applicable.

Informed Consent Statement: Not applicable.

Data Availability Statement: The acquired data are directly presented in this manuscript, and they are original and neither inappropriately selected, manipulated, enhanced, nor fabricated.

Acknowledgments: The materials were supported by Sika AG and Mapei Spain S.A. The fourth author is a Serra Húnter Fellow.

Conflicts of Interest: The authors declare no conflicts of interest.

References

1. Oliveira, T.F.; de Carvalho, J.M.F.; Mendes, J.C.; Souza, G.-B.Z.; Carvalho, V.R.; Peixoto, R.A.F. Precast concrete sandwich panels (PCSP): An analytical review and evaluation of CO_2 equivalent. *Constr. Build. Mater.* **2022**, *358*, 129424. [CrossRef]
2. Oliveira, P.R.; May, M.; Panzera, T.H.; Hiermaier, S. Bio-based/green sandwich structures: A review. *Thin-Walled Struct.* **2022**, *177*, 109426. [CrossRef]
3. Cox, B.; Syndergaard, P.; Al-Rubaye, S.; Pozo-Lora, F.F.; Tawadrous, R.; Maguire, M. Lumped GFRP star connector system for partial composite action in insulated precast concrete sandwich panels. *Compos. Struct.* **2019**, *229*, 111465. [CrossRef]
4. Pantano, A.; Averill, R.C. A 3D Zig-Zag sublaminate model for analysis of thermal stresses in laminated composites and sandwich plates. *J. Sandw. Struct. Mater.* **2000**, *2*, 228–312. [CrossRef]
5. Portal, N.W.; Zandi, K.; Malaga, K.; Wlasak, L. GFRP connectors in textile reinforced concrete sandwich elements. In Proceedings of the IABSE Congress, Stockholm 2016: Challenges in Design and Construction of an Innovative and Sustainable Built Environment, Stockholm, Sweden, 21–23 September 2016; pp. 1336–1343.
6. Hulin, T.; Hodicky, K.; Schmidt, J.W.; Stang, H. Experimental investigations of sandwich panels using high performance concrete thin plates exposed to fire. *Mater. Struct.* **2016**, *49*, 3879–3891. [CrossRef]
7. Tomlinson, D.G.; Teixeira, N.; Fam, A. New Shear Connector Design for Insu-lated Concrete Sandwich Panels Using Basalt Fiber-Reinforced Polymer Bars. *J. Compos. Constr.* **2016**, *20*, 04016003. [CrossRef]
8. Lou, X.; Xue, W.; Bai, H.; Li, Y.; Huang, Q. Shear behavior of stainless-steel plate connectors for insulated precast concrete sandwich panels. *Structures* **2022**, *44*, 1046–1056. [CrossRef]
9. Choi, K.-B.; Choi, W.-C.; Feo, L.; Jang, S.-J.; Yun, H.-D. In-plane shear behavior of insulated precast concrete sandwich panels reinforced with corrugated GFRP shear connectors. *Compos. Part B Eng.* **2015**, *79*, 419–429. [CrossRef]

10. Choi, W.; Jang, S.-J.; Yun, H.-D. Design properties of insulated precast concrete sandwich panels with composite shear connectors. *Compos. Part B Eng.* **2019**, *157*, 36–42. [CrossRef]
11. Sylaj, V.; Fam, A. UHPC sandwich panels with GFRP shear connectors tested under combined bending and axial loads. *Eng. Struct.* **2021**, *248*, 113287. [CrossRef]
12. Hou, H.; Wang, W.; Qu, B.; Dai, C. Testing of insulated sand-wich panels with GFRP shear connectors. *Eng. Struct.* **2020**, *209*, 109954. [CrossRef]
13. Meng, Y.; Wang, L.; Chen, B. Shear resistance and deflection prediction of steel–concrete–steel sandwich panel with headed stud connectors. *Structures* **2023**, *54*, 1690–1704. [CrossRef]
14. Wang, Y.; Chen, J.; Zhai, X.; Zhi, X.; Zhou, H. Static behaviours of steel-concrete-steel sandwich beams with novel interlocked angle connectors: Test and analysis. *J. Constr. Steel Res.* **2023**, *201*, 107723. [CrossRef]
15. Mercedes, L.; Mendizábal, V.; Bernat-Maso, E.; Gil, L. Performance of hemp-FRCM-strengthened beam subjected to cyclic loads. *Mater. Constr.* **2022**, *72*, e270. [CrossRef]
16. Mercedes, L.; Gil, L.; Bernat-Maso, E. Mechanical performance of vegetal fabric reinforced cementitious matrix (FRCM) composites. *Constr. Build. Mater.* **2018**, *175*, 161–173. [CrossRef]
17. *EN 1015-11*; Methods of Test for Mortar for Masonry. Part 11: Determination of Flexural and Compressive Strength of Hardened Mortar. CEN: Brussels, Belgium, 2019.
18. Mercedes, L.; Bernat-Maso, E.; Gil, L. Bending and compression perfor-mance of full-scale sandwich panels of hemp fabric reinforced cementitious matrix. *Eng. Struct.* **2023**, *275 Pt B*, 115241. [CrossRef]
19. Ardanuy, M.; Claramunt, J.; Filho, R.D.T. Cellulosic fiber reinforced cement-based composites: A review of recent research. *Constr. Build. Mater.* **2015**, *79*, 115–128. [CrossRef]
20. Donnini, J.; Corinaldesi, V. Mechanical characterization of different FRCM systems for structural reinforcement. *Constr. Build. Mater.* **2017**, *145*, 565–575. [CrossRef]
21. D'Antino, T.; Papanicolaou, C. Mechanical characterization of textile reinforced inorganic-matrix composites. *Compos. Part B Eng.* **2017**, *127*, 78–91. [CrossRef]
22. *EN ISO 13934-1/2*; Textiles–Tensile Properties of Fabrics—Part 1: Determination of Maximum Force and Elongation at Maximum Force Using the Strip Method. ISO: Geneva, Switzerland, 2013.
23. *AC434-0213-R1(ME/BG)*; Proposed Revisions to the Acceptance Criteria for Masonry and Concrete Strengthening Using Fiber-reinforced Cementitious Matrix(FRCM) Composite Systems. ICC Evaluation Service Inc.: Brea, CA, USA, 2012.

Disclaimer/Publisher's Note: The statements, opinions and data contained in all publications are solely those of the individual author(s) and contributor(s) and not of MDPI and/or the editor(s). MDPI and/or the editor(s) disclaim responsibility for any injury to people or property resulting from any ideas, methods, instructions or products referred to in the content.

Article

A Meta-Analysis of the Effect of Moisture Content of Recycled Concrete Aggregate on the Compressive Strength of Concrete

Sung-Won Cho [1,*], Sung Eun Cho [1] and Alexander S. Brand [1,2,*]

[1] The Charles E. Via, Jr. Department of Civil and Environmental Engineering, Virginia Polytechnic Institute and State University, Blacksburg, VA 24061, USA; csungeun22@vt.edu
[2] Department of Materials Science and Engineering, Virginia Polytechnic Institute and State University, Blacksburg, VA 24061, USA
* Correspondence: authors: csungwon@vt.edu (S.-W.C.); asbrand@vt.edu (A.S.B.)

Abstract: To reduce the environmental impact of concrete, recycled aggregates are of significant interest. Recycled concrete aggregate (RCA) presents a significant resource opportunity, although its performance as an aggregate in concrete is variable. This study presents a meta-analysis of the published literature to refine the understanding of how the moisture content of RCA, as well as other parameters, affects the compressive strength of concrete. Seven machine learning models were used to predict the compressive strength of concrete with RCA, including linear regression, support vector regression (SVR), and k-nearest neighbors (KNN) as single models, and decision tree, random forest, XGBoost, and LightGBM as ensemble models. The results of this study demonstrate that ensemble models, particularly the LightGBM model, exhibited superior prediction accuracy compared to single models. The LightGBM model yielded the highest prediction accuracy with R^2 = 0.94, RMSE = 4.16 MPa, MAE = 3.03 MPa, and Delta RMSE = 1.4 MPa, making it the selected final model. The study, employing feature importance with LightGBM as the final model, identified age, water/cement ratio, and fine RCA aggregate content as key factors influencing compressive strength in concrete with RCA. In an interaction plot analysis using the final model, lowering the water–cement ratio consistently improved compressive strength, especially between 0.3 and 0.4, while increasing the fine RCA ratio decreased compressive strength, particularly in the range of 0.4 to 0.6. Additionally, it was found that maintaining moisture conditions of RCA typically between 0.0 and 0.8 was crucial for maximizing strength, whereas extreme moisture conditions, like fully saturated surface dry (SSD) state, negatively impacted strength.

Keywords: machine learning; recycled concrete aggregate; moisture content

Citation: Cho, S.-W.; Cho, S.E.; Brand, A.S. A Meta-Analysis of the Effect of Moisture Content of Recycled Concrete Aggregate on the Compressive Strength of Concrete. *Appl. Sci.* **2024**, *14*, 3512. https://doi.org/10.3390/app14083512

Academic Editors: Mouhamadou Amar and Nor Edine Abriak

Received: 8 March 2024
Revised: 11 April 2024
Accepted: 15 April 2024
Published: 22 April 2024

Copyright: © 2024 by the authors. Licensee MDPI, Basel, Switzerland. This article is an open access article distributed under the terms and conditions of the Creative Commons Attribution (CC BY) license (https://creativecommons.org/licenses/by/4.0/).

1. Introduction

Concrete recycling can be implemented as a strategy to reduce carbon emissions and promote sustainable development [1–3]. Concrete with recycled aggregates is recognized as one of the most prominent eco-friendly concretes [4–6]. Another concern for using recycled aggregates is that the supply of quality natural aggregates is diminishing in certain regions of the world (e.g., [7]). Of particular concern is increasing amounts of construction and demolition waste [1,8,9], from which recycled concrete aggregates (RCA) are derived.

Consequently, concrete recycling and sustainable development are considered highly important in the construction industry [1,10,11]. Recycled concrete can be utilized by crushing discarded concrete debris and using it as RCA to partially or fully replace natural aggregates in new concrete [1]. Recycled aggregates can be classified depending on the particle size or the type of waste material [12]. Among them, coarse recycled aggregates produced from crushed concrete are the most used in concrete production, and they are referred to as recycled concrete aggregate or RCA [12–18].

The surface of RCA can consist of natural aggregate and adhered mortar, which results in RCA having different physicochemical properties compared to natural aggregates [19–24].

The adhered mortar is porous, resulting in higher water absorption capacity in RCA compared to natural aggregates [19,25]. Additionally, the roughness of the adhered mortar increases the surface area of recycled aggregate particles, requiring more water for achieving consistent workability compared to concrete made with the same natural aggregates [19]. As a result, the mechanical properties of concrete with RCA can be inferior to those of concrete using natural aggregate particles [19–24]. Therefore, a number of researchers have explored methodologies to limit the detrimental effect of RCA, such as through beneficiation methodologies (e.g., [26,27]), mix design approaches (e.g., [28,29]), or alternative mixing approaches (e.g., [13,30–32]).

A significant number of researchers have noted that the moisture content of the RCA at the time of concrete batching can have a significant impact on the properties and performance of the concrete (e.g., [32,33]) in addition to affecting the microstructure development [33–35]. Poon et al. [36] discussed concrete compressive strengths with three different RCA moisture conditions—air dry (AD), oven dry (OD), and saturated surface dry (SSD)—and argued that SSD aggregate would release water, resulting in a weakened interfacial transition zone (ITZ) and a higher water-to-cement ratio [12]. These AD, OD, and SSD moisture states represent different moisture levels in RCA and play a crucial role in concrete performance. AD refers to aggregates dried naturally under atmospheric conditions, while OD indicates aggregates dried in an oven to remove all moisture. SSD signifies aggregates with their surfaces saturated with moisture but not immersed in water. Brand et al. [32,33] found that partially saturated RCA has the potential to have an equivalent concrete strength compared to natural aggregate concrete, and Etxeberria et al. [23] recommended to use partial SSD rather than SSD to secure the compressive strength of concrete with RCA. Mefteh et al. [37] argued that recycled aggregates within an SSD condition have the most negative impact on concrete strength, while AD recycled aggregates optimize concrete strength [37].

The objective of this study is to explore the relationship between the compressive strength of concrete and the moisture content of recycled concrete aggregates (RCA) through the application of machine learning, since there are conflicting conclusions in the literature. Various factors, including the volume fraction of aggregates, aggregate type, aggregate gradation, coarse-to-fine aggregate ratio, aggregate shape and texture, water-to-cement ratio, cement content, type and content of any supplementary cementitious materials, and type and dosage of chemical admixtures [38–42], influence the compressive strength of concrete. Due to the complex and nonlinear interrelationship between these factors and compressive strength, general linear equations are often ineffective [43–45]. In recent years, machine learning algorithms have been increasingly used to predict the performance of concrete with both natural aggregates [46,47] and with recycled aggregates [4]. Studies have focused on predicting various properties of concrete with recycled aggregates, such as strength [43], elastic modulus [19], chloride resistance [1], and durability [48], using machine learning models. However, there remains insufficient established predictive information regarding the relationship between the compressive strength of concrete and the moisture content of RCA. Therefore, this study aims to address this gap by conducting a literature review to investigate the relationship between compressive strength and moisture content in RCA. To achieve this, a database was constructed consisting of 752 entries, considering parameters such as the moisture content, water-to-cement ratio, replacement ratio of recycled aggregates, composition ratio of natural aggregates, curing age, etc. Seven machine learning methods, including linear regression, support vector regression (SVR), k-nearest neighbors (KNN), decision tree, random forest, LightGBM, and XGBoost, were employed to develop a predictive model for the compressive strength of RCA. The models were compared using evaluation metrics, and the final model was selected. Furthermore, feature importance and interaction plots were utilized to analyze the relationship between moisture content and compressive strength. This study contributes to the field by addressing the lack of consideration of RCA moisture content in predicting concrete performance metrics, making it a novel endeavor in the realm of machine learning studies in concrete technology.

2. Experimental Methodology

2.1. Data Collection

In this study, a database was collected to predict the compressive strength of concrete with RCA when considering the moisture condition of the RCA. The database was obtained from published literature (Table 1). The database consists of 752 entries, focusing on studies that explicitly reported the moisture condition of RCA. The output variable is compressive strength (MPa), and there are a total of 13 input variables considered. The input variables represent the mixture materials used in RCA and are expressed as ratios after unifying them in kg/m^3. The coarse RCA ratio represents the ratio of coarse RCA to the total coarse aggregate, and the fine RCA ratio represents the ratio of fine RCA to the total fine aggregate. These ratios are included to understand the influence of RCA proportions on compressive strength. The input variables related to the materials used in RCA are normalized by dividing them by the total material. The reason for dividing each material by the total material is to standardize all input variables on the same scale, allowing the model to consider the influence of each variable equally. This ensures that the model operates consistently even when the quantities or proportions of each material vary, making the results easier to interpret. Also, it can ease replication during future experiments, as following proportions makes it easier to replicate under standardized conditions. These input variables include cement, fly ash, water, superplasticizer, natural coarse aggregate, natural fine aggregate, fine RCA, and coarse RCA. The moisture condition is included as an input variable to investigate its influence on compressive strength. The range for the moisture condition was 0 to 1, where 0 was OD, 1 was SSD, and 0.5 was AD. Additionally, the other input variables are age and water-to-cement ratio.

Table 1. Database source.

No.	Reference	Number of Data	No.	Reference	Number of Data
1	[32]	12	12	[49]	36
2	[50]	10	13	[51]	15
3	[52]	8	14	[53]	10
4	[29]	27	15	[54]	6
5	[55]	50	16	[22]	3
6	[56]	24	17	[57]	14
7	[36]	36	18	[58]	4
8	[59]	54	19	[60]	18
9	[34]	5	20	[61]	20
10	[62]	42	21	[63]	300
11	[64]	12	22	[65]	46
				Total Data: 752	

2.2. Data Analysis

The database containing the input and output data was uploaded to the software as an Excel file, and the database was analyzed using Python code. Table 2 presents the statistical analysis of the database, including the mean, standard deviation, minimum, first quartile to third quartile, and maximum values. From Table 2, it is observed that there are variations among the input variables. Examining the mean values, the curing age is 23.06 days, and the average values for Coarse Aggregate/Total Material and Fine Aggregate/Total Material are 0.21 and 0.26, respectively. The average value for Superplasticizer/Total Material is 0.01, while for Fly Ash/Total Material it is 0.001, indicating significant variations. These variations can affect the performance of the model, hence preprocessing of the database is necessary [66].

Table 2. Database analysis.

Parameters	Unit	Mean	Standard Deviation	Min	25 Percentile	50 Percentile	75 Percentile	Max
Coarse RCA Ratio	-	0.46	0.41	0	0	0.49	1	1
Fine RCA Ratio	-	0.12	0.30	0	0	0	0	1
Cement/Total Material	-	0.18	0.03	0.1	0.16	0.18	0.19	0.28
Fly Ash/Total Material	-	0.001	0.006	0	0	0	0	0.04
Water/Total Material	-	0.09	0.02	0.04	0.08	0.09	0.1	0.18
Superplasticizer/Total Material	-	0.01	0.04	0	0	0	0	0.17
Coarse Aggregate/Total Material	-	0.21	0.17	0	0	0.22	0.37	0.5
Fine Aggregate/Total Material	-	0.26	0.09	0	0.26	0.28	0.3	0.43
Fine RCA/Total Material	-	0.05	0.13	0	0	0	0	0.73
Coarse RCA/Total Material	-	0.2	0.17	0	0	0.19	0.39	0.47
Moisture Condition	-	0.88	0.3	0	1	1	1	1
Age	Day	23.06	25.89	1	7	14	28	90
Water/Cement	-	0.46	0.12	0.27	0.42	0.5	0.55	1.11
Compressive Strength	MPa	36.82	16.20	4.8	24.63	35.4	47.33	85.2

2.3. Data Preprocessing

Scaling is a commonly used data preprocessing technique in machine learning. It is applied to address the issue of significant differences in units or ranges among variables. When variables have different units or ranges, it can make the interpretation of the model difficult. If one variable has a much larger range compared to others, it may have a large impact on the model's predictions. To mitigate this problem, it is necessary to adjust the variables to a consistent scale [67,68].

In this study, before applying the scale, the dataset was divided into training and test sets. The dataset consisting of 752 samples was split into a training set, which accounts for 70% of the data, and a test set, which accounts for the remaining 30%, following the methodology employed in previous studies [19,69,70]. After splitting the data, standard scaling was applied to the input variables. This approach standardizes the variables by adjusting their means to 0 and standard deviations to 1, aligning them with a standard normal distribution [67].

2.4. Cross-Validation and Hyperparameter

Cross-validation (CV) is a technique used in machine learning to evaluate the performance of a model and estimate its generalization ability. It involves dividing the available data into multiple subsets or k-folds [71]. K is a user-specified value, commonly set to 5 or 10 but can be chosen as any other value as well [72,73]. In each iteration, the model is trained on a training set and then evaluated on the validation set. This process is repeated several times, with different subsets of the data serving as the validation set each time. The performance metrics obtained from each iteration are then averaged to provide an overall estimate of the model's performance. CV helps address the issue of overfitting.

Hyperparameters are parameters that are set by the user before training the model. To optimize model performance, hyperparameter tuning is performed by systematically searching for the best combination of hyperparameter values [73]. This is often done in conjunction with CV, where different hyperparameter values are evaluated on different subsets of the data. This helps generalize well across different data subsets, resulting in a more robust and reliable model [74]. In this study, grid search was used to find the optimal CV value and hyperparameter values.

3. Results

3.1. Optimizing the Model

The study employed the grid search method to simultaneously find the optimal CV values and hyperparameter combinations for each model [75,76]. The range of CV fold values was set between a minimum of 2 and a maximum of 10, and various predefined hyperparameter values for each model were explored to find the best combination. The best CV fold values and hyperparameters for each model were selected based on evaluation metrics such as coefficient of determination (R^2), root-mean-square deviation (RMSE), mean absolute error (MAE), and Delta RMSE. Furthermore, to ensure the reproducibility and consistency of the results, the random state parameter was set to '5'.

3.1.1. Linear Regression

Linear regression does not require additional hyperparameter tuning because it does not have many hyperparameters to tune. In linear regression, the focus of model training is to adjust the weights and biases of the input variables to find the best-fitting linear relationship. As a result, for the test dataset, the R^2 is 0.66, RMSE is 9.72 MPa, and MAE is 7.67 MPa. The values for RMSE and MAE are relatively higher than the other six models. The Delta RMSE is 0.31 MPa.

3.1.2. Support Vector Regression (SVR)

The process of optimizing the SVR model involves adjusting the hyperparameters, cost, epsilon, gamma, and kernel values. The cost parameter determines the degree of error tolerance, while epsilon represents the acceptable range of error between predicted and actual values. In this case, the range for the cost parameter was set as 1, 10, and 100, and epsilon was set to 0.01 and 0.1. These values were commonly used and selected as initial choices for the parameters [77,78]. Additionally, gamma plays a role in adjusting the curvature of the decision boundary, and gamma values were set to 0.01 and 0.1. The kernel was considered with options including linear, polynomial, and Gaussian radial basis function (RBF) kernels.

During the evaluation process with varying CV values from 2 to 10, consistent results were observed for the test sets in terms of evaluation metrics, as is visually represented in Figure 1. Figure 1 shows the evaluation metrics for each CV value. The R^2 value for the test set was found to be 0.79, with an RMSE of 7.67 MPa and an MAE of 5.25 MPa. Furthermore, the Delta RMSE was 1.6 MPa. Based on these results, the optimal SVR model was obtained with a CV value of 2 and the following hyperparameter combination: cost value of 100, epsilon value of 0.1, gamma value of 0.1, and the Gaussian kernel. The decision to choose a smaller CV value, such as 2, was that a smaller CV value leads to a simpler model and reduced model complexity, which helps avoid overfitting [79]. Table 3 shows the optimal hyperparameter values and CV fold for each model.

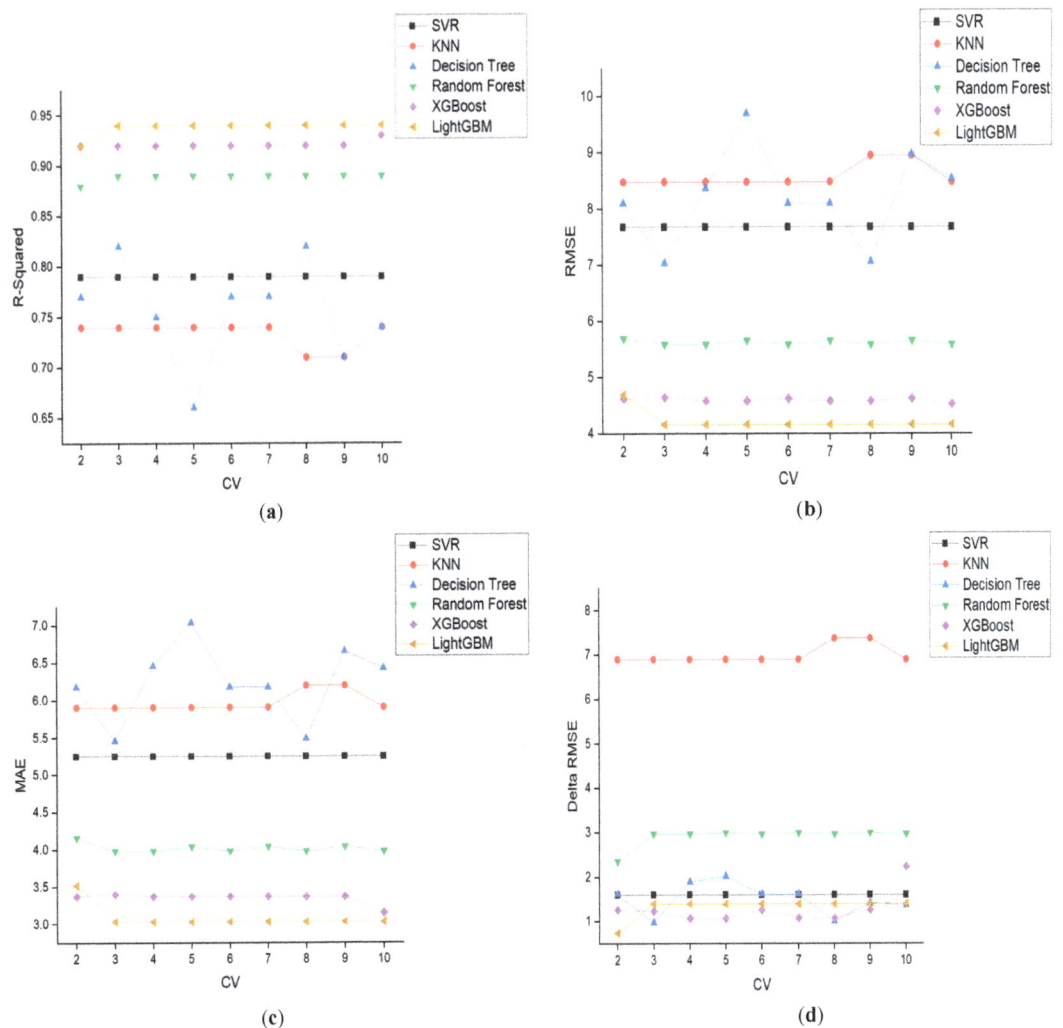

Figure 1. Performance metrics for each cross-validation value of the models, including (**a**) R^2; (**b**) RMSE; (**c**) MAE; and (**d**) Delta RMSE.

Table 3. Optimal hyperparameter values and CV fold.

Model	Best Hyperparameter	CV
SVR	Cost: 100 Epsilon: 0.1 Gamma: 0.1 Kernel: Gaussian	2
KNN	Number of neighbors: 5 Weights: distance Algorithm: auto Power parameter: Euclidean distance	2

Table 3. *Cont.*

Model	Best Hyperparameter	CV
Decision tree	Maximum depth: 7 Minimum number of sample split: 5 Maximum number of features: square root Criterion: mean squared error	3
Random forest	Number of estimators: 200 Maximum depth: None Minimum number of samples split: 2 Number of minimum samples leaf: 1	3
XGBoost	Number of estimators: 200 Learning rate: 0.1 Maximum depth: 5 Number of subsamples: 0.5	10
LightGBM	Number of estimators: 200 Learning rate: 0.1 Maximum depth: 7 Number of subsamples: 0.5	3

3.1.3. K-Nearest Neighbors (KNN)

To optimize the k-nearest neighbors (KNN) model, the values of the number of neighbors, weights, algorithm, and power parameter are adjusted. Number of neighbors is a parameter that specifies the number of nearest neighbors and is usually chosen as an odd value for applying majority voting rule. In this study, the number of neighbors is set to three and five. Weights are a parameter that determines the weight of neighbors, and two options, uniform and distance, are chosen. Uniform assigns equal weight to all neighbors, while distance assigns weights inversely proportional to the distance. Algorithm is used for neighbor search, and auto is used in this study as it automatically selects the most suitable algorithm. Power parameters are methods of distance measurement, where in this study, Manhattan distance and Euclidean distance are used.

The results show that for CV values ranging from 2 to 10, for the test set, the R^2, MAE, and RMSE values are consistent for CV values 2 to 7 and 10, which yielded the highest R^2 and the lowest RMSE, Delta RMSE, and MAE. Therefore, the optimal KNN model is selected with a CV value of 2, 5 neighbors, weights as distance, algorithm is set to auto, and Euclidean distance for power parameter, as summarized in Table 3. CV value as 2 shows an R^2 of 0.74, an RMSE of 8.48 MPa, an MAE of 5.91 MPa with the Delta RMSE as 6.91 MPa.

3.1.4. Decision Tree

The hyperparameters of the decision tree model include maximum depth, minimum number of samples split, maximum number of features, and criterion [25]. Maximum depth represents the maximum depth of the decision tree, and in this study, it was set to 5, 6, and 7. By controlling how deep the tree branches can extend, it helps mitigate the risk of overfitting. The minimum number of samples split refers to the minimum number of samples required to split a node. Nodes with fewer samples than this value will not be split. The default value in this research was set to 3, 4, and 5. The maximum number of features limits the number of features available for splitting and can help control the complexity and overfitting of the model. In this study, a square root of the total number of features and logarithm base 2 of the total number of features was chosen. Criterion is the function used to evaluate the quality of a node's split and mean squared error, Friedman mean squared error, and Poisson loss were applied in this research.

The lowest RMSE and MAE values were achieved when CV values were 3 and 8. For CV = 3, the test set had an R^2 of 0.82, an RMSE of 7.03 MPa, an MAE of 5.46 MPa, and Delta RMSE was 0.98 MPa. For CV = 8, the test set had an R^2 of 0.82, an RMSE of 7.06 MPa, and an MAE of 5.49 MPa with delta RMAE as 1.01 MPa. Since the performance metrics for CV values for 3 and 8 came out very similar, CV value 3 was chosen as the optimal model for the decision tree. The optimal hyperparameters for CV = 3 are as follows: the maximum depth is 7, the minimum number of samples split was set as 5, the maximum number of features as a square root of the total number of features, and the criterion is mean squared error, as summarized in Table 3.

The advantage of the tree model is the ability to visualize the model [80]. Figure 2 represents the optimal decision tree model for CV as 3. Interpreting the figure, the first splitting criterion is 'AGE'. It uses the 'AGE' feature to perform the first split. If the 'AGE' value is less than or equal to 21.0, it branches to the left; otherwise, it branches to the right. 'Squared error' indicates the mean squared error in the split, representing the average squared difference between the predicted and actual values in the split. In the first split, the mean squared error is 254.458. 'Samples' represents the number of data points included in the split, which is 541 in this case. The value of 541 is obtained by multiplying the training set ratio (0.7) by the total number of data points (752). Finally, 'value' denotes the average value of the target variable predicted within the split, which is 36.753 for the first split, representing the average of the target variable values for the data points belonging to the first split.

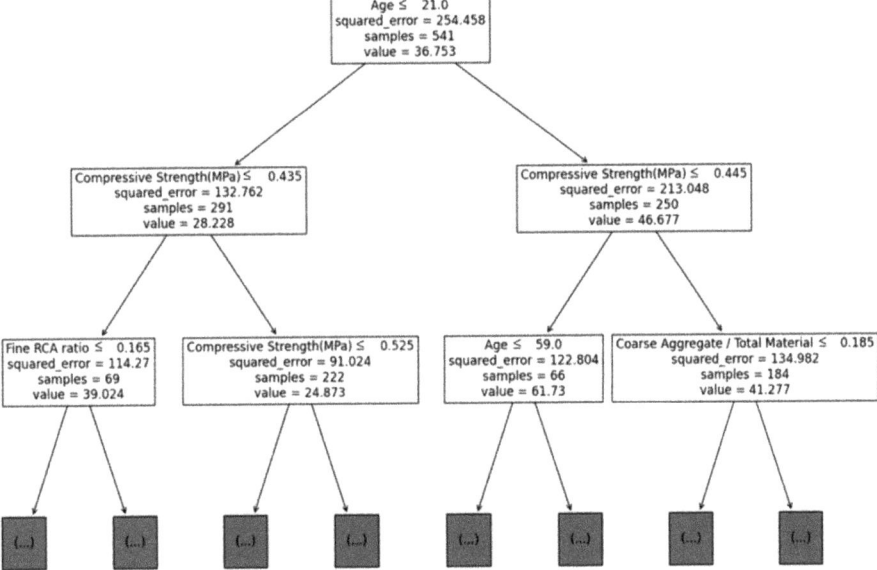

Figure 2. Optimal model of decision tree.

3.1.5. Random Forest

The hyperparameters of the random forest model that were specified in the study are the number of estimators, maximum depth, minimum number of samples split, and minimum number of samples leaf. The number of estimators represents the number of trees to be generated in the random forest. Increasing the number of trees can potentially improve prediction performance, but it can also slow down the model's training and prediction speed [81]. The maximum depth limits the maximum depth of each tree in the random forest [81]. Setting a maximum depth helps control the risk of overfitting, as deeper trees can capture more specific patterns in the training data but may struggle to generalize well to new data. The minimum number of samples split is the minimum number of samples required to split an internal node, while the minimum number samples leaf is the minimum number of samples required to be at a leaf node [81]. These parameters also contribute to controlling the model's complexity and generalization ability. The hyperparameter ranges specified in the study are as follows. The number of estimators is set to 100 and 200; maximum depth as 0, 5, and 10; minimum number of samples split as 2, 5, and 10; and minimum number samples leaf set to 1, 2, and 4.

The analysis of each CV value for the optimal hyperparameter values in terms of R^2, RMSE, MAE, and the Delta RMSE revealed similar results between CV values of 3 and 10. However, there was a noticeable difference when the CV was set to 2. For CV as 2, the test set had an R^2 of 0.88, an RMSE of 5.69 MPa, and an MAE of 4.16 MPa, with a Delta RMSE as 2.35 MPa. On the other hand, for CV values between 3 and 10, similar performance metrics were obtained. Especially for CV values as 3, the R^2 was 0.89, RMSE was 5.59 MPa, and MAE was 3.98 MPa, and Delta RMSE was 2.97 MPa. Comparing the two CV values, 2 and 3, it can be observed that the model performs better when the CV is set to 3. Although the Delta RMSE value was higher for CV = 3 compared to CV = 2, the higher R^2 and lower RMSE and MAE values indicate better overall performance when the CV is set to 3. Therefore, the optimal model for the random forest is obtained when the CV is set to 3, with 200 estimators, no maximum depth, a minimum samples split of 2, and 1 minimum samples leaf, as summarized in Table 3.

3.1.6. XGBoost

The selected hyperparameters for XGBoost in this study are number of estimators as 100 and 200; learning rate as 0.01 and 0.1; maximum depth as 3, 5, and 7; and number of subsamples as 0.5, 0.7, and 0.9 [82,83]. The number of estimators specifies the number of decision trees to be generated. Learning rate determines the contribution of each tree to the final prediction; smaller values result in less contribution from each tree, while larger values increase their contribution. Maximum depth limits the maximum depth of each tree, as deeper trees can lead to overfitting. Subsample specifies the proportion of samples used to train each tree.

When the CV value is 10, the highest R^2 and lowest RMSE and MAE were observed. For CV as 10, the test set achieved an R^2 of 0.93, RMSE of 4.52 MPa, and MAE of 3.15 MPa. The Delta RMSE was the highest at 2.24 MPa, suggesting that the model may be slightly overfitting to the training data. Despite this, the model still demonstrates superior performance in terms of R^2, RMSE, and MAE. The CV as 10 yielded the best results. Thus, based on the higher R^2, lower RMSE, and MAE, the model with the CV as 10 was chosen as the optimal model, even though the delta RMSE is higher compared to other CV values. Therefore, the optimal model is achieved with a CV of 10, and the corresponding optimal hyperparameters are 200 number estimators, a learning rate of 0.1, a maximum depth of 5, and subsamples of 0.5, as summarized in Table 3.

3.1.7. Light GBM

The hyperparameters for LightGBM were set with the same conditions as XGBoost [84]. Among the different CV values, the CV as 2 resulted in the lowest Delta RMSE of 0.74 MPa. However, when considering other evaluation metrics, CV values ranging from 3 to 10 showed better performance. Specifically, CV values between 3 and 10 achieved the highest R^2 value of 0.94, along with the lowest RMSE of 4.16 MPa and MAE of 3.03 MPa. Furthermore, the Delta RMSE was the second lowest at 1.4 MPa. Taking all these factors into account, the optimal LightGBM model was selected with a CV value of 3. Consequently, the optimal hyperparameters for the LightGBM model are 200 estimators, a learning rate of 0.1, a maximum depth of 7, and a subsample of 0.5, as summarized in Table 3.

3.2. Final Model Selection

Figure 3 compares the performance of the seven models and visualizes the reliability of their predictions. By examining the scatter plots of the optimal models on the test set, both the model's performance and the reliability of its predictions can be evaluated. The distribution of the actual values and predicted values is displayed visually, and the regression line and error range of ±10% show how well the predicted values fall within the acceptable range. The data points of XGBoost and LightGBM models are concentrated within the error range of 10%, indicating a better fit compared to the linear regression, SVR, KNN, decision tree, and random forest models. This suggests that the predictions of the XGBoost and LightGBM models can be considered more reliable compared to other models. Previous studies have also demonstrated similar findings. For instance, Cakiroglu et al. [85] used machine learning to study fiber-reinforced concrete and found that both model data samples remained within the ±10% deviation lines, while a study by Abdulalim Alabdullah et al. [71] on high-strength concrete prediction using LightGBM and XGBoost found a strong correlation between experimental and predicted results for both models.

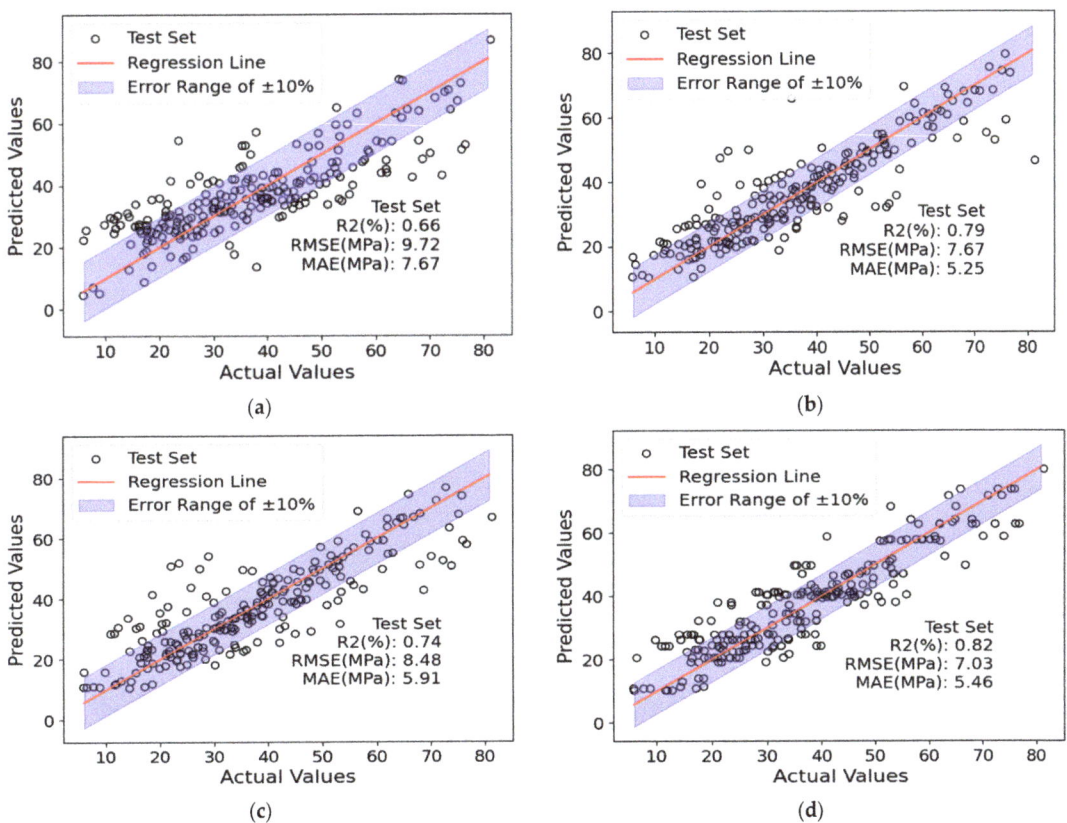

Figure 3. Cont.

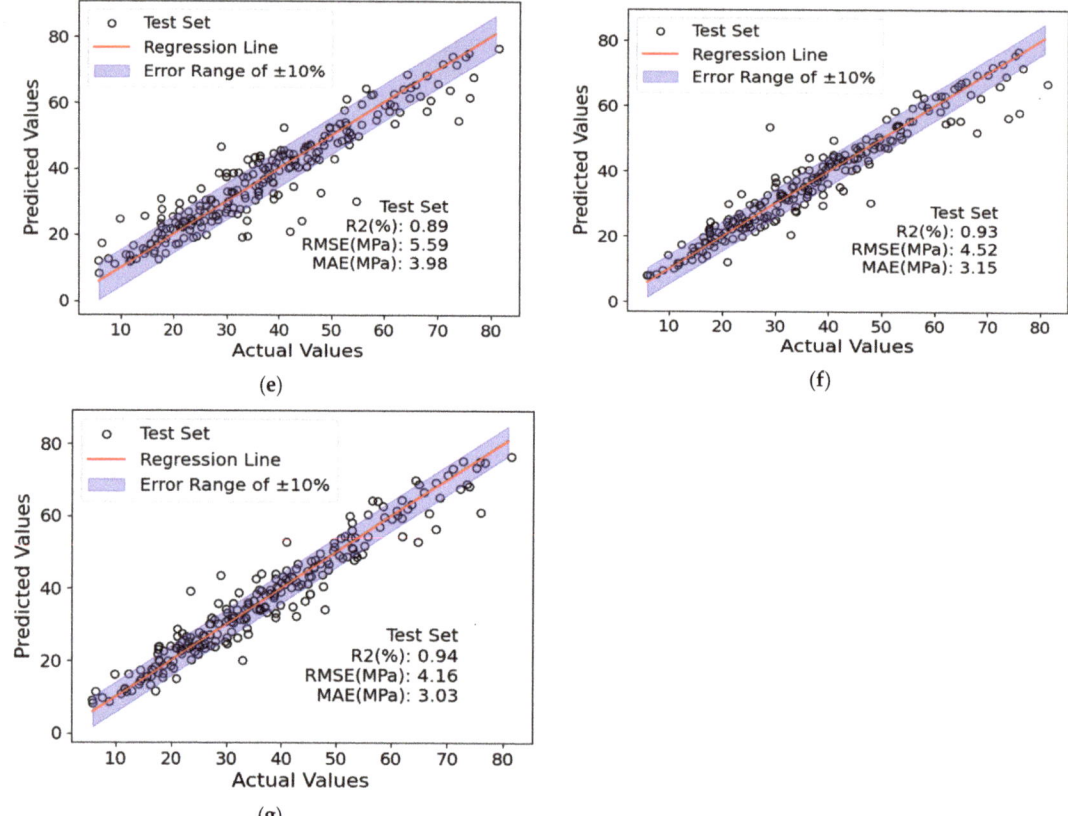

Figure 3. Evaluation of the optimal model's performance of (**a**) linear regression; (**b**) SVR; (**c**) KNN; (**d**) decision tree; (**e**) random forest; (**f**) XGBoost; and (**g**) LightGBM.

Figure 4 compares the actual values and predicted values from the test dataset of seven optimal models. The horizontal axis represents the index of the test data, and the vertical axis represents the compressive strength. The data points connected by the red line represent the predicted values, while the data points connected by the blue line represent the actual values. A larger distance between the two data points on the y-axis indicates a lower accuracy of the model's predictions. From the analysis, it can be observed that both the XGBoost and LightGBM models fit the original data very well. Both models show high prediction accuracy between the data indices 50 and 100. Following that, the random forest and decision tree models also exhibit good alignment with similar results. The model with the highest error rate is the linear regression model, with errors evenly distributed across the entire data index. Based on these results, XGBoost and LightGBM models demonstrate high reliability, while the linear regression model shows the least reliable predictions.

In Table 4, the values comparing the performance of the seven optimal models are presented, and Figure 5 illustrates the performance of the models based on Table 4. Through the comparison, the final model was selected based on performance metrics. Firstly, considering the R^2 values, the linear regression model had the lowest R^2 of 0.66 on the test set. In contrast, XGBoost and LightGBM models demonstrated strong predictive performance, achieving test set R^2 values of 0.93 and 0.94, respectively. This aligns with the findings of [85], who reported R^2 values of 0.93 for XGBoost and 0.94 for LightGBM. Secondly, looking at the RMSE and MAE values on the test set, the linear regression model had the highest values, while LightGBM had the lowest values of 4.16 MPa and 3.03 MPa, respectively, among the seven models. The RMSE and MAE values for the XGBoost and LightGBM models were also similarly low. Lastly, considering the Delta RMSE, the XGBoost model had a relatively low RMSE difference of 2.24 MPa, while the LightGBM model showed an even lower difference of 1.4 MPa. This indicates better generalization performance on the model. LightGBM exhibits high R^2 values on the test set along with low RMSE, MAE, and Delta RMSE values. Based on the provided information, LightGBM exhibits better performance than XGBoost in terms of training time, with LightGBM taking 2.92 s compared to XGBoost 37.26 s. The results from Wang [86] support the superior prediction accuracy of the LightGBM model compared to other models. Similarly, Amin [74] observed that LightGBM exhibited the highest reliability among the XGBoost and random forest models, as indicated by R^2, RMSE, and MAE values of 0.865, 3.56 MPa, and 1.3 MPa, respectively. Considering these findings and the evaluation of performance metrics, the LightGBM model was chosen as the optimized and final model for this study.

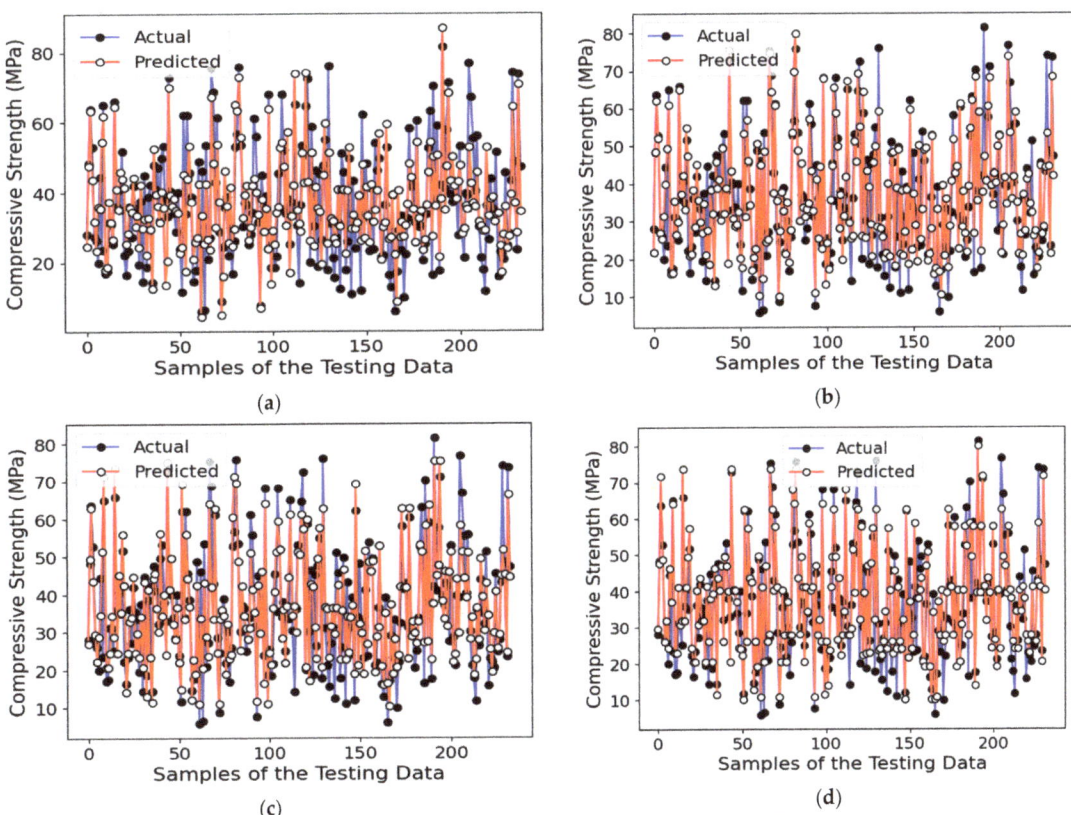

Figure 4. *Cont.*

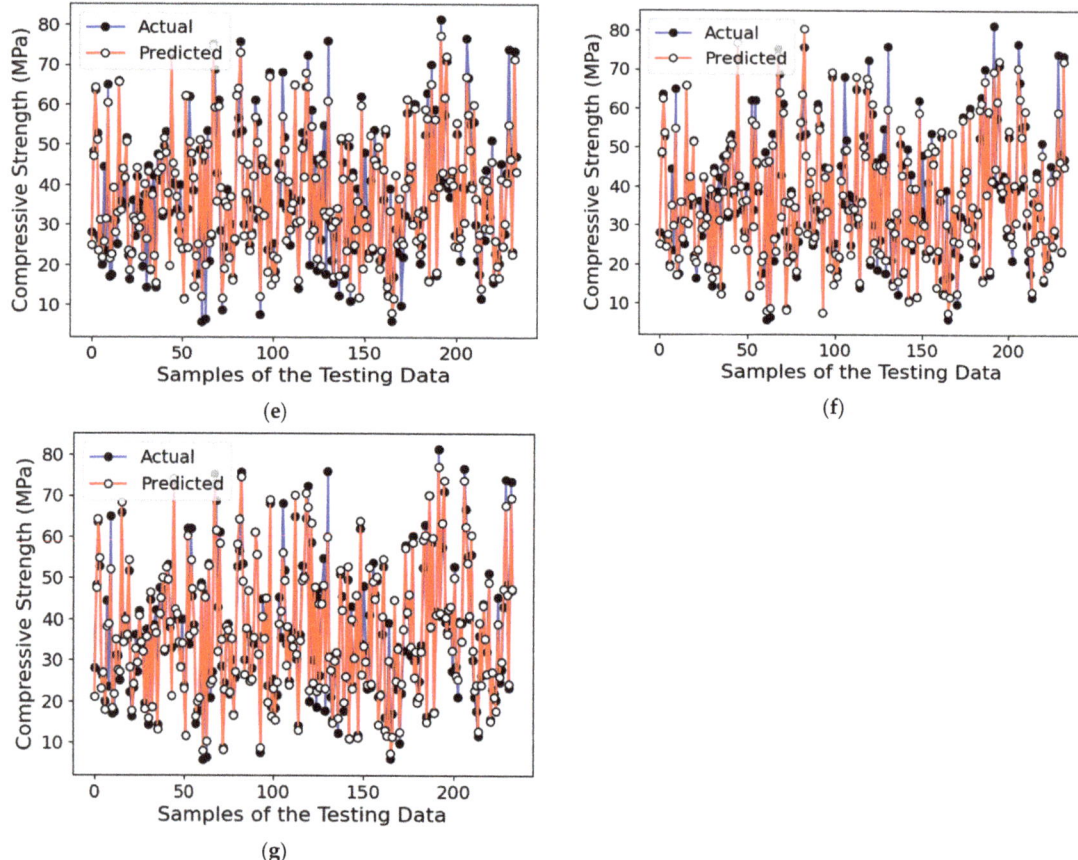

Figure 4. Actual and predicted values for the test data using the optimal models of (**a**) linear regression; (**b**) SVR; (**c**) KNN; (**d**) decision tree; (**e**) random forest; (**f**) XGBoost; and (**g**) LightGBM.

Table 4. Comparing the performance indicators of optimal models.

Database	Indicators	Models						
		Linear Regression	SVR	KNN	Decision Tree	Random Forest	XGBoost	LightGBM
Training set	R^2	0.65	0.86	0.99	0.86	0.97	0.98	0.97
	RMSE (MPa)	9.41	6.07	1.57	6.05	2.62	2.28	2.76
	MAE (MPa)	7.41	3.75	0.38	4.52	1.8	1.54	1.98
Test set	R^2	0.66	0.79	0.74	0.82	0.89	0.93	0.94
	RMSE (MPa)	9.72	7.67	8.48	7.03	5.59	4.52	4.16
	MAE (MPa)	7.67	5.25	5.91	5.46	3.98	3.15	3.03
Delta RMSE (MPa)		0.31	1.6	6.91	0.98	2.97	2.24	1.4
Training time (seconds)		0.03	5.01	0.07	0.37	43.37	37.26	2.92

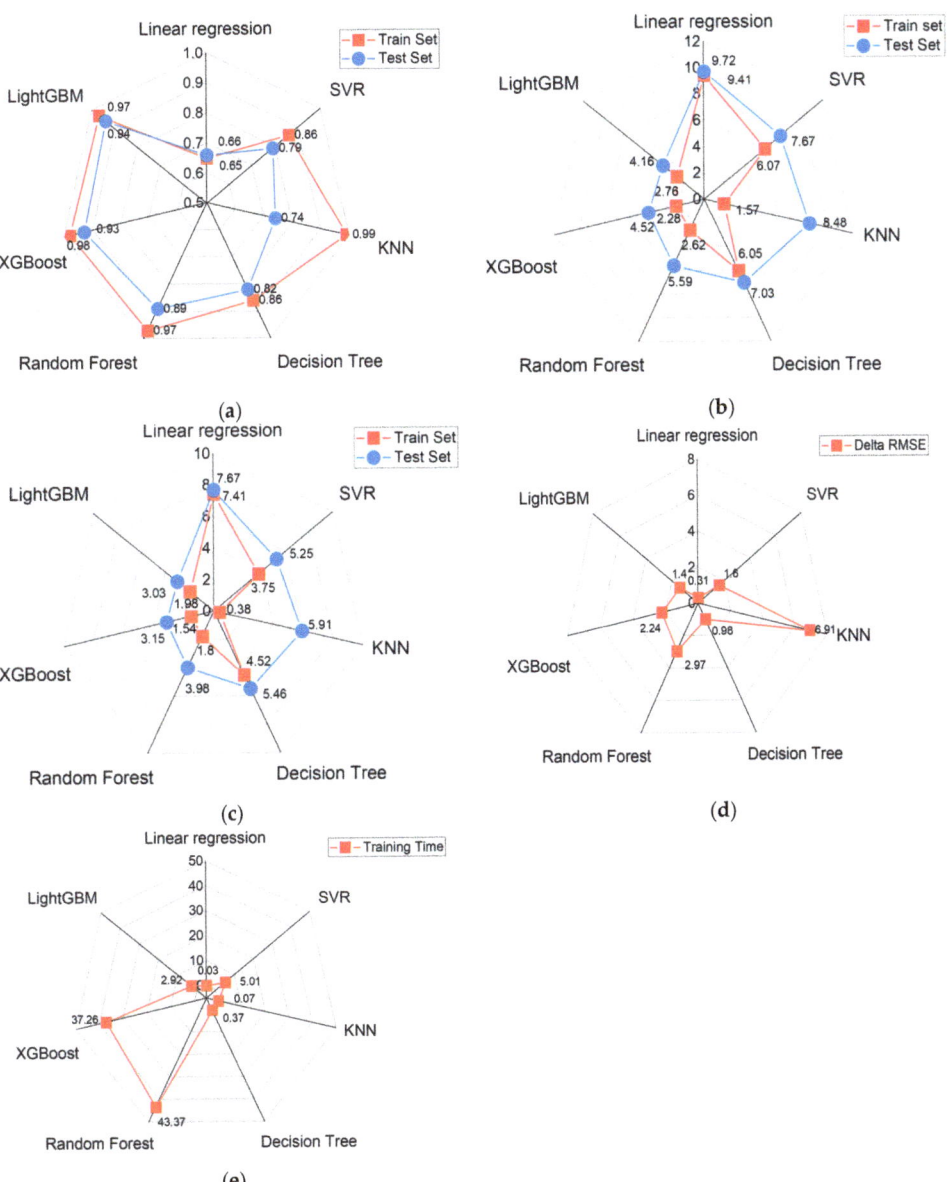

Figure 5. Comparing the optimal models using (**a**) R^2; (**b**) RMSE; (**c**) MSE; (**d**) Delta RMSE; and (**e**) training time.

4. Discussion

4.1. Feature Importance Analysis

Figure 6 shows the feature importance and represents the relative importance of each feature in the final model. Feature importance is a metric used in machine learning models to evaluate the importance of each feature in predicting the outcome. It helps to understand which features have the most significant influence on the model's predictions. In general, a higher feature importance value indicates a greater impact of that feature on the model's predictions [87]. The statement regarding the importance of the "Fine Aggregate/Total Material" feature and its influence on compressive strength is consistent with previous studies (e.g., [88]). Similarly, the "Water/Cement" feature is identified as the second most important, aligning with the general understanding that reducing the water–cement ratio can improve the compressive strength of concrete [1,43,75,89]. The third-largest impact is attributed to the "Age" feature, suggesting that curing time or the age of the concrete influences compressive strength, which of course is well known. However, the "Fly Ash/Total Material" feature is reported to have a value of '0', indicating no impact on compressive strength, which is attributed to limited data availability given that only 12 data in the 752 total dataset included fly ash.

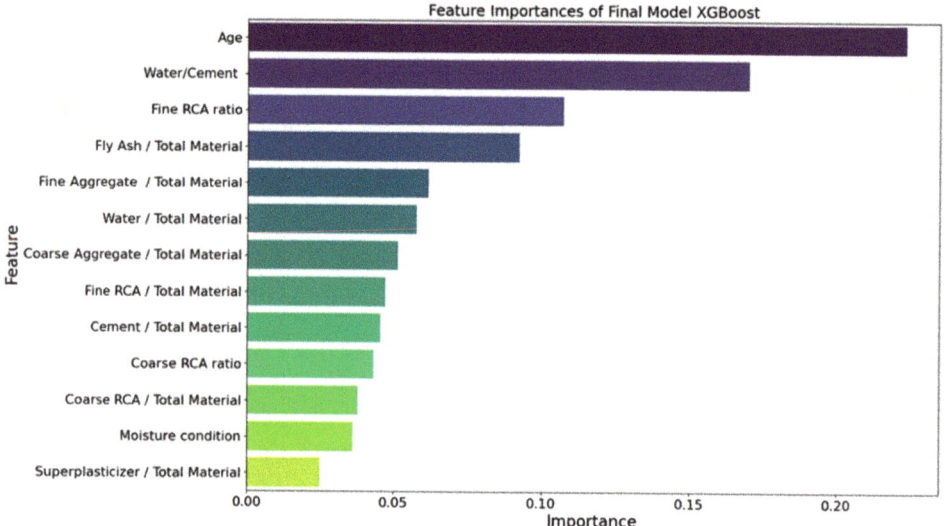

Figure 6. Feature importance of the final model.

As the main motivator for this study, it can be seen that the moisture condition does not rank very high on the feature importance plot. Moisture condition of the RCA is the 8th ranked feature, indicating that other features, including water-to-cement ratio, total cement content, and coarse RCA content, have a greater impact on the compressive strength of concrete.

4.2. Interaction Plot Analysis

To gain a detailed understanding of the relationships between variables, an interaction plot was utilized to explore the interaction effects among variables. The interaction plot aids in visually comprehending the interplay between variables and comparing their effects at different levels. In this study, based on feature importance, water/cement ratio was chosen as the variable with a significant influence on compressive strength, and an interaction plot was generated. Figure 7 presents the interaction relationship between water/cement ratio, fine aggregate/total material, and compressive strength. The analysis reveals that from a water/cement ratio of 0.6 onwards, there is a sharp decrease in compressive strength; however, when water/cement is fixed at 0.3 and 0.4, the average compressive strength is 55.87 MPa and 50.73 MPa, respectively.

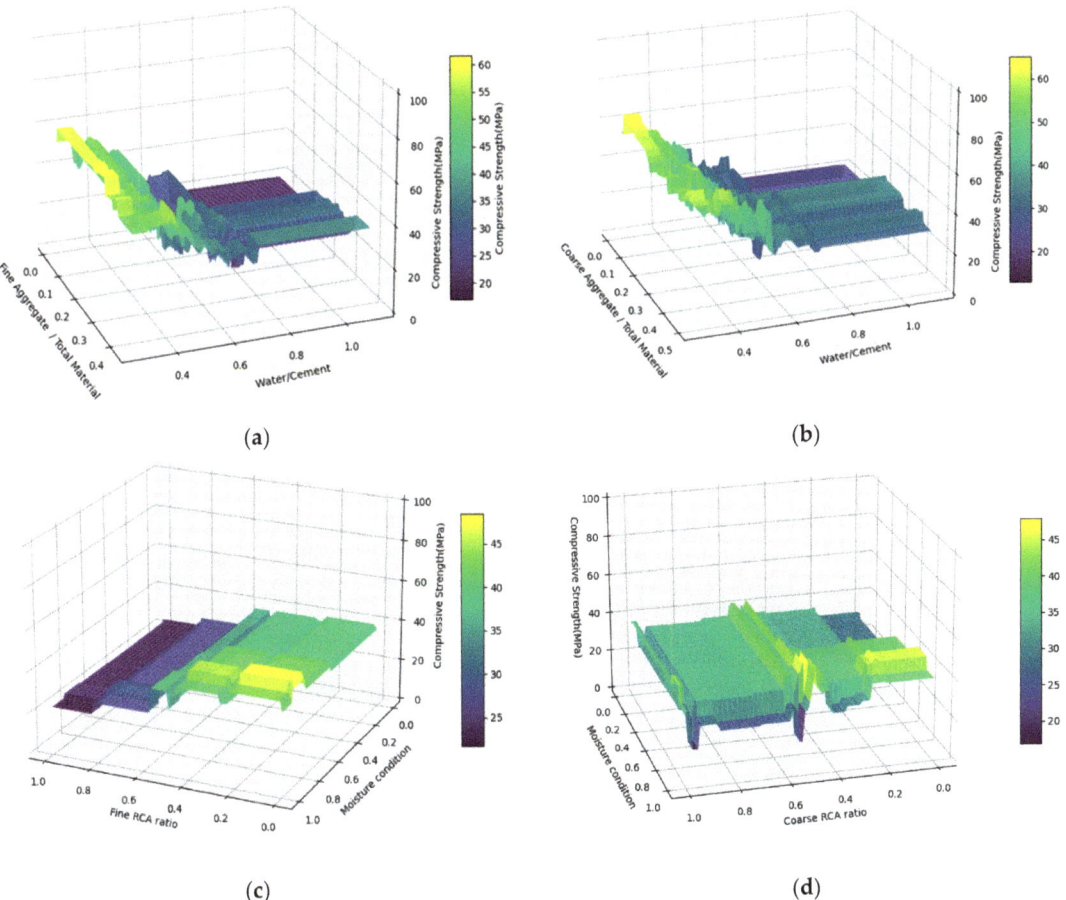

Figure 7. Result of interaction plot (**a**) relationship between fine aggregate/total material, water/cement, and compressive strength; (**b**) relationship between coarse aggregate/total material, water/Cement, and compressive strength; (**c**) relationship between fine RCA ratio, moisture condition, and compressive Strength; and (**d**) relationship between coarse RCA, moisture condition, and compressive strength.

Figure 7 represents an interaction plot among coarse aggregate/total material, water/cement ratio, and compressive strength (MPa). It shows that when the water/cement ratio increases from 0.4 to 0.6, the compressive strength decreases rapidly by at least 17 MPa. Additionally, within the coarse aggregate/total material range of 0 to 0.5, at a water/cement ratio value of 0.3, the average compressive strength is 56.47 MPa, and at a water/cement ratio value of 0.4, the average compressive strength is 48.66 MPa. In conclusion, the research shows that reducing the water–cement ratio in cement mortar leads to higher compressive strength, particularly within the range of 0.3 to 0.4 water–cement ratio. These results are in line with findings from other studies [90]. Similarly, Zhou et al. [91] observed that the dynamic compressive strength of cement mortar increases as water content decreases.

Figure 7 shows the interaction between fine RCA ratio, moisture condition, and compressive strength (MPa). Increasing the fine RCA ratio leads to a decrease in compressive strength, especially in the range of 0.4 to 0.6. This aligns with Kou and Poon [92], who also found reduced strength with higher fine RCA content. They suggested using 25%

to 50% fine RCA for maximum compressive and tensile strength. Checking the moisture condition ranging from 0 to 1.0, it was observed that as the moisture condition increased, the compressive strength also increased. A moisture condition at 0.8 exhibited the highest compressive strength within the entire fine RCA ratio range. However, when the moisture condition reaches the fully saturated surface dry (SSD) state, the compressive strength decreases slightly. This indicates that extreme moisture conditions can have a negative impact on the compressive strength of the material. In summary, the research suggests that an ideal fine RCA ratio could be around 0.4 to achieve higher compressive strength in concrete mixtures containing fine recycled aggregates. It is also recommended to maintain a moisture condition in the range of 0.0 to 0.8 to optimize compressive strength properties.

Based on the observations from Figure 7, it is evident that the interaction between coarse RCA ratio, moisture condition, and compressive strength follows certain trends. The highest compressive strength is achieved when the moisture condition is 0.8, and as the moisture condition increases, the compressive strength tends to increase, except when the moisture condition reaches 1.0, where there is a slight decrease in compressive strength. This was observed by Mefteh et al. [37] as well. The use of SSD recycled aggregates had the most significant adverse effect on concrete strength. Also, in the study by Poon et al. [36], the compressive strength decreased when recycled fine and coarse aggregates were in the SSD moisture condition. This decrease in strength can be attributed to the "bleeding" phenomenon, where water from the concrete mixture migrates to the surface of the aggregate particles, and then evaporates from the surface of the concrete mixture. As a result, the water–cement ratio increases during construction, leading to a reduction in the compressive strength of the concrete. Therefore, caution should be exercised when using recycled aggregates in the SSD state to avoid compromising the strength of the concrete.

In summary, the study reveals that for the coarse RCA ratio, values within the range of 0.0 to 0.2 lead to an increase in compressive strength, with the highest strength observed at a coarse RCA ratio of 0.2, irrespective of the moisture condition. However, for coarse aggregate/total material values exceeding 0.4 to 1.0, compressive strength decreases, especially when the coarse RCA ratio falls within the range of 0.4 to 0.6. Additionally, Etxeberria et al. [23] found that incorporating 25% recycled coarse aggregates can achieve mechanical properties equivalent to conventional concrete using the same cement quantity and water-to-cement ratio. Andal et al. [93] also recommended that incorporating 30% recycled coarse aggregates of preserved quality results in concrete of comparable quality to that made with natural aggregates. Based on these findings, the ideal moisture condition would be to avoid reaching SSD for coarse RCA, and the optimum coarse RCA ratio for achieving the highest compressive strength in the concrete mixture is around 0.2.

5. Conclusions and Future Recommendations

This study investigated the relationship between the compressive strength of concrete and the moisture content of RCA using machine learning techniques. Furthermore, various machine learning models were employed to comprehensively understand the impact of RCA moisture content on predicting concrete performance metrics. A literature review was conducted to explore the relationship between RCA moisture content and concrete compressive strength, based on which a database consisting of 752 items was constructed. Subsequently, a predictive model for RCA compressive strength was developed using seven machine learning models, and evaluation metrics were used to assess its performance.

Through the comprehensive construction of the database and the application of seven machine learning models, including XGBoost and LightGBM, the study developed a predictive model for RCA compressive strength. Evaluation results demonstrated that the LightGBM model outperformed other models in terms of R^2 values, RMSE, MAE, and generalization performance, thereby proving to be the optimal choice for this study.

Feature importance and interaction plot analyses were conducted to investigate how moisture content affects compressive strength. The analysis revealed that "Age", "Water/Cement" and "Fine RCA Ratio" were the most influential features, in line with prior research. Surprisingly, the moisture condition of the RCA ranked relatively low in importance, indicating that factors like water-to-cement ratio and aggregate content had a greater influence. Interaction plot analysis highlighted the significance of water-to-cement ratio, aggregate ratios, and moisture conditions on compressive strength. Notably, reducing the water-to-cement ratio consistently led to higher compressive strength. Maintaining optimal aggregate ratios, both fine and coarse, proved crucial for enhancing compressive strength. Additionally, controlling moisture within specific ranges, particularly avoiding extremes like fully saturated surface dry (SSD) states, was crucial for maximizing compressive strength. These findings support established research, emphasizing the importance of these factors in concrete mixture design for achieving desired mechanical properties.

Overall, this study fills the gap in predicting concrete performance metrics considering RCA moisture content and provides valuable insights for optimizing concrete mixtures containing recycled aggregates. These findings underscore the importance of comprehensive parameter consideration and the use of machine learning techniques in enhancing predictive models for concrete technology.

From a recommendation perspective, further research is needed to understand the impact of various recycled materials on concrete performance and to develop optimal mixtures. Among these, slag, generated during the steel manufacturing process, stands out as a valuable recycled material for concrete production. Properly processed, slag can enhance concrete quality and provide environmental benefits. Leveraging innovative technologies like machine learning to optimize the utilization of recycled materials holds significant promise in promoting sustainable construction and material production. These efforts are expected to contribute to achieving sustainable architecture and material production by enhancing environmental protection and resource efficiency.

Supplementary Materials: The following supporting information can be downloaded at: https://www.mdpi.com/article/10.3390/app14083512/s1, The Supplementary Materials for this article include Python Code and Database Source.

Author Contributions: S.-W.C.: Investigation, Writing—original draft, Writing—review and editing. S.E.C.: Data curation, Software, Validation, Visualization, Writing—original draft. A.S.B.: Methodology, Project Administration, Supervision, Writing—original draft, Writing—review and editing. All authors have read and agreed to the published version of the manuscript.

Funding: This research received no external funding.

Institutional Review Board Statement: Not applicable.

Informed Consent Statement: Not applicable.

Data Availability Statement: The data presented in this study are openly available as Supplementary Materials to this manuscript at https://www.mdpi.com/article/10.3390/app14083512/s1.

Acknowledgments: Thank you to the Peer reviewers for elevating the quality of the paper through their feedback, thereby enhancing its credibility and verifiability.

Conflicts of Interest: The authors declare no conflict of interest.

References

1. Liu, K.-H.; Zheng, J.-K.; Pacheco-Torgal, F.; Zhao, X.-Y. Innovative modeling framework of chloride resistance of recycled aggregate concrete using ensemble-machine-learning methods. *Constr. Build. Mater.* **2022**, *337*, 127613. [CrossRef]
2. Cavalline, T.; Snyder, M.B.; Taylor, P. *Tech Brief: Use of Recycled Concrete Aggregate in Concrete Paving Mixtures*; FHWA-HIF-22-020; Federal Highway Administration: Washington, DC, USA, 2022.
3. Crackler, T. *Recycled Concrete Aggregate Usage in the US*; National Concrete Pavement Technology Center, Iowa State University: Ames, IA, USA, 2018.
4. Feng, J.; Zhang, H.; Gao, K.; Liao, Y.; Gao, W.; Wu, G. Efficient creep prediction of recycled aggregate concrete via machine learning algorithms. *Constr. Build. Mater.* **2022**, *360*, 129497. [CrossRef]
5. Xie, T.; Gholampour, A.; Ozbakkaloglu, T. Toward the development of sustainable concretes with recycled concrete aggregates: Comprehensive review of studies on mechanical properties. *J. Mater. Civ. Eng.* **2018**, *30*, 04018211. [CrossRef]

6. Torgal, F.; Miraldo, S.; Labrincha, J.A.; Brito, J. An overview on concrete carbonation in the context of eco-efficient construction: Evaluation, use of SCMs and/or RAC. *Constr. Build. Mater.* **2012**, *36*, 141–150. [CrossRef]
7. Pavlů, T.; Kočí, V.; Hájek, P. Environmental assessment of two use cycles of recycled aggregate concrete. *Sustainability* **2019**, *11*, 6185. [CrossRef]
8. Torres, A.; Brandt, J.; Lear, K.; Liu, J. A looming tragedy of the sand commons. *Science* **2017**, *357*, 970–971. [CrossRef] [PubMed]
9. Zhang, C.; Hu, M.; Di Maio, F.; Sprecher, B.; Yang, X.; Tukker, A. An overview of the waste hierarchy framework for analyzing the circularity in construction and demolition waste management in Europe. *Sci. Total Environ.* **2022**, *803*, 149892. [CrossRef] [PubMed]
10. Pacheco-Torgal, F.; Ding, Y.; Colangelo, F.; Tuladhar, R.; Koutamanis, A. *Advances in Construction and Demolition Waste Recycling: Management, Processing and Environmental Assessment*; Woodhead Publishing: Cambridge, UK, 2020.
11. Tam, V.W.Y.; Soomro, M.; Evangelista, A.C.J. A review of recycled aggregate in concrete applications (2000–2017). *Constr. Build. Mater.* **2018**, *172*, 272–292. [CrossRef]
12. Zhang, H.; Xu, X.; Liu, W.; Zhao, B.; Wang, Q. Influence of the moisture states of aggregate recycled from waste concrete on the performance of the prepared recycled aggregate concrete (RAC)—A review. *Constr. Build. Mater.* **2022**, *326*, 126891. [CrossRef]
13. Kong, D.; Lei, T.; Zheng, J.; Ma, C.; Jiang, J.; Jiang, J. Effect and mechanism of surface-coating pozzalanics materials around aggregate on properties and ITZ microstructure of recycled aggregate concrete. *Constr. Build. Mater.* **2010**, *24*, 701–708. [CrossRef]
14. Singh, N.; Singh, S.P. Carbonation and electrical resistance of self compacting concrete made with recycled concrete aggregates and metakaolin. *Constr. Build. Mater.* **2016**, *121*, 400–409. [CrossRef]
15. Zhang, H.; Zhao, Y. Cracking of reinforced recycled aggregate concrete beams subjected to loads and steel corrosion. *Constr. Build. Mater.* **2019**, *210*, 364–379. [CrossRef]
16. Zhang, H.; Xu, X.; Su, X.; Zeng, W. To improve the resistance of recycled aggregate concrete (RAC) to the internal steel corrosion by the pre-treatment of aggregate. *Constr. Build. Mater.* **2021**, *306*, 124911. [CrossRef]
17. Montero, J.; Laserna, S. Influence of effective mixing water in recycled concrete. *Constr. Build. Mater.* **2017**, *132*, 343–352. [CrossRef]
18. Marinković, S.; Radonjanin, V.; Malešev, M.; Ignjatović, I. Comparative environmental assessment of natural and recycled aggregate concrete. *Waste Manag.* **2010**, *30*, 2255–2264. [CrossRef]
19. Han, T.; Siddique, A.; Khayat, K.; Huang, J.; Kumar, A. An ensemble machine learning approach for prediction and optimization of modulus of elasticity of recycled aggregate concrete. *Constr. Build. Mater.* **2020**, *244*, 118271. [CrossRef]
20. Li, C.; Geng, H.; Deng, C.; Li, B.; Zhao, S. Experimental investigation on columns of steel fiber reinforced concrete with recycled aggregates under large eccentric compression load. *Materials* **2019**, *12*, 445. [CrossRef]
21. Corinaldesi, V. Mechanical and elastic behaviour of concretes made of recycled-concrete coarse aggregates. *Constr. Build. Mater.* **2010**, *24*, 1616–1620. [CrossRef]
22. Evangelista, L.; de Brito, J. Durability performance of concrete made with fine recycled concrete aggregates. *Cem. Concr. Compos.* **2010**, *32*, 9–14. [CrossRef]
23. Etxeberria, M.; Vázquez, E.; Marí, A.; Barra, M. Influence of amount of recycled coarse aggregates and production process on properties of recycled aggregate concrete. *Cem. Concr. Res.* **2007**, *37*, 735–742. [CrossRef]
24. Xiao, J.; Li, J.; Zhang, C. Mechanical properties of recycled aggregate concrete under uniaxial loading. *Cem. Concr. Res.* **2005**, *35*, 1187–1194. [CrossRef]
25. Behnood, A.; Olek, J.; Glinicki, M. Predicting modulus elasticity of recycled aggregate concrete using M5′ model tree algorithm. *Constr. Build. Mater.* **2015**, *94*, 137–147. [CrossRef]
26. Prajapati, R.; Gettu, R.; Singh, S. Thermomechanical beneficiation of recycled concrete aggregates (RCA). *Constr. Build. Mater.* **2021**, *310*, 125200. [CrossRef]
27. Kim, J. Properties of recycled aggregate concrete designed with equivalent mortar volume mix design. *Constr. Build. Mater.* **2021**, *301*, 124091. [CrossRef]
28. Bidabadi, M.S.; Akbari, M.; Panahi, O. Optimum mix design of recycled concrete based on the fresh and hardened properties of concrete. *J. Build. Eng.* **2020**, *32*, 101483. [CrossRef]
29. Jang, H.; Kim, J.; Sicakova, A. Effect of aggregate size on recycled aggregate concrete under equivalent mortar volume mix design. *Appl. Sci.* **2021**, *11*, 11274. [CrossRef]
30. Tam, Y.; Gao, X.; Tam, C. Comparing performance of modified two-stage mixing approach for producing recycled aggregate concrete. *Mag. Concr. Res.* **2006**, *58*, 477–484. [CrossRef]
31. Tam, V.W.Y.; Tam, C.M.; Wang, Y. Optimization on proportion for recycled aggregate in concrete using two-stage mixing approach. *Constr. Build. Mater.* **2007**, *21*, 1928–1939. [CrossRef]
32. Brand, A.; Roesler, J.; Salas, A. Initial moisture and mixing effects on higher quality recycled coarse aggregate concrete. *Constr. Build. Mater.* **2015**, *79*, 83–89. [CrossRef]
33. Brand, A.; Roesler, J. Interfacial transition zone of cement composites with recycled concrete aggregate of different moisture states. *Adv. Civ. Eng. Mater.* **2018**, *7*, 87–102. [CrossRef]
34. Leite, M.B.; Monteiro, P.J.M. Microstructural analysis of recycled concrete using X-ray microtomography. *Cem. Concr. Res.* **2016**, *81*, 38–48. [CrossRef]
35. Le, T.; Le Saout, G.; Garcia-Diaz, E.; Damien, B.; Rémond, S. Hardened behavior of mortar based on recycled aggregate: Influence of saturation state at macro- and microscopic scales. *Constr. Build. Mater.* **2017**, *141*, 479–490. [CrossRef]

36. Poon, C.S.; Shui, Z.H.; Lam, L.; Fok, H.; Kou, S.C. Influence of moisture states of natural and recycled aggregates on the slump and compressive strength of concrete. *Cem. Concr. Res.* **2004**, *34*, 31–36. [CrossRef]
37. Mefteh, H.; Kebaïli, O.; Oucief, H.; Berredjem, L.; Arabi, N. Influence of moisture conditioning of recycled aggregates on the properties of fresh and hardened concrete. *J. Clean. Prod.* **2013**, *54*, 282–288. [CrossRef]
38. McNeil, K.; Kang, T.H.K. Recycled concrete aggregates: A review. *Int. J. Concr. Struct. Mater.* **2013**, *7*, 61–69. [CrossRef]
39. Montgomery, D. Workability and compressive strength properties of concrete containing recycled concrete aggregate. In *Sustainable Construction: Use of Recycled Concrete Aggregate*; Thomas Telford: London, UK, 1998; pp. 287–296.
40. Bui, N.K.; Satomi, T.; Takahashi, H. Effect of mineral admixtures on properties of recycled aggregate concrete at high temperature. *Constr. Build. Mater.* **2018**, *184*, 361–373. [CrossRef]
41. Dosho, Y. Effect of mineral admixtures on the performance of low-quality recycled aggregate concrete. *Crystals* **2021**, *11*, 596. [CrossRef]
42. Ju, M.; Jeong, J.G.; Palou, M.; Park, K. Mechanical behavior of fine recycled concrete aggregate concrete with the mineral admixtures. *Materials* **2020**, *13*, 2264. [CrossRef] [PubMed]
43. Tran, V.Q.; Dang, V.Q.; Ho, L.S. Evaluating compressive strength of concrete made with recycled concrete aggregates using machine learning approach. *Constr. Build. Mater.* **2022**, *323*, 126578. [CrossRef]
44. Choi, W.-C.; Yun, H.-D. Compressive behavior of reinforced concrete columns with recycled aggregate under uniaxial loading. *Eng. Struct.* **2012**, *41*, 285–293. [CrossRef]
45. Cree, D.; Green, M.; Noumowé, A. Residual strength of concrete containing recycled materials after exposure to fire: A review. *Constr. Build. Mater.* **2013**, *45*, 208–223. [CrossRef]
46. Gamil, Y. Machine learning in concrete technology: A review of current researches, trends, and applications. *Front. Built Environ.* **2023**, *9*, 1145591. [CrossRef]
47. Dabiri, H.; Kioumarsi, M.; Kheyroddin, A.; Kandiri, A.; Sartipi, F. Compressive strength of concrete with recycled aggregate; a machine learning-based evaluation. *Clean. Mater.* **2022**, *3*, 100044. [CrossRef]
48. Ozbakkaloglu, T.; Gholampour, A.; Xie, T. Mechanical and durability properties of recycled aggregate concrete: Effect of recycled aggregate properties and content. *J. Mater. Civ. Eng.* **2018**, *30*, 04017275. [CrossRef]
49. Li, L.G.; Lin, C.J.; Chen, G.M.; Kwan, A.K.H.; Jiang, T. Effects of packing on compressive behaviour of recycled aggregate concrete. *Constr. Build. Mater.* **2017**, *157*, 757–777. [CrossRef]
50. Pepe, M.; Toledo Filho, R.D.; Koenders, E.A.B.; Martinelli, E. A novel mix design methodology for Recycled Aggregate Concrete. *Constr. Build. Mater.* **2016**, *122*, 362–372. [CrossRef]
51. Berredjem, L.; Arabi, N.; Molez, L. Mechanical and durability properties of concrete based on recycled coarse and fine aggregates produced from demolished concrete. *Constr. Build. Mater.* **2020**, *246*, 118421. [CrossRef]
52. Butler, L.; West, J.S.; Tighe, S.L. Effect of recycled concrete coarse aggregate from multiple sources on the hardened properties of concrete with equivalent compressive strength. *Constr. Build. Mater.* **2013**, *47*, 1292–1301. [CrossRef]
53. Puente de Andrade, G.; de Castro Polisseni, G.; Pepe, M.; Toledo Filho, R.D. Design of structural concrete mixtures containing fine recycled concrete aggregate using packing model. *Constr. Build. Mater.* **2020**, *252*, 119091. [CrossRef]
54. Chen, X.; Zhang, H.; Geng, Y.; Wang, Y.-Y. Tests and prediction model for time-dependent internal relative humidity of recycled aggregate concrete due to self-desiccation. *Constr. Build. Mater.* **2022**, *352*, 129024. [CrossRef]
55. Brandes, M.R.; Kurama, Y. Effect of recycled concrete aggregates on strength and stiffness gain of concrete and on bond strength of steel prestressing strand. *PCI J.* **2018**, *63*, 87–105. [CrossRef]
56. Gumede, M.T.; Franklin, S.O. Studies on strength and related properties of concrete incorporating aggregates from demolished wastes: Part 2—Compressive and flexural strengths. *Open J. Civ. Eng.* **2015**, *05*, 175. [CrossRef]
57. Sun, D.; Huang, W.; Liu, K.; Ma, R.; Wang, A.; Guan, Y.; Shen, S. Effect of the moisture content of recycled aggregate on the mechanical performance and durability of concrete. *Materials* **2022**, *15*, 6299. [CrossRef] [PubMed]
58. Garcia-Gonzalez, J.; Rodriguez-Robles, D.; Juan-Valdes, A.; Moran-Del Pozo, J.M.; Guerra-Romero, M.I. Pre-saturation technique of the recycled aggregates: Solution to the water absorption drawback in the recycled concrete manufacture. *Materials* **2014**, *7*, 6224–6236. [CrossRef] [PubMed]
59. Yildirim, S.T.; Meyer, C.; Herfellner, S. Effects of internal curing on the strength, drying shrinkage and freeze–thaw resistance of concrete containing recycled concrete aggregates. *Constr. Build. Mater.* **2015**, *91*, 288–296. [CrossRef]
60. Ismail, S. Effect of different moisture states of surface-treated recycled coarse aggregate on properties of fresh and hardened concrete. *Int. J. Mater. Metall. Eng.* **2014**, *8*, 65–71. [CrossRef]
61. Koenders, E.A.B.; Pepe, M.; Martinelli, E. Compressive strength and hydration processes of concrete with recycled aggregates. *Cem. Concr. Res.* **2014**, *56*, 203–212. [CrossRef]
62. Fan, C.-C.; Huang, R.; Hwang, H.; Chao, S.-J. Properties of concrete incorporating fine recycled aggregates from crushed concrete wastes. *Constr. Build. Mater.* **2016**, *112*, 708–715. [CrossRef]
63. Shicong, K. Reusing Recycled Aggregates in Structural Concrete. Ph.D. Thesis, Hong Kong Polytechnic University, Hong Kong, 2006.
64. Pradhan, S.; Kumar, S.; Barai, S.V. Recycled aggregate concrete: Particle Packing Method (PPM) of mix design approach. *Constr. Build. Mater.* **2017**, *152*, 269–284. [CrossRef]
65. Lin, Y.-H.; Tyan, Y.-Y.; Chang, T.-P.; Chang, C.-Y. An assessment of optimal mixture for concrete made with recycled concrete aggregates. *Cem. Concr. Res.* **2004**, *34*, 1373–1380. [CrossRef]

66. Zhang, C.; Zhu, Z.; Liu, F.; Yang, Y.; Wan, Y.; Huo, W.; Yang, L. Efficient machine learning method for evaluating compressive strength of cement stabilized soft soil. *Constr. Build. Mater.* **2023**, *392*, 131887. [CrossRef]
67. Shahriyari, L. Effect of normalization methods on the performance of supervised learning algorithms applied to HTSeq-FPKM-UQ data sets: 7SK RNA expression as a predictor of survival in patients with colon adenocarcinoma. *Brief. Bioinform.* **2019**, *20*, 985–994. [CrossRef] [PubMed]
68. Ahsan, M.M.; Mahmud, M.A.P.; Saha, P.K.; Gupta, K.D.; Siddique, Z. Effect of data scaling methods on machine learning algorithms and model performance. *Technologies* **2021**, *9*, 52. [CrossRef]
69. Young, B.A.; Hall, A.; Pilon, L.; Gupta, P.; Sant, G. Can the compressive strength of concrete be estimated from knowledge of the mixture proportions?: New insights from statistical analysis and machine learning methods. *Cem. Concr. Res.* **2019**, *115*, 379–388. [CrossRef]
70. Cook, R.; Lapeyre, J.; Ma, H.; Kumar, A. Prediction of Compressive Strength of Concrete: Critical Comparison of Performance of a Hybrid Machine Learning Model with Standalone Models. *J. Mater. Civ. Eng.* **2019**, *31*, 1–15. [CrossRef]
71. Abdulalim Alabdullah, A.; Iqbal, M.; Zahid, M.; Khan, K.; Nasir Amin, M.; Jalal, F.E. Prediction of rapid chloride penetration resistance of metakaolin based high strength concrete using light GBM and XGBoost models by incorporating SHAP analysis. *Constr. Build. Mater.* **2022**, *345*, 128296. [CrossRef]
72. Ayaz, Y.; Kocamaz, A.F.; Karakoç, M.B. Modeling of compressive strength and UPV of high-volume mineral-admixtured concrete using rule-based M5 rule and tree model M5P classifiers. *Constr. Build. Mater.* **2015**, *94*, 235–240. [CrossRef]
73. Zhang, J.; Ma, G.; Huang, Y.; Sun, J.; Aslani, F.; Nener, B. Modelling uniaxial compressive strength of lightweight self-compacting concrete using random forest regression. *Constr. Build. Mater.* **2019**, *210*, 713–719. [CrossRef]
74. Amin, M.N.; Salami, B.A.; Zahid, M.; Iqbal, M.; Khan, K.; Abu-Arab, A.M.; Alabdullah, A.A.; Jalal, F.E. Investigating the bond strength of FRP laminates with concrete using LIGHT GBM and SHAPASH analysis. *Polymers* **2022**, *14*, 4717. [CrossRef]
75. Zhang, X.; Dai, C.; Li, W.; Chen, Y. Prediction of compressive strength of recycled aggregate concrete using machine learning and Bayesian optimization methods. *Front. Earth Sci.* **2023**, *11*, 1112105. [CrossRef]
76. Bergstra, J.; Bardenet, R.; Kégl, B.; Bengio, Y. Algorithms for Hyper-Parameter Optimization. *Adv. Neural Inf. Process. Syst.* **2011**, *24*, 2546–2554.
77. Cortes, C.; Vapnik, V. Support-vector networks. *Mach. Learn.* **1995**, *20*, 273–297. [CrossRef]
78. Montesinos López, O.A.; Montesinos López, A.; Crossa, J. Support vector machines and support vector regression. In *Multivariate Statistical Machine Learning Methods for Genomic Prediction*; Montesinos López, O.A., Montesinos López, A., Crossa, J., Eds.; Springer International Publishing: Cham, Switzerland, 2022; pp. 337–378. [CrossRef]
79. Escalante, H.J.; Montes, M.; Sucar, L. Particle swarm model selection. *J. Mach. Learn. Res.* **2009**, *10*, 405–440.
80. Song, Y.-Y.; Lu, Y. Decision tree methods: Applications for classification and prediction. *Shanghai Arch. Psychiatry* **2015**, *27*, 130–135. [CrossRef] [PubMed]
81. Probst, P.; Boulesteix, A.-L.; Wright, M. Hyperparameters and Tuning Strategies for Random Forest. *Wiley Interdiscip. Rev. Data Min. Knowl. Discov.* **2018**, *9*, e1301. [CrossRef]
82. Sommer, J.; Sarigiannis, D.; Parnell, T. Learning to tune XGBoost with XGBoost. *arXiv* **2019**, arXiv:1909.07218.
83. Xiong, X.; Guo, X.; Zeng, P.; Zou, R.; Wang, X. A short-term wind power forecast method via xgboost hyper-parameters optimization. *Front. Energy Res.* **2022**, *10*, 905155. [CrossRef]
84. Gan, M.; Pan, S.; Chen, Y.; Cheng, C.; Pan, H.; Zhu, X. Application of the machine learning lightgbm model to the prediction of the water levels of the lower columbia river. *J. Mar. Sci. Eng.* **2021**, *9*, 496. [CrossRef]
85. Cakiroglu, C.; Aydın, Y.; Bekdaş, G.; Geem, Z.W. Interpretable predictive modelling of basalt fiber reinforced concrete splitting tensile strength using ensemble machine learning methods and SHAP approach. *Materials* **2023**, *16*, 4578. [CrossRef]
86. Wang, X.; Chen, A.; Liu, Y. Explainable ensemble learning model for predicting steel section-concrete bond strength. *Constr. Build. Mater.* **2022**, *356*, 129239. [CrossRef]
87. Liu, L.; Chen, J. Effect of concretion wastes recycled aggregate on concrete performance. In Proceedings of the 2017 3rd International Forum on Energy, Environment Science and Materials (IFEESM 2017), Shenzhen, China, 25–26 November 2017. [CrossRef]
88. Peng, Y.; Unluer, C. Modeling the mechanical properties of recycled aggregate concrete using hybrid machine learning algorithms. *Resour. Conserv. Recycl.* **2023**, *190*, 106812. [CrossRef]
89. Liu, J.; Han, X.; Pan, Y.; Cui, K.; Xiao, Q. Physics-assisted machine learning methods for predicting the splitting tensile strength of recycled aggregate concrete. *Sci. Rep.* **2023**, *13*, 9078. [CrossRef] [PubMed]
90. Haach, V.G.; Vasconcelos, G.; Lourenço, P.B. Influence of aggregates grading and water/cement ratio in workability and hardened properties of mortars. *Constr. Build. Mater.* **2011**, *25*, 2980–2987. [CrossRef]
91. Zhou, J.; Chen, X.; Wu, L.; Kan, X. Influence of free water content on the compressive mechanical behaviour of cement mortar under high strain rate. *Sadhana* **2011**, *36*, 357–369. [CrossRef]

92. Kou, S.C.; Poon, C.S. Properties of self-compacting concrete prepared with coarse and fine recycled concrete aggregates. *Cem. Concr. Compos.* **2009**, *31*, 622–627. [CrossRef]
93. Andal, J.; Shehata, M.; Zacarias, P. Properties of concrete containing recycled concrete aggregate of preserved quality. *Constr. Build. Mater.* **2016**, *125*, 842–855. [CrossRef]

Disclaimer/Publisher's Note: The statements, opinions and data contained in all publications are solely those of the individual author(s) and contributor(s) and not of MDPI and/or the editor(s). MDPI and/or the editor(s) disclaim responsibility for any injury to people or property resulting from any ideas, methods, instructions or products referred to in the content.

MDPI AG
Grosspeteranlage 5
4052 Basel
Switzerland
Tel.: +41 61 683 77 34

Applied Sciences Editorial Office
E-mail: applsci@mdpi.com
www.mdpi.com/journal/applsci

Disclaimer/Publisher's Note: The statements, opinions and data contained in all publications are solely those of the individual author(s) and contributor(s) and not of MDPI and/or the editor(s). MDPI and/or the editor(s) disclaim responsibility for any injury to people or property resulting from any ideas, methods, instructions or products referred to in the content.

www.ingramcontent.com/pod-product-compliance
Lightning Source LLC
LaVergne TN
LVHW070359100526
838202LV00014B/1350